STRUCTURES ON MANIFOLDS

SERIES IN PURE MATHEMATICS

Part I. Monographs and Textbooks

Volume 1: Total Mean Curvature and Submanifolds of Finite Type
B Y Chen

Volume 3: Structures on Manifolds
K Yano & M Kon

Volume 4: Goldbach Conjecture
Wang Yuan (editor)

Part II. Lecture Notes

Volume 2: A Survey of Trace Forms of Algebraic Number Fields
P E Conner & R Perlis

Series in Pure Mathematics — Volume 3

STRUCTURES ON MANIFOLDS

KENTARO YANO
Tokyo Institute of Technology
Tokyo, Japan

MASAHIRO KON
Hirosaki University
Hirosaki, Japan

World Scientific

Published by
World Scientific Publishing Co. Pte. Ltd.
P. O. Box 128, Farrer Road, Singapore 9128

STRUCTURES ON MANIFOLDS

ISBN 9971-966-15-8
9971-966-16-6 pbk

Printed in Singapore by Kim Hup Lee Printing Co Pte Ltd.

PREFACE

The theory of structures on manifolds is a very interesting topic of modern differential geometry and the differential geometric aspects of submanifolds of manifolds with certain structures are vast and very fruitful fields for Riemannian geometry.

The purpose of this book is to provide an introduction to the theory of various differential geometric structures on manifolds and to gather and arrange the results on submanifolds of Riemannian manifolds with certain structures.

In Chapter I we have given a brief survey of differentiable manifolds, tensor fields, connections, fibre bundles and Riemannian curvature tensors. Chapter II contains the theory of submanifolds of Riemannian manifolds. We first of all state general formulas on submanifolds and the formulas of Laplacians of the second fundamental forms which will play very important roles in the following discussions. We also prove various theorems of submanifolds of Riemannian space forms. In Chapter III we provide differential geometric foundations for almost complex manifolds and Hermitian metrics, in particular, complex manifolds and Kaehlerian metrics. We also discuss the theory of nearly Kaehlerian manifolds and quaternion Kaehlerian manifolds. Chapter IV is devoted to the study of submanifolds of Kaehlerian manifolds. We study complex submanifolds, anti-invariant submanifolds and CR submanifolds of Kaehlerian manifolds. In Chapter V we give the basic results of almost contact manifolds and contact manifolds. The main purpose of this chapter is to introduce the fundamental properties of Sasakian manifolds. We also consider the Boothby-Wang fiberation and the Brieskorn manifolds. In Chapter VI we present the theory of submanifolds of Sasakian manifolds which include invariant submanifolds, anti-invariant submanifolds tangent to the structure vector field, anti-invariant submanifolds normal to the structure vector field and contact

CR submanifolds. Chapter VII is devoted to the study of f-structures on Riemannian manifolds. We also consider the hypersurfaces of framed manifolds. In Chapter VIII, we study the theory of product manifolds. We give the fundamental formulas for product manifolds, and consider its submanifolds. We also discuss the Kaehlerian product manifolds and its submanifolds. In the last Chapter IX, we first give the fundamental formulas for submersions, and present some results of almost Hermitian submersions. In the last section of this chapter, we discuss the relations of Sasakian manifolds and Kaehlerian manifolds and submanifolds of these manifolds by using the theory of submersions.

The Exercises are intended to introduce the intereting results which are concerned with the subject of this book.

The authors wish to express here their deep gratitude to Professor C. C. Hsiung who suggested that we include this book in "Series in Pure Mathematics". It is a pleasant duty for us to acknowledge that World Scientific Publishing took all possible care in the production of the book.

November 13, 1984

Kentaro Yano
Masahiro Kon

CONTENTS

CHAPTER I
RIEMANNIAN MANIFOLDS

1. Manifolds and tensor fields 1
2. Connections and covariant differentiations 18
3. Sectional curvature 31
4. Transformations 40
5. Fibre bundles and covering spaces 46
Exercises 54

CHAPTER II
SUBMANIFOLDS OF RIEMANNIAN MANIFOLDS

1. Induced connection and second fundamental form 61
2. Equations of Gauss, Codazzi and Ricci 67
3. Laplacian of the second fundamental form 73
4. Submanifolds of space forms 78
5. Minimal submanifolds 89
Exercises 99

CHAPTER III
COMPLEX MANIFOLDS

1. Almost complex manifolds and complex manifolds 104
2. Examples of complex manifolds and almost complex manifolds . 118
3. Hermitian manifolds 124
4. Kaehlerian manifolds 129
5. Nearly Kaehlerian manifolds 144
6. Quaternion Kaehlerian manifolds 158
Exercises 174

CHAPTER IV

SUBMANIFOLDS OF KAEHLERIAN MANIFOLDS

1. Kaehlerian submanifolds 180
2. Anti-invariant submanifolds of Kaehlerian manifolds 199
3. CR submanifolds of Kaehlerian manifolds 214
Exercises 244

CHAPTER V

CONTACT MANIFOLDS

1. Almost contact manifolds 252
2. Contact manifolds 255
3. Torsion tensor of almost contact manifolds 263
4. Contact distribution 269
5. Sasakian manifolds 272
6. Regular contact manifolds 286
7. Brieskorn manifolds 291
Exercises 306

CHAPTER VI

SUBMANIFOLDS OF SASAKIAN MANIFOLDS

1. Invariant submanifolds of Sasakian manifolds 312
2. Anti-invariant submanifolds tangent to the structure vector field of Sasakian manifolds 329
3. Anti-invariant submanifolds normal to the structure vector field of Sasakian manifolds 344
4. Contact CR submanifolds 351
5. Induced structures on submanifolds 366
Exercises 372

CHAPTER VII

f-STRUCTURES

1. f-structure on manifolds 379
2. Normal f-structure 392
3. Framed f-structure 402

4. Hypersurfaces of framed manifolds 408
Exercises 412

CHAPTER VIII

PRODUCT MANIFOLDS

1. Locally product manifolds 414
2. Locally decomposable Riemannian manifolds 418
3. Submanifolds of product manifolds 424
4. Submanifolds of Kaehlerian product manifolds 429
Exercises 436

CHAPTER IX

SUBMERSIONS

1. Fundamental equations of submersions 439
2. Almost Hermitian submersions 448
3. Submersions and submanifolds 455
Exercises 467

BIBLIOGRAPHY 473

INDEX 495

CHAPTER I

RIEMANNIAN MANIFOLDS

In this chapter, we have given a brief survey of Riemannian geometry. In §1, we give the basic definitions and formulas on differentiable manifolds. §2 is devoted to the study of connections and covariant differentiation on manifolds. We introduce the torsion tensor field and the curvature tensor field and derive the structure equation of Cartan. Moreover, we define the Riemannian connection on a manifold. In §3, we define the sectional curvature of a Riemannian manifold and give examples of space forms, that is, Riemannian manifolds of constant curvature. In §4, we discuss transformations on a Riemannian manifold and give some integral formulas (cf. Yano [6]). In the last §5, we prepare some results of fibre bundles for later use.

In §§1, 2 and 3 we followed Chapter I of Helgason [1] and Kobayashi-Nomizu [1]. In §5, we followed fairly closely Kobayashi-Nomizu [1]. We also refer to Matsushima [1].

1. MANIFOLDS AND TENSOR FIELDS

Let M be a topological space. We assume that M satisfies the Hausdorff separation axiom which states that any two different points in M can be separated by disjoint open sets. An open *chart* on M is a pair (U,ϕ) where U is an open subset of M and ϕ is a homeomorphism of U onto an open subset of R^n, where R^n is an n-dimensional Euclidean space.

Definition. Let M be a Hausdorff space. A *differentiable structure* on M of dimension n is a collection of open charts $(U_i, \phi_i)_{i\varepsilon\Lambda}$ on M where $\phi_i(U_i)$ is an open subset of R^n such that the following conditions are satisfied:

(a) $M = \bigcup_{i\varepsilon\Lambda} U_i$;

(b) For each pair $i,j \varepsilon \Lambda$ the mapping $\phi_j \cdot \phi_i^{-1}$ is a differentiable mapping of $\phi_i(U_i \cap U_j)$ onto $\phi_j(U_i \cap U_j)$;

(c) The collection $(U_i, \phi_i)_{i\varepsilon\Lambda}$ is a maximal family of open charts for which (a) and (b) hold.

A *differentiable manifold* (or C^∞*-manifold* or simply *manifold*) of dimension n is a Hausdorff space with differentiable structure of dimension n. If M is a manifold, a *local coordinate system* on M (or a *local chart* on M) is by definition a pair (U_i, ϕ_i). If p is a point in U_i and $\phi_i(p) = (x^1(p),\ldots,x^n(p))$, the set U_i is called a *coordinate neighborhood* of p and the numbers $x_j(p)$ are called *local coordinates* of p. The mapping $\phi_i : q \longrightarrow (x^1(q),\ldots,x^n(q))$, $q \varepsilon U_i$, is often denoted by $\{x^1,\ldots,x^n\}$.

We notice that the condition (c) is not essential in the definition of a manifold. In fact, if only (a) and (b) are satisfied, the family $(U_i, \phi_i)_{i\varepsilon\Lambda}$ can be extended in a unique way to a larger family of open charts such that (a), (b) and (c) are all fulfilled. This is easily seen by defining the larger family as the set of all open charts (V,ϕ) on M satisfying: (i) $\phi(V)$ is an open set in R^n; (ii) for each $i \varepsilon \Lambda$, $\phi_i \cdot \phi^{-1}$ is a diffeomorphism of $\phi(V \cap U_i)$ onto $\phi_i(V \cap U_i)$.

An *analytic structure* of dimension n is defined in a similar fashion. In (b) we just replace "differentiable" by "analytic". In this case M is called an *analytic manifold.*

In order to define a complex manifold of (complex) dimension n we replace R^n in the definition of differentiable manifold by n-dimensional complex number space C^n. The condition (b) is replaced by the condition that the n coordinates of $\phi_j \cdot \phi_i^{-1}(p)$ should be holomorphic functions of the coordinates of p.

Given two manifolds M and M', a mapping $f : M \longrightarrow M'$ is said to be *differentiable* (or C^∞ differentiable), if for every chart

(U_i,ϕ_i) of M and every chart (V_j,ψ_j) of M' such that $f(U_i) \subset V_j$, the mapping $\psi_j \cdot f \cdot \phi_i^{-1}$ of $\phi_i(U_i)$ into $\psi_j(V_j)$ is differentiable. A *differentiable function* on M is a differentiable mapping of M into R. If M is analytic, the function f on M is said to be *analytic* if for every chart (U_i,ϕ_i), the function $f \cdot \phi_i^{-1}$ is an analytic function on $\phi_i(U_i)$. The definition of a *holomorphic* (or *complex analytic*) mapping or function is similar.

Let M and N be two manifolds of dimension n and m, respectively. Let $(U_i,\phi_i)_{i\in\Lambda}$ and $(V_a,\psi_a)_{a\in\Lambda'}$ be collections of open charts on M and N, respectively. For $i \in \Lambda$, $a \in \Lambda'$, let $\phi_i \times \psi_a$ denote the mapping $(p,q) \longrightarrow (\phi_i(p),\psi_a(q))$ of the product set $U_i \times V_a$ into R^{n+m}. Then the collection $(U_i \times V_a, \phi_i \times \psi_a)_{i\in\Lambda, a\in\Lambda'}$ of open charts on the product space $M \times N$ satisfies (a) and (b), and hence $M \times N$ can be turned into a manifold the *product* of M and N.

By a *differentiable curve* in a manifold M, we shall mean a differentiable mapping of a closed interval [a,b] of R into M. We shall now define a *tangent vector* (or simply a *vector*) at a point p of M. Let $\mathfrak{F}(p)$ be the algebra of differentiable functions defined in a neighborhood of p. Let $\tau(t)$ $(a \leq t \leq b)$ be a curve such that $\tau(t_0)=p$. The vector tangent to the curve $\tau(t)$ at p is a mapping $X : \mathfrak{F}(p) \longrightarrow R$ defined by

$$Xf = (df(\tau(t))/dt)_{t_0}.$$

In other words, Xf is the derivative of f in the direction of the curve $\tau(t)$ at $t = t_0$. The vector X satisfies the following conditions:

(1) X is a linear mapping of $\mathfrak{F}(p)$ into R;

(2) $X(fg) = (Xf)g(p) + f(p)(Xg)$ for $f,g \in \mathfrak{F}(P)$.

The set of mappings X of $\mathfrak{F}(p)$ into R satisfying the preceding two conditions forms a real vector space. Let $x^1,\ldots,x^n$ be local coordinates in a coordinate neighborhood U of p. For each i, $(\partial/\partial x^i)_p$ is a mapping of $\mathfrak{F}(p)$ into R which satisfies (1) and (2). Given any curve $\tau(t)$ with $p = \tau(t_0)$, let $x^i = \tau^i(t)$, $i = 1,\ldots,n$, be its equations in terms of the local coordinates $x^1,\ldots,x^n$. Then

$$(df(\tau(t))/dt)_{t_0} = \sum_i (\partial f/\partial x^i)_p (d\tau^i(t)/dt)_{t_0},$$

which proves that every vector at p is a linear combination of $(\partial/\partial x^1)_p, \ldots, (\partial/\partial x^n)_p$. Conversely, given a linear combination $\Sigma\xi^i(\partial/\partial x^i)_p$, consider the curve defined by

$$x^i = x^i(p) + \xi^i t, \quad i = 1,\ldots,n.$$

Then the vector tangent to this curve at $t = 0$ is $\Sigma\xi^i(\partial/\partial x^i)_p$. If we assume $\Sigma\xi^i(\partial/\partial x^i)_p = 0$, then $0 = \Sigma\xi^i(\partial x^j/\partial x^i)_p = \xi^j$ for $j = 1,\ldots,n$. Therefore, $(\partial/\partial x^1)_p, \ldots, (\partial/\partial x^n)_p$ are linearly independent and hence these form a basis of the set of vectors at p. The set of tangent vectors at p, denoted by $T_p(M)$, is called the *tangent space* of M at p. The n-tuple of numbers $\xi^1,\ldots,\xi^n$ are *components* of the vectors $\Sigma\xi^i(\partial/\partial x^i)_p$ with respect to the local coordinates $x^1,\ldots,x^n$.

We notice that on a C^∞ differentiable manifold M the tangent space $T_p(M)$ coincides with the space of $X : \mathfrak{F}(p) \longrightarrow R$ satisfying the conditions (1) and (2) above.

A *vector field* X on a manifold M is an assignment of a vector X_p to each point p of M. If f is a differentiable function on M, then Xf is a function on M defined by $(Xf)(p) = X_p f$. A vector field X is said to be *differentiable* if Xf is differentiable for every differentiable function f. In terms of local coordinates $x^1,\ldots,x^n$, X may be expressed by $X = \Sigma\xi^i(\partial/\partial x^i)$, where ξ^i are functions defined in the coordinate neighborhood, called the components of X with respect to $x^1,\ldots,x^n$. X is differentiable if and only if its components ξ^i are differentiable.

We denote by $\mathfrak{X}(M)$ the set of all differentiable vector fields on M. From now on we shall consider mainly manifolds of class C^∞, mappings of class C^∞ and vector fields of class C^∞.

If X and Y are vector fields, define the bracket [X,Y] as a mapping from the ring of functions on M into itself by

$$[X,Y]f = X(Yf) - Y(Xf).$$

Let $X = \Sigma\xi^i(\partial/\partial x^i)$ and $Y = \Sigma\xi^j(\partial/\partial x^j)$. Then

$$[X,Y]f = \sum_{i,j}(\xi^j(\partial\eta^i/\partial x^j) - \eta^j(\partial\xi^i/\partial x^j)(\partial f/\partial x^i).$$

This means that $[X,Y]$ is a vector field with components $\Sigma_j(\xi^j(\partial\eta^i/\partial x^j) - \eta^j(\partial\xi^i/\partial x^j))$, $i = 1,\dots,n$. With respect to this bracket operation, $\mathfrak{X}(M)$ is a Lie algebra over R. For any vector fields X, Y and Z, we have the Jacobi identity:

$$[[X,Y],Z] + [[Y,Z],X] + [[Z,X],Y] = 0.$$

We may also regard $\mathfrak{X}(M)$ as a module over the algebra $\mathfrak{F}(M)$ of differentiable functions on M as follows. If f is a function and X is a vector field on M, then fX is a vector field on M defined by $(fX)_p = f(p)X_p$ for $p \in M$. We also have

$$[fX,gY] = fg[X,Y] + f(Xg)Y - g(Yf)X.$$

Let $T^*_p(M)$ be the dual space of the tangent space $T_p(M)$ of M at p. An element of $T^*_p(M)$ is called a *covector* at p. An assignment of a covector at each point p is called a 1-*form* (*differential form of degree* 1). For each function for M, the *total differential* $(df)_p$ of f at p is defined by

$$\langle (df)_p, X\rangle = Xf \qquad \text{for } X \in T_p(M),$$

where $\langle\ ,\ \rangle$ denotes the value of the first entry on the second entry as a linear functional on $T_p(M)$. Let $(U;x^i)$ be a local coordinate system at p. Then $(dx^1)_p,\dots,(dx^n)_p$ form a basis for $T^*_p(M)$. They form the dual basis of the basis $(\partial/\partial x^1)_p,\dots,(\partial/\partial x^n)_p$ for $T_p(M)$. In a neighborhood of p, every 1-form ω can be uniquely written as

$$\omega = \sum_j f_j dx^j,$$

where f_j are functions in U and are called the *components* of ω with respect to $x^1,\ldots,x^n$. The 1-form ω is called *differentiable* if f_j are differentiable. This condition is independent of the choice of a local coordinate system. We shall only consider differentiable 1-forms.

A 1-form ω can be defined also as an $\mathfrak{F}(M)$-linear mapping of the $\mathfrak{F}(M)$-module $\mathfrak{X}(M)$ into $\mathfrak{F}(M)$. The two definitions are related by

$$(\omega(X))_p = \langle\omega_p, X_p\rangle, \quad X \in \mathfrak{X}(M).$$

Let A be a commutative ring with identity element and $E_1,\ldots,E_s$ be A-modules. Then $E_1 \times \ldots \times E_a$ is also an A-module. A mapping $f : E_1 \times \ldots \times E_s \longrightarrow F$, where F is an A-module, is said to be A-*maltilinear* if it is A-linear in each argument. The set of all A-multilinear mappings of $E_1 \times \ldots \times E_s$ into F is again A-module. Suppose that all the facters E_i coincide. The A-multilinear mapping is called *alternate* if $f(X_1,\ldots,X_s) = 0$ whenever at least two X_i coincide.

Let T^r_s denote the $\mathfrak{F}(M)$-module of all $\mathfrak{F}(M)$-multilinear mapping of

$$\mathfrak{X}(M)^* \times \ldots \times \mathfrak{X}(M)^* \times \mathfrak{X}(M) \times \ldots \times \mathfrak{X}(M)$$
$$(\mathfrak{X}(M)^* \text{ r times, } \mathfrak{X}(M) \text{ s times})$$

into $\mathfrak{F}(M)$, where $\mathfrak{X}(M)^*$ is the dual $\mathfrak{F}(M)$-module of $\mathfrak{X}(M)$. We put $T^r_0 = T^r$, $T^0_s = T_s$ and $T^0_0 = \mathfrak{F}(M)$. A *tensor field* K on M of type (r,s) is an element of T^r_s. This tensor field K is said to be *contravariant* of degree r, *covariant* of degree s. In particular, the tensor fields of type (0,0), (1,0) and (0,1) on M are just the differentiable functions on M, the vector fields and the 1-forms on M, respectively. If p is a point in M, we define $T^r_s(p)$ as the set of all R-multilinear mappings of

$$T_p^*(M) \times \ldots \times T_p^*(M) \times T_p(M) \times \ldots \times T_p(M)$$
$$(T_p^*(M) \text{ r times, } T_p(M) \text{ s times})$$

into R. The set $T_s^r(p)$ is a vector spece over R and is nothing but the tensor product

$$T_p(M) \otimes \ldots \otimes T_p(M) \otimes T_p^*(M) \otimes \ldots \otimes T_p^*(M)$$
$$(T_p(M) \text{ r times, } T_p^*(M) \text{ s times})$$

or otherwise written

$$T_s^r(p) = \otimes^r T_p(M) \otimes {}^s\otimes T_p^*(M).$$

We also put $T_0^0(p) = R$.

Let $\{x^1,\ldots,x^n\}$ be a system of coordinates valid on an open neighborhood U of p. Then there exist vector fields $X_1,\ldots,X_n$ and 1-forms $\omega_1,\ldots,\omega_n$ on M and an open neighborhood V of p, $p \in V \in U$ such that on V

$$X_i = (\partial/\partial x^i), \quad \omega^j(X_i) = \delta_{ij} \quad (1 \leq i,j \leq n).$$

On V we put

$$Z_i = \sum_k f_{ik} X_k, \qquad \theta_j = \sum_l g_{jl}\omega^l,$$

where f_{ik}, $g_{jl} \in \mathfrak{F}(V)$. Then we have, for $q \in V$,

$$K(\theta^1,\ldots,\theta^r,Z_1,\ldots,Z_s)(q)$$
$$= \sum_{l_j=1,k_i=1}^{n} g_{1l_1}\cdots g_{rl_r} f_{1k_1}\cdots f_{sk_s} K(\omega^{l_1},\ldots,\omega^{l_r},X_{k_1},\ldots,X_{k_s})(q).$$

This shows that $K(\theta^1,\ldots,\theta^r,Z_1,\ldots,Z_s)(p) = 0$ if some θ^j or some Z_i

vanishes at p. We can therefore define an element $K_p \in T^r_s(p)$ by

$$K_p((\theta^1)_p,\dots,(\theta^r)_p,(Z_1)_p,\dots,(Z_s)_p) = K(\theta^1,\dots,\theta^r,Z_1,\dots,Z_s)(p).$$

Thus the tensor field K gives rise to a family K_p, $p \in M$, where $K_p \in T^r_s(p)$. If $K_p = 0$ for all p, then K=0. The element K_p depends differentiably on p in the sense that if V is a coordinate neighborhood of p and K_q $(q \in V)$ is expressed as above in terms of basis for $\mathfrak{X}(V)^*$ and $\mathfrak{X}(V)$, then the coefficients are differentiable functions on V. On the other hand, if there is a rule $p \longrightarrow K_p$ which to each $p \in M$ assigns a member K(p) of $T^r_s(p)$ in a differentiable manner, then there exists a tensor field K of type (r,s) such that $K_p = K(p)$ for all $p \in M$. In the case when M is analytic it is clear how to define analyticity of a tensor field K.

Let T denote the direct sum of the $\mathfrak{F}(M)$-modules T^r_s,

$$T = \sum_{r,s=0}^{\infty} T^r_s.$$

Similarly, if $p \in M$ we consider

$$T(p) = \sum_{r,s=0}^{\infty} T^r_s(p).$$

An element of T is of the form $\Sigma_{r,s} K^r_s$, where $K^r_s \in T^r_s$ are zero except for a finite number of them at each $p \in M$. The vector spece T(p) can be turned into an associative algebra over R as follows: Let $a = e_1 \otimes \dots \otimes e_r \otimes f^1 \otimes \dots \otimes f^s$, $b = e'_1 \otimes \dots \otimes e'_c \otimes f'^1 \otimes \dots \otimes f'^d$, where e_i, e'_j are members of a basis for $T_p(M)$, f^k, f'^l are members of a dual basis for $T^*_p(M)$. Then $a \otimes b$ is defined by

$$a \otimes b = e_1 \otimes \dots \otimes e_r \otimes e'_1 \otimes \dots \otimes e'_c \otimes f^1 \otimes \dots \otimes f^s \otimes f'^1 \otimes \dots \otimes f'^d.$$

We put $a \otimes 1 = a$, $1 \otimes b = b$ and extend the operation $(a,b) \longrightarrow a \otimes b$ to a bilinear mapping of $T(p) \times T(p)$ into T(p). Then T(p) is an associative algebra over R. The multiplication in T(p) is independent of the choice of basis.

The tensor product ⊗ in T is defined as the $\mathcal{F}(M)$-bilinear mapping $(K,L) \longrightarrow K \otimes L$ of $T \times T$ into T such that

$$(K \otimes L)_p = K_p \otimes L_p, \qquad K \in T^r_s,\ L \in T^c_d.$$

This turnes the $\mathcal{F}(M)$-module T into a ring satisfying

$$f(K \otimes L) = fK \otimes L = K \otimes fL$$

for $f \in \mathcal{F}(M)$, $K,L \in T$. In other words, T is an associative algebra over $\mathcal{F}(M)$. The algebras T and T(p) are called the *mixed tensor algebras* over M and $T_p(M)$ respectively. The submodules

$$T^* = \sum_{r=0}^{\infty} T^r, \qquad T_* = \sum_{s=0}^{\infty} T_s$$

are subalgebras of T and the subspaces

$$T^*(p) = \sum_{r=0}^{\infty} T^r(p), \qquad T_*(p) = \sum_{s=0}^{\infty} T_s(p)$$

are subalgebras of T(p).

We now define the notion of *contraction*. Now let r, s be two integers ≥ 1, and let i, j be integers such that $1 \leq i \leq r$, $1 \leq j \leq s$. Consider the R-linear mapping $C^i_j : T^r_s(p) \longrightarrow T^{r-1}_{s-1}(p)$ defined by

$$C^i_j(e_1 \otimes \ldots \otimes e_r \otimes f^1 \otimes \ldots \otimes f^s) = \langle e_i, f_j \rangle (e_1 \otimes \ldots \hat{e}_i \ldots \otimes e_r \otimes f^1 \ldots \hat{f}^j \ldots \otimes f^s),$$

where $\{e_k\}$ is a basis of $T_p(M)$, $\{f^l\}$ is the dual basis of $T^*_p(M)$. (The symbol ^ over a letter means that the letter is missing.) We note that C^i_j is independent of the choice of basis. There exists now a unique $\mathcal{F}(M)$-linear mapping $C^i_j : T^r_s \longrightarrow T^{r-1}_{s-1}$ such that

$$(C^i_j(K))_p = C^i_j(K_p)$$

for all $K \in T^r_s$ and all $p \in M$. This mapping satisfies the relation

$$C^i_j(X_1\otimes\ldots\otimes X_r\otimes\omega^1\otimes\ldots\otimes\omega^s)$$
$$= \langle X_i,\omega_j\rangle(X_1\otimes..\hat{X}_i..\otimes X_r\otimes\omega^1\otimes..\hat{\omega}^j..\otimes\omega^s)$$

for all $X_1,\ldots,X_r \in T^1$, $\omega^1,\ldots,\omega^s \in T_1$. The mapping C^i_j is called the contraction of the i-th contravariant index and the j-th covariant index.

For the basis $\{e_i\}$ for $T_p(M)$ and the dual basis $\{f_s\}$ for $T^*_p(M)$, every tensor K of type (r,s) can be expressed uniquely as

$$K = \sum_{i_1,..,i_r,j_1,..,j_s} K^{i_1..i_r}_{j_1..j_s} e_{i_1}\otimes\ldots\otimes e_{i_r}\otimes f^{j_1}\otimes\ldots\otimes f^{j_s},$$

where $K^{i_1..i_r}_{j_1..j_s}$ are called the components of K with respect to the above basis. In terms of components, the contraction C^i_j is represented by

$$(C^i_jK)^{i_1..i_{r-1}}_{j_1..j_{s-1}} = \sum_k K^{i_1..k..i_{r-1}}_{j_1..k..j_{s-1}},$$

where the superscript k appears at the i-th position and the subscript k appears at the j-th position.

Let $\Lambda T^*_p(M)$ be the exterior algebra over $T^*_p(M)$. An r-*form* ω is an assignment of an element of degree r in $\Lambda T^*_p(M)$ to each point p of M. In terms of a local coordinate system $x^1,\ldots,x^n$, r-form ω can be expressed uniquely as

$$\omega = \sum_{i_1<..<i_r} f_{i_1..i_r} dx^{i_1} \quad\ldots\quad dx^{i_r}.$$

The r-form ω is said to be *differentiable* if the components $f_{i_1..i_r}$ are all differentiable. By an r-form we shall mean a differentiable r-form. An r-form ω can be defined also as a skew-symmetric r-linear mapping over $\mathfrak{F}(M)$ of $\mathfrak{X}(M) \times \ldots \times \mathfrak{X}(M)$ (r times) into $\mathfrak{F}(M)$. The two definitions are related as follows. If $\omega_1,\ldots,\omega_r$ are 1-forms and

$X_1,\ldots,X_r$ are vector fields, then $(\omega_1 \wedge \ldots \wedge \omega_r)(X_1,\ldots,X_r)$ is $1/r!$ times the determinant of the matrix $(\omega_j(X_k))_{j,k=1,\ldots,r}$ of degree r.

Let $D^r = D^r(M)$ be the totality of r-forms on M for each $r = 0,1,\ldots,n$. Then $D^0 = \mathfrak{F}(M)$. Each D^r is a real vector space and can be also considered as an $\mathfrak{F}(M)$-module: for $f \in \mathfrak{F}(M)$ and $\omega \in D^r$, $f\omega$ is an r-form defined by $(f\omega)_p = f(p)\omega_p$, $p \in M$. We set

$$D = D(M) = \sum_{r=0}^{n} D^r(M).$$

With respect to the exterior product, D(M) forms an algebra over R. *Exterior differentiation* d can be characterized as follows:

(1) d is an R-linear mapping of D(M) into itself such that $d(D^r) \subset D^{r+1}$;

(2) For $f \in D^0$, df is the total differential;

(3) If $\omega_1 \in D^r$ and $\omega_2 \in D^s$, then

$$d(\omega_1 \wedge \omega_2) = d\omega_1 \wedge \omega_2 + (-1)^r \omega_1 \wedge d\omega_2;$$

(4) $d^2 = 0$.

Let V be an m-dimensional real vector space. We define a V-valued r-form ω on M as an assignment to each point p of M a skew-symmetric r-linear mapping of $T_p(M) \times \ldots \times T_p(M)$ (r times) into V. If we take a basis $e_1,\ldots,e_m$ for V, we can write ω uniquely as $\omega = \Sigma_j \omega^j e_j$, where ω^j are usual r-forms on M. If ω^j are all differentiable, ω is said to be differentiable. The exterior derivative $d\omega$ is defined to be $\Sigma_j d\omega^j e_j$, which is a V-valued (r+1)-form.

Let f be a mapping of a manifold M into another manifold M'. We define the *differential* $f_* : T_p(M) \longrightarrow T_{f(p)}(M')$ at p of f as follows: For each $X \in T_p(M)$, choose a curve $\tau(t)$ in M such that X is the vector tangent to $\tau(t)$ at $p = \tau(t_0)$. Then $f_*(X)$ is the vector tangent to the curve $f(\tau(t))$ at $f(p) = f(\tau(t_0))$. If g is a function in a neighborhood of f(p), then $(f_*(X))g = X(g \cdot f)$. When it is necessary to specify the point p, we write $(f_*)_p$. When there is no danger of confusion, we may simply write f instead of f_*. The transpose of $(f_*)_p$ is a linear mapping of $T^*_{f(p)}(M')$ into $T^*_p(M)$. For any r-form ω' on M', we define an r-form $f^*\omega'$ on M by

$$(f^*\omega')(X_1,\ldots,X_r) = \omega'(f_*X_1,\ldots,f_*X_r), \quad X_1,\ldots,X_r \in T_p(M).$$

We also have $d(f^*\omega') = f^*(d\omega')$, that is, d commutes with f^*.

A mapping f of M into M' is said to be of *rank* r at $p \in M$ if the dimension of $f_*(T_p(M))$ is r. If the rank of f at p is equal to n = dim M, $(f_*)_p$ is injective and dim M $\leq$ dim M'. If the rank of f at p is equal to n' = dim M', $(f_*)_p$ is surjective and dim M $\geq$ dim M'. We notice that the following proposition is hold (see Chevalley [;pp. 79-80]).

PROPOSITION 1.1. Let f be a mapping of M into M' and p be a point of M.

(1) If $(f_*)_p$ is injective, there exist a local coordinate system $\{U;x^i\}$ of p and a local coordinate system $\{U';y^a\}$ of f(p) such that

$$y^i(f(q)) = x^i(q) \quad \text{for } q \in U,\ i = 1,\ldots,n.$$

In particular, f is a homeomorphism of U onto f(U);

(2) If $(f_*)_p$ is surjective, there exist a local coordinate system $\{U;x^i\}$ of p and a local coordinate system $\{U';y^a\}$ of f(p) such that

$$y^a(f(q)) = x^a(q) \quad \text{for } q \in U,\ a = 1,\ldots,n'.$$

In particular, $f : U \longrightarrow M'$ is open;

(3) If $(f_*)_p$ is a linear isomorphism of $T_p(M)$ onto $T_{f(p)}(M')$, then f defines a homeomorphism of a neighborhood U of p onto a neighborhood U' of f(p) and its inverse $f^{-1} : U' \longrightarrow U$ is also differentiable.

A mapping f of M into M' is called an *immersion* if $(f_*)_p$ is injective for every point p of M. Then M is called an *immersed submanifold* of M'. If the immersion f is injective, it is called an *imbedding* of M into M'. We say then that M (or the image f(M)) is an *imbedded submanifold* of M'. When there is no danger of confusion, we say simply that M is a submanifold of M' instead of that M is an immersed submanifold of M' or an imbedded submanifold of M'.

A homeomorphism f of M onto M' is called a *diffeomorphism* if both f and f^{-1} are differentiable. A diffeomorphism of M onto itself is called a *differentiable transformation* (or simply a *transformation*) of M. A transformation ϕ of M induces an automorphism ϕ^* of the algebra D(M) of differential forms on M and, in particular, an automorphism of the algebra $\mathfrak{F}(M)$: $(\phi^*f)(p) = f(\phi(p))$ for $f \in \mathfrak{F}(M)$, $p \in M$. From this we have an automorphism ϕ_* of the Lie algebra $\mathfrak{X}(M)$ by $(\phi_*X)_p = (\phi_*)_q(X_q)$, where $X \in \mathfrak{X}(M)$, $p = \phi(q)$. They are related by $\phi^*((\phi_*X)f) = X(\phi^*f)$. Although any mapping ϕ of M into M' carries a differential form ω' on M' into a differential form $\phi^*(\omega')$ on M, ϕ does not send a vector field on M into a vector field on M' in general. We say that a vector field X on M is ϕ-*related* to X' on M' if $(\phi_*)_p X_p = X'_{\phi(p)}$ for all p. If X and Y are ϕ-related to X' and Y' respectively, then [X,Y] is ϕ-related to [X',Y'].

A 1-*parameter group of* (*differentiable*) *transformations* of M is a mapping of $R \times M$ into M, $(t,p) \in R \times M \longrightarrow \phi_t(p) \in M$, which satisfies the following conditions:

(1) For each $t \in R$, $\phi_t : p \longrightarrow \phi_t(p)$ is a transformation of M;
(2) For all t, $s \in R$ and $p \in M$, $\phi_{t+s}(p) = \phi_t(\phi_s(p))$.

A curve $\tau(t)$ in M is called an *integral curve* of a vector field X if the vector $X_{\tau(t)}$ is tangent to $\tau(t)$ for every t. For any point p of M, there is a unique integral curve $\tau(t)$ of a vector field X, defined for $|t| < \varepsilon$ for some $\varepsilon > 0$, such that $p = \tau(0)$.

Each 1-parameter group of transformations ϕ_t induces a vector field X as follows. For any point $p \in M$, X_p is the vector tangent to the curve $\tau(t) = \phi_t(p)$, called the *orbit* of p, at $p = \phi_0(p)$. The orbit $\phi_t(p)$ is an integral curve of X starting at p. A *local* 1-*parameter group of local transformations* can be defined in the same way, except that $\phi_t(p)$ is defined only for t in a neighborhood of 0 and p in an open set of M. A local 1-parameter group of local transformations defined on $I_\varepsilon \times U$ is a mapping of $I_\varepsilon \times U$ into M which satisfies the following conditions:

(1') For each $t \in I_\varepsilon$, $\phi_t : p \longrightarrow \phi_t(p)$ is a diffeomorphism of U onto the open set $\phi_t(U)$ of M;

(2') If $t, s, t+s \in I_\varepsilon$ and if $p, \phi_s(p) \in U$, then $\phi_{t+s}(t)=\phi_t(\phi_s(p))$.

As in the case of a 1-parameter group of transformations, ϕ_t induces a vector field X defined on U. We also have the converse, that is, we can prove the following: Let X be a vector field on M. For each point p_0 of M, there exist a neighborhood U of p_0, a positive number ε and a local 1-parameter group of transformations $\phi_t : U \longrightarrow M$, $t \in I_\varepsilon$, which induces the given X (see Kobayashi-Nomizu [1;p.13]). We say that X generates a local 1-parameter group of local transformations ϕ_t in a neighborhood of p_0. If there exists a (global) 1-parameter group of transformations of M which induces X, then we say that X is *complete*. If $\phi_t(p)$ is defined on $I_\varepsilon \times M$ for some ε, then X is complete. Thus, if M is compact, every vector field X is complete.

We notice here that the differential ϕ_* of a transformation ϕ of M gives a linear isomorphism of $T_{\phi^{-1}(x)}(M)$ onto $T_x(M)$. This linear isomorphism can be extended to an isomorphism of the tensor algebra $T(\phi^{-1}(x))$ onto the tensor algebra $T(x)$, which we denote by the same ϕ. Given a tensor field K, we define a tensor field K by

$$(\phi K)_x = \phi(K_{\phi^{-1}(x)}).$$

In this way, every transformation ϕ of M induces an algebra automorphism of T which preserves type and commutes with contractions.

Let X be a vector field on M and ϕ_t a global 1-parameter group of transformations of M. For each t, ϕ_t is an automorphism of the algebra T. For any tensor field K on M, we set

$$(L_X K)_x = \lim_{t \to 0} \frac{1}{t}[K_x - (\phi_t K)_x].$$

The mapping L_X of T into itself which sends K into $L_X K$ is called the *Lie differentiation* with respect to X. It will be no difficulty in modifying the definition of Lie differentiation when X is not complete, that is, when ϕ_t is a local 1-parameter group of transformations generated by X. For the Lie differentiation we have

PROPOSITION 1.2. Lie differentiation L_X with respect to a vector field X satisfies the following conditions:

(a) L_X is a derivative of T, that is, it is linear and satisfies

$$L_X(K \otimes K') = (L_X K) \otimes K' + K \otimes (L_X K'), \quad K, K' \in T;$$

(b) L_X is type-preserving: $L_X(T^r_s) \subset T^r_s$;

(c) L_X commutes with every contraction of a tensor field;

(d) $L_X f = Xf$, $f \in \mathcal{F}(M)$;

(e) $L_X Y = [X,Y]$, $Y \in \mathfrak{X}(M)$.

Proof. It is clear L_X is linear. Moreover, we have

$$\begin{aligned} L_X(K \otimes K') &= \lim_{t\to 0} \frac{1}{t}[K \otimes K' - \phi_t(K \otimes K')] \\ &= \lim_{t\to 0} \frac{1}{t}[K \otimes K' - (\phi_t K) \otimes K'] \\ &\qquad + \lim_{t\to 0} \frac{1}{t}[(\phi_t K) \otimes K' - (\phi_t K) \otimes (\phi_t K')] \\ &= (\lim_{t\to 0} \frac{1}{t}[K - (\phi_t K)]) \otimes K' \\ &\qquad + \lim_{t\to 0}(\phi_t K) \otimes (\frac{1}{t}[K' - (\phi_t K')]) \\ &= (L_X K) \otimes K' + K \otimes (L_X K'). \end{aligned}$$

Since ϕ_t preserves type and commutes with contractions, so does L_X. If f is a function on M, then

$$(L_X f)(x) = \lim_{t\to 0} \frac{1}{t}[f(x) - f(\phi_t^{-1}(x)] = -\lim_{t\to 0} \frac{1}{t}[f(\phi_t^{-1}x) - f(x)].$$

We see that $\phi_t^{-1} = \phi_{-t}$ is a local 1-parameter group of local transformations generated by -X, and hence we have $L_X f = -(-X)f = Xf$.

Let f be a function on M. We can take a function g_t such that $f \cdot \phi_t = f + t \cdot g_t$ and $g_0 = Xf$ (see Kobayashi-Nomizu [1;p.15]). We put $p(t) = \phi_t^{-1}(p)$. Then

$$((\phi_t)_* Y)_p f = (Y(f\cdot\phi_t))_{p(t)} = (Yf)_{p(t)} + t(Yg_t)_{p(t)}$$

and

$$\lim_{t\to 0}\frac{1}{t}[Y - (\phi_t)_* Y]_p f = \lim_{t\to 0}\frac{1}{t}[(Yf)_p - (Yf)_{p(t)}] - \lim_{t\to 0}(Yg_t)_{p(t)}$$
$$= X_p(Yf) - Y_p g_0 = [X,Y]_p f.$$

Therefore we have (e). QED.

By a *derivative* of T, we shall mean a mapping of T into itself which satisfies conditions (a), (b) and (c) of Proposition 1.2.

We now prepare some basic formulas for latter use. For the proof see Kobayashi-Nomizu [1].

Let X and Y be vector fields on M. Then

$$L_{[X,Y]} = [L_X, L_Y].$$

Let K be a tensor field of type (1,r). Then we have

$$(L_X K)(Y_1,\ldots,Y_r) = [X, K(Y_1,\ldots,Y_r)] - \sum_{i=1}^{r} K(Y_1,\ldots,[X,Y_i],\ldots,Y_r).$$

We notice that a tensor field K is invariant by ϕ_t for every t if and only if $L_X K = 0$.

A *derivation* (resp. *skew-derivation*) of D(M) is a linear mapping A of D(M) into itself which satisfies

$$A(\omega\wedge\omega') = A\omega\wedge\omega' + \omega\wedge A\omega', \quad \omega, \omega' \in D(M)$$

(resp. $A(\omega\wedge\omega') = A\omega\wedge\omega' + (-1)^r\omega\wedge A\omega'$, $\omega\in D^r(M)$, $\omega'\in D(M)$).
A derivation or a skew-derivation A of D(M) is said to be of degree k if it maps $D^r(M)$ into $D^{r+k}(M)$ for every r. The exterior differentiation d is a skew-derivation of degree 1 and the Lie differentiation L_X is a derivation of degree 0. Indeed, the formula

$$(L_X\omega)(Y_1,\dots,Y_r) = X(\omega(Y_1,\dots,Y_r)) - \sum_{i=1}^{r} \omega(Y_1,\dots,[X,Y_i],\dots,Y_r)$$

implies that $L_X(D^r(M)) \subset D^r(M)$ and, for $\omega, \omega' \in D(M)$,

$$L_X(\omega \wedge \omega') = L_X\omega \wedge \omega' + \omega \wedge L_X\omega'.$$

Moreover, L_X commutes with d. If ω is an r-form, then

$$(d\omega)(X_0,X_1,\dots,X_r) = \frac{1}{r+1}\sum_{i=0}^{r} (-1)^i X_i(\omega(X_0,\dots,\hat{X}_i,\dots,X_r))$$

$$+ \frac{1}{r+1}\sum_{0\le i<j\le r} (-1)^{i+j}\omega([X_i,X_j],X_0,\dots,\hat{X}_i,\dots,\hat{X}_j,\dots,X_r),$$

where ^ means that the term is omitted. If ω is a 1-form, then

$$2(d\omega)(X,Y) = X(\omega(Y)) - Y(\omega(X)) - \omega([X,Y]).$$

If ω is a 2-form, then

$$3(d\omega)(X,Y,Z) = X(\omega(Y,Z)) + Y(\omega(Z,X)) + Z(\omega(X,Y))$$

$$- \omega([X,Y],Z) - \omega([Y,Z],X) - \omega([Z,X],Y).$$

Remark. Sometimes, we define $d\omega$ without the coefficients $1/(r+1)$ of the right hand side of the above equation to simplify the notation.

In the next place, we give the definition of a *distribution* on a manifold. A distribution S of dimension r on M is an assignment to each point p of M an r-dimensional subspace S_p of $T_p(M)$. S is called *differentiable* if every point p has a neighborhood U and r differentiable vector fields, say, $X_1,\dots,X_r$, which form a basis of S_q at every $q \in U$. The set $X_1,\dots,X_r$ is called a local basis for S in U. A vector field X is said to belong to S if $X_p \in S_p$ for all $p \in M$. S is called *involutive* if $[X,Y] \in S$ for any $X, Y \in S$. By a distribution we shall always mean a differentiable distribution.

A connected submanifold N of M is called an *integral manifold* of S if $f_*(T_p(N)) = S_p$ for all $p \in N$, f being the imbedding of N into M. If there is no other integral manifold of S which contains N, N is called a *maximal integral manifold* of S. The classical theorem of Frobenius can be formulated as follows (cf. Chevalley [1;p.94]).

PROPOSITION 1.3. Let S be an involutive distribution on a manifold M. Through every point $p \in M$, there passes a unique maximal integral manifold N(p) of S. Any integral manifold through p is an open submanifold of N(p).

2. CONNECTIONS AND COVARIANT DIFFERENTIATIONS

First of all we give the definition of affine connection on a manifold M.

Definition. An *affine connection* on a manifold M is a rule ∇ which assigns to each $X \in$ (M) a linear mapping ∇_X of the vector space $\mathfrak{X}(M)$ into itself satisfying the following two conditions:

(1) $\quad \nabla_{fX+gY} = f\nabla_X + g\nabla_Y$;

(2) $\quad \nabla_X(fY) = f\nabla_X Y + (Xf)Y$

for f, $g \in \mathfrak{F}(M)$, X, $Y \in \mathfrak{X}(M)$. The operator ∇_X is called the *covariant differentiation* with respect to X.

LEMMA 2.1. Suppose M has the affine connection ∇ and let U be an open submanifold of M. Let X, $Y \in \mathfrak{X}(M)$. If X or Y vanishes identically on U, then so does $\nabla_X Y$.

Proof. Suppose Y vanishes identically on U. Let $p \in U$ and $g \in \mathfrak{F}(M)$. To prove that $((\nabla_X Y)g)(p) = 0$, we select $f \in \mathfrak{F}(M)$ such that $f(p) = 0$ and $f = 1$ outside U. Then $fY = Y$ and

$$(\nabla_X Y)g = (\nabla_X(fY))g = (Xf)(Yg) + f(\nabla_X Y)g$$

which vanishes at p. The statement about X follows similarly. QED.

By Lemma 2.1, an affine connection ∇ on M induces an affine connection on an arbitrary open submanifold U of M, which will be denoted by the same ∇. In fact, let X, Y be two vector fields on U. For each $p \in U$ there exist vector fields X', Y' on M which agree with X and Y in an open neighborhood V of p. We then put $(\nabla_X Y)_q = (\nabla_{X'} Y')_q$ for $q \in V$. By Lemma 2.1, the right-hand side of this equation is independent of the choice of X', Y'. It follows immediately that the rule $\nabla : X \longrightarrow \nabla_X$ $(X \in \mathfrak{X}(U))$ is an affine connection on U.

In particular, suppose U is a coordinate neughborhood with coordinate system $\{x^1,\ldots,x^n\}$. To simplyfy the notation, we write ∇_i instead of $\nabla_{\frac{\partial}{\partial x^i}}$. We define the functions Γ^k_{ij} on U bu

$$(2.1) \qquad \nabla_i\left(\frac{\partial}{\partial x^j}\right) = \sum_k \Gamma^k_{ij}\frac{\partial}{\partial x^k}.$$

If $\{y^1,\ldots,y^n\}$ is another coordinate system valid on U, we get another set of functions $\bar{\Gamma}^c_{ab}$ by

$$\nabla_a\left(\frac{\partial}{\partial y^b}\right) = \sum_c \bar{\Gamma}^c_{ab}\frac{\partial}{\partial y^c}.$$

From (1) and (2) in the definition of affine connection we find easily

$$(2.2) \qquad \bar{\Gamma}^c_{ab} = \sum_{i,j,k} \frac{\partial x^i}{\partial y^a}\frac{\partial x^j}{\partial y^b}\frac{\partial y^c}{\partial x^k}\Gamma^k_{ij} + \sum_j \frac{\partial^2 x^j}{\partial y^a \partial y^b}\frac{\partial y^c}{\partial x^j}.$$

On the other hand, suppose that there is given an open covering of M by open coordinate neighborhoods U and in each U a system of functions Γ^k_{ij} such that (2.2) holds whenever two of these neighborhoods overlap. Then we can define ∇_i by (2.1) and thus we get an affine connection ∇ in each U. We define an affine connection $\bar{\nabla}$ on M as follows: Let $X, Y \in \mathfrak{X}(M)$ and $p \in M$. If U is a coordinate neighborhood containing p, let

$$(\bar{\nabla}_X Y)_p = (\nabla_{X'} Y')_p$$

if X' and Y' are the vector fields on U induced by X and Y, respectively. Then $\bar{\nabla}$ is an affine connection on M which on each U induces the connection .

Let $\{x^1,\ldots,x^n\}$ be a coordinate system valid on U of p. On the set U we have $X = \Sigma f_i(\partial/\partial x^i)$ where $f_i \in \mathfrak{F}(U)$ and $f_i(p) = 0$ $(1 \leq i \leq n)$. Using Lemma 2.1 we find $(\nabla_X Y)_p = \Sigma f_i(p)(\nabla_i Y)_p = 0$.

LEMMA 2.2. Let $X, Y \in \mathfrak{X}(M)$. If X vanishes at a point p of M, then so does $\nabla_X Y$.

Definition. Suppose ∇ is an affine connection on M and f is a diffeomorphism of M. A new affine connection ∇' can be defined on M by

$$\nabla'_X Y = f_*^{-1}(\nabla_{f_*X} f_* Y), \quad X, Y \in \mathfrak{X}(M).$$

The affine connection ∇ is called *invariant* under f if $\nabla' = \nabla$. In this case f is called an *affine transformation* of M. Similarly we can define an affine transformation of one manifold onto another.

Let $\gamma : t \longrightarrow \gamma(t)$ $(t \in I)$ be a curve in M defined on an open interval $I \in R$. Differentiation with respect to the parameter will often denoted by a dot $(\dot{})$. Let J be a closed subinterval of I such that $\gamma_J : t \longrightarrow \gamma(t)$ $(t \in J)$ has no double points and such that $\gamma(J)$ is contained in a coordinate neighborhood U. We need the following

LEMMA 2.3. Let g(t) be a differentiable function on an open interval containing J. Then there exists a function $G \in \mathfrak{F}(M)$ such that

$$G(\gamma(t)) = g(t) \quad (t \in J).$$

We put $X(t) = \dot{\gamma}(t)$ $(t \in I)$. By Lemma 2.3 it is easy to see that there exist vector fields $X, Y \in \mathfrak{X}(M)$ such that ($Y(t)$ being an associated vector to $t \in I$)

$$X_{\gamma(t)} = X(t), \quad Y_{\gamma(t)} = Y(t) \quad (t \in J).$$

The family $Y(t)$ $(t \in J)$ is said to be *parallel* with respect to γ_J

(or parallel along γ_J) if

$$(\nabla_X Y)_{\gamma(t)} = 0 \quad \text{for all } t \in J. \tag{2.3}$$

We show that this definition is independent of the choice of X and Y. We express (2.3) in the coordinates $\{x^1,\ldots,x^n\}$. Then X and Y are given by

$$X = \sum_i X^i \frac{\partial}{\partial x^i}, \qquad Y = \sum_j Y^j \frac{\partial}{\partial x^j} \qquad \text{on } U.$$

For simplicity we put $x^i(t) = x^i(\gamma(t))$, $X^i(t) = X^i(\gamma(t))$ and $Y^i(t) = Y^i(\gamma(t))$, $t \in J$, $1 \leq i \leq n$. Then $X^i(t) = \dot{x}^i(t)$. Since

$$\nabla_X Y = \sum_k \Big(\sum_i X^i \frac{\partial Y^k}{\partial x^i} + \sum_{i,j} X^i Y^j \Gamma^k_{ij}\Big) \frac{\partial}{\partial x^k} \qquad \text{on } U,$$

we obtain

$$\frac{dY^k}{dt} + \sum_{i,j} \Gamma^k_{ij} \frac{dx^i}{dt} Y^j = 0 \quad (t \in J). \tag{2.4}$$

This equation involves X and Y only through their values on the curvae. Consequently, condition (2.3) for parallelism is independent of the choice of X and Y. It is well now obvious how to define parallelism with respect to any finite curve segment γ_J and finally with respect to the entire curve γ.

Definition. Let $\gamma : t \longrightarrow \gamma(t)$ $(t \in I)$ be a curve in M. The curve γ is called a *geodesic* if the vector field $X(t) = \dot{\gamma}(t)$ defined along γ is parallel with respect to γ, that is, $\nabla_X X = 0$ for all t. A geodesic γ is called maximal if it is not a proper restriction of any geodesic.

From (2.4), if γ_J is a geodesic segment, then

$$\frac{d^2x^k}{dt^2} + \sum_{i,j} \Gamma^k_{ij} \frac{dx^i}{dt} \frac{dx^j}{dt} = 0 \quad (t \in J). \tag{2.5}$$

If we change the parameter on the geodesic and put $t = f(s)$, $(f'(s) \neq 0)$, then we get a new curve $s \longrightarrow \gamma_J(f(s))$. This curve is a geodesic if and only if f is a linear function, as (2.5) shows.

We notice that the two following propositions hold.

PROPOSITION 2.1. Let p and q be two points in M and γ a curve segment from p to q. The parallelism τ with respect to γ induces an isomorphism $T_p(M)$ onto $T_q(M)$.

PROPOSITION 2.2. Let M be a manifold with an affine connection. Let p be any point in M and let $X \neq 0$ in $T_p(M)$. Then there exists a unique maximal geodesic $\gamma : t \longrightarrow \gamma(t)$ in M such that $\gamma(0) = p$ and $\dot{\gamma}(0) = X$.

The geodesic with properties in Proposition 2.2 will be denoted by γ_X. If $X = 0$, we put $\gamma_X(t) = p$ for all $t \in R$. We then have (cf. Helgason [1])

THEOREM 2.1. Let M be a manifold with an affine connection. Let p be any point in M. Then there exists an open neighborhood N_0 of 0 in $T_p(M)$ and an open neighborhood N_p of p in M such that the mapping $X \longrightarrow \gamma_X(1)$ is a diffeomorphism of N_0 onto N_p.

The mapping $X \longrightarrow \gamma_X(1)$ descrived in Theorem 2.1 is called the *exponential mapping* at p and will be denoted by Exp (or Exp_p).

Let M be a manifold with an affine connection and p be a point in M. An open neighborhood N_0 of the origin $T_p(M)$ is said to be *normal* if : (1) Exp is a diffeomorphism of N_0 onto an open neighborhood N_p of p in M; (2) if $X \in N_0$, $0 \leq t \leq 1$, then $tX \in N_0$.

A neighborhood N_p of p in M is called a *normal beighborhood* of p if $N_p = \mathrm{Exp} N_0$, where N_0 is a normal neighborhood of 0 in $T_p(M)$. Assuming this to be the case, and letting $X_1,\ldots,X_n$ denote some basis of $T_p(M)$, the inverse mapping

$$\mathrm{Exp}(a^1X_1 + \ldots + a^nX_n) \longrightarrow (a^1,\ldots,a^n)$$

of N_p into R^n is called a system of normal coordinates at p.

We now give a useful refinement of Theorem 2.1 (cf. Helgason [1]).

THEOREM 2.2. Let M be a manifold with an affine connection. Then each point $p \in M$ has a normal neighborhood N_p which is a normal neighborhood of each of its points.

In the next place, we shall define covariant derivatives of arbitrary tensor fields. We first prove

THEOREM 2.3. Let M be a manifold with an affine connection. Let $p \in M$ and let X, Y be two vector fields on M. Assume $X_p \neq 0$. Let $s \longrightarrow \phi(s)$ be an integral curve of X through $p = \phi(0)$ and τ_t the parallel translation from p to $\phi(t)$ with respect to the curve ϕ. Then

$$(\nabla_X Y)_p = \lim_{s\to 0} \frac{1}{s}(\tau_s^{-1} Y_{\phi(s)} - Y_p).$$

Proof. Consider a fixed $s > 0$ and the family $Z_{\phi(t)}$ $(0 \leq t \leq s)$ which is parallel with respect to the curve ϕ such that $Z_{\phi(0)} = \tau_s^{-1} Y_{\phi(s)}$. We can put

$$Z_{\phi(t)} = \sum_i Z^i(t)\left(\frac{\partial}{\partial x^i}\right)_{\phi(t)}, \quad Y_{\phi(t)} = \sum_i Y^i(t)\left(\frac{\partial}{\partial x^i}\right)_{\phi(t)}$$

and we have the relations

$$\dot{Z}^k(t) + \sum_{i,j} \Gamma^k_{ij} \dot{x}^i(t) Z^j(t) = 0 \quad (0 \leq t \leq s),$$

$$Z^k(s) = Y^k(s) \quad (1 \leq k \leq n).$$

By the mean value theorem $Z^k(s) = Z^k(0) + s\dot{Z}^k(t^*)$ for a suitable number t^* between 0 and s. Hence the k-th component of $(1/s)(\tau_s^{-1}Y_{(s)} - Y_p)$ is

$$\frac{1}{s}(Z^k(0) - Y^k(0)) = \frac{1}{s}(Z^k(s) - s\dot{Z}^k(t^*) - Y^k(0))$$

$$= \sum_{i,j} \Gamma^k_{ij}(\phi(t^*))\dot{x}^i(t^*)Z^j(t^*) + \frac{1}{s}(Y^k(s) - Y^k(0)).$$

As s ——> 0 this expression has the limit

$$\frac{dY^k}{ds} + \sum_{i,j} \Gamma^k_{ij} \frac{dx^i}{ds} Y^j.$$

Let this last expression be denoted by A_k. It was shown earlier that

$$(\nabla_X Y)_p = \sum_k A_k \left(\frac{\partial}{\partial x^k}\right)_p.$$

This proves the theorem. QED.

By using Theorem 2.3 it is now possible to define covariant derivatives of arbitrary tensor fields.

Let p and q be two points of M and γ a curve segment in M from p to q. Let τ be the parallel translation along γ. If $F \in T^*_p(M)$, we define $\tau \cdot F \in T^*_q(M)$ by $(\tau \cdot F)(A) = F(\tau^{-1} \cdot A)$ for each $A \in T_q(M)$. If K is a tensor field on M of type (r,s), we define $\tau \cdot K_q \in T^r_s(q)$ by

$$(\tau \cdot K_p)(F_1, \ldots, F_r, A_1, \ldots, A_s) = K_p(\tau^{-1}F_1, \ldots, \tau^{-1}F_r, \tau^{-1}A_1, \ldots, \tau^{-1}A_s)$$

for $A_i \in T_q(M)$ and $F_j \in T^*_q(M)$. Let $X \in \mathfrak{X}(M)$ and p be any point in M where $X_p \neq 0$. With the notation of Theorem 2.3 we put

$$(2.6) \qquad (\nabla_X K)_p = \lim_{s \to 0} \frac{1}{s}(\tau_s^{-1} K_{\phi(s)} - K_p).$$

For each point $q \in M$ where $X_q = 0$ we put $(\nabla_X K)_q = 0$ in accordance with Lemma 2.2. For a functionf on M we put

$$(\nabla_X f)_p = \lim_{s \to 0} \frac{1}{s}(f(\phi(s)) - f(p)),$$

if $X_p \neq 0$, otherwise we put $(\nabla_X f)_p = 0$. Then we have $\nabla_X f = Xf$. Finally ∇_X is extended to a linear mapping of T into itself.

THEOREM 2.4. The operator ∇_X has the following properties:

(1) ∇_X is a derivation of the mixed tensor algebra T;

(2) ∇_X preserves type of tensors;

(3) ∇_X commutes with contractions.

We now give the structure equations of Cartan. For this purpose we define the torsion tensor field and the curvature tensor field.

On a manifold M with an affine connection, we put

$$T(X,Y) = \nabla_X Y - \nabla_Y X - [X,Y],$$

$$R(X,Y) = \nabla_X \nabla_Y - \nabla_Y \nabla_X - \nabla_{[X,Y]},$$

where X, Y are vector fields on M. Note that $T(X,Y) = -T(Y,X)$ and $R(X,Y) = -R(Y,X)$. It is easy to verify that $T(fX,gY) = fgT(X,Y)$ and $R(fX,gY)hZ = fghR(X,Y)Z$ for all $f, g, h \in \mathfrak{F}(M)$, $X, Y, Z \in \mathfrak{X}(M)$. The mapping $(\omega,X,Y) \longrightarrow \omega(T(X,Y))$ is an $\mathfrak{F}(M)$-multilinear mapping $\mathfrak{X}(M)^* \times \mathfrak{X}(M) \times \mathfrak{X}(M)$ into $\mathfrak{F}(M)$ and is an element of $T^1_2(M)$. This element is called the*torsion tensor field* and is also denoted by T. Similarly, the mapping $(\omega,Z,X,Y) \longrightarrow \omega(R(X,Y)Z)$ is an $\mathfrak{F}(M)$-multilinear mapping $\mathfrak{X}(M)^* \times \mathfrak{X}(M) \times \mathfrak{X}(M) \times \mathfrak{X}(M)$ into $\mathfrak{F}(M)$ and therefore is an element of $T^1_3(M)$. This element is called the *curvature tensor field* and is denoted by R. The tensor fields T and R is of type (1,2) and (1,3) respectively.

Let p be a point of M and suppose $X_1,\ldots,X_n$ is a basis for the vector fields in some open neighborhood N_p of p, that is, each vector field X on N_p can be written as $X = \Sigma_i f_i X_i$ where $f_i \in \mathfrak{F}(N_p)$. We define the functions Γ^k_{ij}, T^k_{ij}, R^k_{lij} on N_p by the formulas

$$\nabla_{X_i} X_j = \sum_k \Gamma^k_{ij} X_k,$$

$$T(X_i,X_j) = \sum_k T^k_{ij} X_k,$$

$$R(X_i,X_j)X_k = \sum_l R^l_{kij} X_l.$$

Let ω^i, ω^i_j $(1 \leq i,j \leq n)$ be the 1-forms on N_p determined by

$$\omega^i(X_j) = \delta^i_j, \qquad \omega^i_j = \sum_k \Gamma^i_{kj}\omega^k.$$

That is, we put

$$\omega^i_j(X_k) = \Gamma^i_{kj}.$$

Thus the forms ω^i_j determine the functions Γ^i_{kj} on N_p and thereby the connection ∇. On the other hand, as the next theorem shows, the forms ω^i_j are descrived by the torsion and curvature tensor fields.

THEOREM 2.5 (the structure equations of Cartan).

$$d\omega^i = -\sum_j \omega^i_j \wedge \omega^j + \tfrac{1}{2}\sum_{j,k} T^i_{jk}\omega^j \wedge \omega^k, \tag{2.7}$$

$$d\omega^i_j = -\sum_k \omega^i_k \wedge \omega^k_j + \tfrac{1}{2}\sum_{k,l} R^i_{jkl}\omega^k \wedge \omega^l. \tag{2.8}$$

Proof. If we define the functions c^i_{jk} by $[X_j,X_k] = \Sigma_i c^i_{jk}X_i$, we obtain

$$d\omega^i(X_j,X_k) = \tfrac{1}{2}[X_j\omega^i(X_k)-X_k\omega^i(X_j)-\omega^i([X_j,X_k])] = -\tfrac{1}{2}c^i_{jk}.$$

As for the right-hand side of this equation, we have

$$\begin{aligned} T(X_j,X_k) &= \nabla_{X_j}X_k - \nabla_{X_k}X_j - [X_j,X_k] \\ &= \sum_i(\Gamma^i_{jk} - \Gamma^i_{kj} - c^i_{jk})X_i, \end{aligned}$$

from which

$$T^i_{jk} = \Gamma^i_{jk} - \Gamma^i_{kj} - c^i_{jk}.$$

We also have

$$\begin{aligned} (-\sum_l \omega^i_l \wedge \omega^l)(X_j,X_k) &= -\tfrac{1}{2}\sum_l[\omega^i_l(X_j)\omega^l(X_k) - \omega^l(X_j)\omega^i_l(X_k)] \\ &= \tfrac{1}{2}(\Gamma^i_{kj} - \Gamma^i_{jk}). \end{aligned}$$

From these equations we have (2.7).

Similarly, we find

$$R^i_{jkl} = \sum_p (\Gamma^p_{lj}\Gamma^i_{kp} - \Gamma^p_{kj}\Gamma^i_{lp}) + X_k\Gamma^i_{lj} - X_l\Gamma^i_{kj} - c^p_{kl}\Gamma^i_{pj},$$

$$d\omega^i_j(X_k,X_l) = \tfrac{1}{2}(X_k\Gamma^i_{lj} - X_l\Gamma^i_{kj} - c^p_{kl}\Gamma^i_{pj}),$$

$$-\omega^i_p \wedge \omega^p_j(X_k,X_l) = -\tfrac{1}{2}(\Gamma^p_{lj}\Gamma^i_{kp} - \Gamma^p_{kj}\Gamma^i_{lp}).$$

From these equations we have (2.8). QED.

We now define a *Riemannian metric* on M. It is a tensor field g of type (0,2) which satisfies the following two conditions:

(i) It is symmetric: $g(X,Y) = g(Y,X)$ for any $X,Y \in \mathfrak{X}(M)$;

(ii) It is positive-definite: $g(X,X) \geq 0$ for every $X \in \mathfrak{X}(M)$ and $g(X,X) = 0$ if and only if $X = 0$.

A manifold M with Riemannian metric g is called a *Riemannian manifold*. A Riemannian metric gives rise to an inner product on each tangent space $T_x(M)$ to M at x. Let $\{x^1,\ldots,x^n\}$ be a local coordinate system in M. The components g_{ij} of g with respect to this local coordinate system are given by

$$g_{ij} = g(\partial/\partial x^i, \partial/\partial x^j), \qquad i,j = 1,\ldots,n.$$

We call g_{ij} the *covariant components* of g. The *contravariant components* g^{ij} of g are defined by

$$g^{ij} = g(dx^i, dx^j), \qquad i,j = 1,\ldots,n.$$

We have then

$$g_{ij}g^{jk} = \delta^k_i, \qquad \delta^k_i = \begin{cases} 1, & k = i \\ 0, & k \neq i \end{cases},$$

where, here and in the sequel, if we need, we use the *Einstein* convention, that is, repeated indices, one upper index and the other lower index, denotes summation over its range. If X^i are components of a vector field X with respect to $\{x^1,\ldots,x^n\}$, that is, $X = X^i(\partial/\partial x^i)$, then the components X_i of the corresponding covector or the corresponding 1-form are related to X^i by

$$X^i = g^{ij}X_j, \qquad X_i = g_{ij}X^j.$$

The inner product g in the tangent space $T_x(M)$ and in its dual space $T^*_x(M)$ can be extended to an inner product, denoted also by g, in the tensor space T^r_s at x for each type (r,s). If K and L are tensors at x of type (r,s) with components

$$K^{i_1\ldots i_r}_{j_1\ldots j_s} \quad \text{and} \quad L^{i_1\ldots i_r}_{j_1\ldots j_s}$$

with respect to $\{x^1,\ldots,x^n\}$, then the inner product g(K,L) of K and L is defined to be

$$g(K,L) = g_{i_1k_1}\cdots g_{i_rk_r}g^{j_1t_1}\cdots g^{j_st_s}K^{i_1\ldots i_r}_{j_1\ldots j_s}L^{k_1\ldots k_r}_{t_1\ldots t_s}.$$

We put

$$|K| = g(K,K)^{1/2},$$

which is called the length of the tensor field K with respect to g.

On a Riemannian manifold M, the arc length of a differentiable curve $\gamma: t \longrightarrow \gamma(t)$, $a \leq t \leq b$, is defined by

$$L(\gamma) = \int_b^a g(\dot{\gamma}(t),\dot{\gamma}(t))^{1/2}dt.$$

This definition can be generalized to a piecewise differentiable curve in an obvious manner. The *distance* d(x,y) between two points x and y of M is defined by the infinimum of the lengths of all piecewise differentiable curves joining x and y. Then we have

$$d(x,y) = d(y,x), \quad d(x,y) + d(y,z) \geq d(x,z), \quad d(x,y) \geq 0$$

$$d(x,y) = 0 \text{ if and only if } x = y.$$

We can see that the topology defined by the distance function (metric) d is the same as the manifold topology of M.

THEOREM 2.6. On a Riemannian manifold M there exists one and only one affine connection satisfying the following two conditions:

(1) The torsion tensor T vanishes, i.e.,

$$T(X,Y) = \nabla_X Y - \nabla_Y X - [X,Y] = 0;$$

(2) g is parallel, i.e., $\nabla_X g = 0$.

Proof. Existence: Given vector fields X and Y on M, we define $\nabla_X Y$ by setting

$$\text{(2.9)} \qquad 2g(\nabla_X Y,Z) = Xg(Y,Z) + Yg(X,Z) - Zg(X,Y) + g([X,Y],Z) + g([Z,X],Y) + g(X,[Z,Y])$$

for any vector field Z on M. Then the mapping $(X,Y) \longrightarrow \nabla_X Y$ defines an affine connection on M. From the above definition of $\nabla_X Y$, we have $T(X,Y) = 0$ and

$$Xg(Y,Z) = g(\nabla_X Y,Z) + g(Y,\nabla_X Z),$$

which shows that $\nabla_X g = 0$, that is, ∇ is a *metric connection* on M.

Uniqueness: By a straightforward computation, we can see that, if $\nabla_X Y$ satisfies $\nabla_X g = 0$ and $T(X,Y) = 0$, then it satisfies the equation which defines $\nabla_X Y$. QED.

The connection ∇ given by (2.9) is called the *Riemannian connection* (sometimes called the *Levi-Civita connection*).

Putting $X = \partial/\partial x^j$, $Y = \partial/\partial x^i$ and $Z = \partial/\partial x^k$ in (2.9), the components Γ^i_{jk} of the Riemannian connection with respect to a local coordinate system $\{x^1,\ldots,x^n\}$ are given by

$$\sum_l g_{lk}\Gamma^l_{ji} = \tfrac{1}{2}\left(\frac{\partial g_{ki}}{\partial x^j} + \frac{\partial g_{jk}}{\partial x^i} - \frac{\partial g_{ji}}{\partial x^k}\right).$$

Let M and M' be Riemannian manifolds with Riemannian metrices g and g' respectively. A mapping $f : M \longrightarrow M'$ is called *isometric* at a point x of M if $g(X,Y) = g'(f_*X,f_*Y)$ for all $X,Y \in T_x(M)$. In this case, f_* is injective at x, because $f_*X = 0$ implies $X = 0$. A mapping f which is isometric at every point of M is thus an immersion, which we call an *isometric immersion*. If, moreover, f is 1 : 1, then it is called an *isometric imbedding* of M into M'. If f maps M 1 : 1 onto M', then f is called an *isometry* of M onto M'. In this case, the differential of the isometry f commutes with the parallel translation. Moreover, if f is an isometric immersion of M into M' and if f(M) is open in M', then the differential of f commutes the parallel translation and every geodesic of M is mapped by f into a geodesic of M'.

A Riemannian manifold M or a Riemannian metric g on M is said to be *complete* if the metric function d is complete, that is, all Cauchy sequences converge. It is well-known that the following conditions on M are equivalent:

(1) M is complete;

(2) Every bounded subset of M with respect to d is relatively compact;

(3) All geodesic arc can be extended in two directions indifinitely with respect to the arc length.

It is also well-known that any two points x and y in M can be joined by a geodesic arc whose length is equal to d(x,y) (Hopf-Rinow [1], de Rham [2]). We can also see that every compact Riemannian manifold is complete.

3. SECTIONAL CURVATURE

Let M be an n-dimensional Riemannian manifold with metric tensor field g. We see that

$$R(X,Y) = \nabla_X \nabla_Y - \nabla_Y \nabla_X - \nabla_{[X,Y]}$$

is an endomorphism of $T_x(M)$. We call this R(X,Y) the *(Riemannian) curvature transformation* of $T_x(M)$ determined by X and Y. It follows that R is a tensor field of type (1,3) such that $R(X,Y) = -R(Y,X)$, which will be called a *(Riemannian) curvature tensor field* (or simply, *curvature*) of type (1,3).

For any vector fields X, Y and Z, using T = 0, we obtain

$$\begin{aligned} &R(X,Y)Z + R(Y,Z)X + R(Z,X)Y \\ &\quad = \nabla_X[Y,Z] - \nabla_{[Y,Z]}X + \nabla_Y[Z,X] - \nabla_{[Z,X]}Y + \nabla_Z[X,Y] - \nabla_{[X,Y]}Z \\ &\quad = [X,[Y,Z]] + [Y,[Z,X]] + [Z,[X,Y]] = 0 \end{aligned}$$

by the Jacobi identity. Thus we have

$$R(X,Y)Z + R(Y,Z)X + R(Z,X)Y = 0. \tag{3.1}$$

We call (3.1), *Bianchi's* 1*st identity*. We have then, by a straightforward computation, the following *Bianchi's* 2*nd identity*:

$$(\nabla_X R)(Y,Z) + (\nabla_Y R)(Z,X) + (\nabla_Z R)(X,Y) = 0. \tag{3.2}$$

We now give algebraic preliminaries for a quadrilinear mapping on a real vector space.

Let V be an n-dimensional real vector space and $B : V \times V \times V \times V \longrightarrow R$ a quadrilinear mapping with the following three properties:

(a) $B(v_1,v_2,v_3,v_4) = -B(v_2,v_1,v_3,v_4)$,

(b) $B(v_1,v_2,v_4,v_4) = -B(v_1,v_2,v_4,v_3)$,

(c) $B(v_1,v_2,v_3,v_4) + B(v_1,v_3,v_4,v_2) + B(v_1,v_4,v_2,v_3) = 0$.

LEMMA 3.1. If B possesses the above three properties, the it possesses also the following fourth property:

(d) $B(v_1,v_2,v_3,v_4) = B(v_3,v_4,v_1,v_2)$.

Proof. Putting the left hand side of (c) by $S(v_1,v_2,v_3,v_4)$, we obtain

$$\begin{aligned} 0 &= S(v_1,v_2,v_3,v_4) - S(v_2,v_3,v_4,v_1) - S(v_3,v_4,v_1,v_2) \\ &\quad + S(v_4,v_1,v_2,v_3) \\ &= B(v_1,v_2,v_3,v_4) - B(v_2,v_1,v_3,v_4) - B(v_3,v_4,v_1,v_2) \\ &\quad + B(v_4,v_3,v_1,v_2). \end{aligned}$$

By applying (a) and (b), we have (d). QED.

LEMMA 3.2. Let B and T be two quadrilinear mappings with the properties (a), (b) and (c). If

$$B(v_1,v_2,v_1,v_2) = T(v_1,v_2,v_1,v_2) \text{ for all } v_1,\ v_2 \in V,$$

then B = T.

Proof. We may assume that T = 0; consider B − T and 0 instead of B and T. We assume therefore that $B(v_1,v_2,v_1,v_2) = 0$ for all v_1, v_2 of V. We have

$$0 = \tfrac{1}{2}B(v_1,v_2+v_4,v_1,v_2+v_4) = B(v_1,v_2,v_1,v_4).$$

From this we obtain

$$0 = B(v_1+v_3,v_2,v_1+v_3,v_4) = B(v_1,v_2,v_3,v_4) + B(v_3,v_2,v_1,v_4).$$

Now, by applying (d) and then (b), we have

$$0 = B(v_1,v_2,v_3,v_4) + B(v_1,v_4,v_3,v_2)$$

$$= B(v_1,v_2,v_3,v_4) - B(v_1,v_4,v_2,v_3).$$

Hence,

$$B(v_1,v_2,v_3,v_4) = B(v_1,v_4,v_2,v_3).$$

Replacing v_2, v_3, v_4 by v_3, v_4, v_2, respectively, we obtain

$$B(v_1,v_2,v_3,v_4) = B(v_1,v_3,v_4,v_2).$$

From these equations we obtain

$$3B(v_1,v_2,v_3,v_4) = B(v_1,v_2,v_3,v_4) + B(v_1,v_3,v_4,v_2) + B(v_1,v_4,v_2,v_3),$$

where the right hand side vanishes by (c). Thus we have

$$B(v_1,v_2,v_3,v_4) = 0.$$

This proves our assertion. QED.

Besides a quadrilinear mapping B, we consider an inner product on V, which will be denoted by (,). Let p be a plane, that is, 2-dimensional subspace in V and let v_1 and v_2 be an orthonormal basis for p.

We put

$$K(p) = B(v_1,v_2,v_1,v_2),$$

which is independent of the choice of an orthonormal basis for p.

PROPOSITION 3.1. If v_1, v_2 is a basis (not necessarily orthonormal) of a plane p in V, then

$$K(p) = B(v_1,v_2,v_1,v_2)/[(v_1,v_1)(v_2,v_2)-(v_1,v_2)^2].$$

Proof. We obtain the formula making use of the following orthonormal basis for p:

$$\frac{v_1}{(v_1,v_1)^{1/2}}, \qquad \frac{1}{a}[(v_1,v_1)v_2 - (v_1,v_2)v_1],$$

where we have put

$$a = [(v_1,v_1)((v_1,v_1)(v_2,v_2)-(v_1,v_2)^2)]^{1/2}. \qquad \text{QED.}$$

We put

$$R_1(v_1,v_2,v_3,v_4) = (v_1,v_3)(v_2,v_4) - (v_2,v_3)(v_4,v_1).$$

Then R_1 is a quadrilinear mapping having the properties (a), (b) and (c). If v_1, v_2 is an orthonormal basis for p, we obtain

$$K_1(p) = R_1(v_1,v_2,v_1,v_2) = 1.$$

PROPOSITION 3.2. Let B be a quadrilinear mapping with properties (a), (b) and (c). If K(p) = c for all planes p, then $B = cR_1$.

Proof. By Proposition 3.1 we have

$$B(v_1,v_2,v_1,v_2) = cR_1(v_1,v_2,v_1,v_2)$$

for all v_1, $v_2 \in V$. Applying Lemma 3.2 to B and cR_1, we conclude $B = cR_1$. QED.

We apply these results to the tangent space $T_x(M)$ of an n-dimensional Riemannian manifold M.

The *Riemannian curvature tensor field* of covariant degree 4 (of type (0,4)) of M, which denoted also R, is defined by

$$R(X_1,X_2,X_3,X_4) = g(R(X_3,X_4)X_2,X_1),\ X_i \in T_x(M),\ i=1,..,4.$$

Obviously, R is a quadrilinear mapping $T_x(M) \times T_x(M) \times T_x(M) \times T_x(M) \longrightarrow R$ at each $x \in M$ with properties (a), (b) and (c).

For each plane p in the tangent space $T_x(M)$, the *sectional curvature* K(p) for p is defined by

$$K(p) = R(X_1,X_2,X_1,X_2) = g(R(X_1,X_2)X_2,X_1),$$

where X_1, X_2 is an orthonormal basis for p. K(p) is independent of the choice of an orthonormal basis X_1, X_2. Lemma 3.2 implies that the set of values of K(p) for all plane p in $T_x(M)$ determines the Riemannian curvature tensor at x.

If K(p) is a constant for all plane p in $T_x(M)$ and for all points x of M, then M is called a *space of constant curvature*. A Riemannian manifold of constant curvature is called a *space form*. Sometimes, a space form is defined as a complete simply connected Riemannian manifold of constant curvature. The following theorem due to Schur [1] is well-known.

THEOREM 3.1. Let M be a connected Riemannian manifold of dimension > 2. If the sectional curvature K(p) depends only on the point x, then M is a space of constant curvature.

Proof. For any W, Z, X, Y $\in T_x(M)$, we put

$$R_1(W,Z,X,Y) = g(W,X)g(Z,Y) - g(Z,X)g(Y,W).$$

From Proposition 3.2 we have $R = kR_1$, where k is a function on M. Since g is parallel, so R_1. Hence

$$(\nabla_U R)(W,Z,X,Y) = (\nabla_U k)R_1(W,Z,X,Y)$$

for any $U \in T_x(M)$. Thus we have

$$((\nabla_U R)(X,Y))Z = (Uk)[g(Z,Y)X - g(Z,X)Y].$$

Therefore, Bianchi's second identity implies

$$0 = (Uk)[g(Z,Y)X - g(Z,X)Y] + (Xk)[g(Z,U)Y - g(Z,Y)U] + (Yk)[g(Z,X)U - g(Z,U)X].$$

For an arbitrary X, we choose Y, Z and U in such a way that X, Y and Z are mutually orthogonal and that U = Z with g(Z,Z) = 1. This is possible since n > 2. Then we obtain

$$(Xk)Y - (Yk)X = 0.$$

Since X and Y are linearly independent, we have Xk = Yk = 0. This shows that k is a constant. QED.

In the course of the proof of Theorem 3.1 we have

THEOREM 3.2. For a space of constant curvature k, we have

$$R(X,Y)Z = k[g(Y,Z)X - g(X,Z)Y].$$

If R^i_{jkl} and g_{ij} are the components of the curvature tensor and the metric tensor with respect to a local coordinate system, then the components R_{ijkl} of the Riemannian curvature tensor of type (0,4) are given by

$$R_{ijkl} = g_{im}R^m_{jkl}.$$

If M is a space of constant curvature k, then

$$R_{ijkl} = k(g_{ik}g_{jl} - g_{il}g_{jk}) \text{ or } R^i_{jkl} = k(\delta^i_k g_{jl} - \delta^i_l g_{jk}).$$

If we take an orthonormal frame field, the we have $g_{ij} = \delta_{ij}$ at x of M, and hence

$$R_{ijkl} = R^i_{jkl} = k(\delta_{ik}\delta_{jl} - \delta_{il}\delta_{jk}).$$

In view of a local coordinate system, we obtain

$$R_{ijkl} + R_{jikl} = 0, \quad R_{ijkl} + R_{ijlk} = 0,$$

$$R_{ijkl} + R_{iklj} + R_{iljk} = 0, \quad R_{ijkl} - R_{klij} = 0.$$

If $E_1,\ldots,E_n$ are local orthonormal vector fields, then

$$(3.3) \qquad S(X,Y) = \sum_{i=1}^{n} g(R(E_i,X)Y,E_i)$$

defines a global tensor field S of type (0,2) with local components

$$R_{jl} = R^i_{jil} = g^{ik}R_{jkil}.$$

From the tensor field S we define a global scalar field

$$(3.4) \qquad r = \sum_{i=1}^{n} S(E_i,E_i)$$

with local components

$$r = g^{ij}R_{ij}.$$

The tensor field S and the scalar function r are called the *Ricci tensor* and the *scalar curvature* of M respectively. If n = 2, then $G = \frac{1}{2}r$ is called the *Gaussian curvature*.

If the Ricci tensor S is of the form

$$S = ag, \qquad R_{ij} = ag_{ij},$$

where a is a constant, then M is called an *Einstein manifold*.

We have the following

THEOREM 3.3. Let M be a connected Riemannian manifold. If $S = ag$, where a is a function on M, then a is necessarily a constant provided that $n = \dim M > 2$.

Proof. From the Bianchi's second identity, we have

$$R_{ijkl;m} + R_{ijlm;k} + R_{ijmk;l} = 0.$$

Multiplying by g^{ik} and g^{jl}, summing up with respect to i, j, k and l and finally using the following formulas

$$R_{ijkl} = -R_{jikl} = -R_{ijlk}, \quad g^{ik}R_{ijkl} = R_{jl} = ag_{jl},$$

we obtain

$$(n-2)a_{;m} = 0.$$

Hence a is a constant. QED.

The following proposition is due to Schouten-Struik [1].

PROPOSITION 3.3. If M is a 3-dimensional Einstein manifold, then M is a space of constant curvature.

If the curvature R vanishes, that is, M is a space of zero curvature, then we call such a Riemannian manifold M a *locally flat space*. A Riemannian manifold M is called a *locally symmetric space* if its curvature tensor is parallel, that is, $\nabla R = 0$. A complete locally symmetric space is called a *symmetric space*.

We define the Ricci operator Q of a Riemannian manifold M by setting

$$g(QX,Y) = S(X,Y)$$

for any vector fields X and Y on M. Q is a tensor field of type (1,1).

We now give examples of simply connected complete Riemannian manifolds of constant curvature.

It is well-known that any two simply connected complete Riemannian manifolds of constant curvature k are isometric to each other (cf. Kobayashi-Nomizu [1;p.265]).

In the following we shall construct, for each constant k, a simply connected complete space form with curvature k.

Let R^n be the affine space of dimension n with cartesian coordinates $x^1,\ldots,x^n$ and let g be the Euclidean metric on R^n, that is,

$$g = (dx^1)^2 + \ldots + (dx^n)^2.$$

Then R^n with this metric g forms a space form of zero curvature. We call it the *Euclidean* n-*space* and we denote it by the same R^n.

We put

$$M^n(k) = \{(x^1,\ldots,x^{n+1}) \in R^{n+1}: \sqrt{|k|}((x^1)^2+\ldots+(x^n)^2+\mathrm{sgn}(k)(x^{n+1})^2) - 2x^{n+1} = 0,\ x^{n+1} \geq 0\},$$

where $\mathrm{sgn}(k) = 1$ or -1 according as $k \geq 0$ or $k < 0$. Then the Riemannian connection induced by

$$g = (dx^1)^2 + \ldots + (dx^n)^2 + \mathrm{sgn}(k)(dx^{n+1})^2$$

on R^{n+1} is the ordinaly Euclidean connection for each value k. In each case the metric tensor induced on $M^n(k)$ is complete and of constant curvature k. Moreover, each $M^n(k)$ is simply connected. Thus $M^n(k)$ gives a model space of a simply connected complete space form of curvature k.

A Riemannian manifold of constant curvature is said to be *elliptic*, *hyperbolic* or *flat* (or *locally Euclidean*) according as the sectional curvature is positive, negative or zero.

The *hyperspheres* in $M^n(k)$ are those hypersurfaces given by quadratic equations of the form

$$(x^1-a^1)^2 + \ldots + (x^n-a^n)^2 + \operatorname{sgn}(k)(x^{n+1}-a^{n+1})^2 = \text{constants},$$

where $a = (a^1,\ldots,a^{n+1})$ is an arbitrary fixed vector in R^{n+1}. In $M^n(0)$, there are just the usual hyperspheres. Among these hyperspheres the *great hypersphere* are those sections of hyperplanes which pass through the center $(0,\ldots,0,\operatorname{sgn}(k)/\sqrt{|k|})$ of $M^n(k)$ in R^{n+1}, $k \neq 0$. For $k = 0$, we consider the point at infinite on the X^{n+1}-axis as the center in R^{n+1}. The intersection of a hyperplane through the center in R^{n+1} is just a hyperplane in $M^n(0)$. All other hyperspheres in $M^n(k)$ are called *small hyperspheres*.

4. TRANSFORMATIONS

Let M be an n-dimensional Riemannian manifold with metric tensor g. Let ρ be a positive function on M. Then $g^* = \rho^2 g$ defines a change of metric on M which does not change the angle between two vectors at a point. Hence it is a *conformal change* of the metric. In particular, if the function ρ is a constant, the conformal transformation is said to be *homothetic*. If ρ is identically equal to 1, the transformation is nothing but an isometry. If a Riemannian metric g is conformally related to a Riemannian metric g* which is locally flat, then the Riemannian manifold is said to be *conformally flat* or *conformally Euclidean*.

The *Weyl conformal curvature tensor field* of M is the tensor field C of type (1,3) defined by

$$(4.1)\qquad C(X,Y)Z = R(X,Y)Z + \frac{1}{n-2}[S(X,Z)Y - S(Y,Z)X + g(X,Z)QY - g(Y,Z)QX] - \frac{r}{(n-1)(n-2)}[g(X,Z)Y - g(Y,Z)X]$$

for any vector fields X, Y and Z on M. Moreover, we put

$$(4.2)\qquad c(X,Y) = (\nabla_X Q)Y - (\nabla_Y Q)X - \frac{1}{2(n-2)}[(\nabla_X r)Y - (\nabla_Y r)X].$$

The tensor field C of type (1,3) vanishes identically for $n = 3$. The Weyl conformal curvature tensor C is invariant under any conformal change of the metric. The following is a well-known theorem of Weyl [1], [2].

THEOREM 4.1. A necessary and sufficient condition for a Riemannian manifold M to be conformally flat is that $C = 0$ for $n > 3$ and $c = 0$ for $n = 3$.

It should be noted that if M is conformally flat and of dimension $n > 3$, then $C = 0$ implies $c = 0$.

Let Γ be an affine connection on M. A vector field X on M is called an *infinitesimal affine transformation* of M if, for each $x \in M$, a local 1-parameter group of local transformations ϕ_t of a neighborhood U of x into M preserves the connection Γ.

An infinitesimal transformation X of a Riemannian manifold M is said to be *conformal* if $L_X g = \rho g$, where ρ is a function on M. In this case, the vector field X is called a *conformal Killing vector field*. The local 1-parameter group of local transformations generated by an infinitesimal transformation X is conformal if and only if X is conformal. If ρ is constant, then X is homothetic, and if $\rho = 0$, then X is isometric.

A vector field X on M is called an *infinitesimal isometry* (or, a *Killing vector field*) if the local 1-parameter group of local transformations generated by X in a neighborhood of each point of M consists of local isometries. An infinitesimal isometry is necessarily an infinitesimal affine transformation. X is an infinitesimal isometry if and only if $L_X g = 0$. The condition $L_X g = \rho g$ and $L_X g = 0$ can be rewritten as, respectively:

$$(L_X g)(Y,Z) = g(\nabla_Y X, Z) + g(\nabla_Z X, Y) = g(Y,Z),$$

$$(L_X g)(Y,Z) = g(\nabla_Y X, Z) + g(\nabla_Z X, Y) = 0,$$

for any vector fields Y and Z on M.

We shall give some integral formulas. When a Riemannian manifold M is orientable, we can define the volume element of M by

$$*1 = \sqrt{g}\,dx^1 \wedge dx^2 \wedge \dots \wedge dx^n,$$

where $g = |g_{ji}|$. Then we can consider the integral of a scalar function f, $\int_D f(x)*1$, over a domain D of M.

We now consider a p-form

$$\omega = \frac{1}{p!}\omega_{i_1 i_2 \dots i_p} dx^{i_1} \wedge dx^{i_2} \wedge \dots \wedge dx^{i_p}.$$

Then the exterior differential $d\omega$ of ω is the (p+1)-form given by

$$d\omega = \frac{1}{(p+1)!}(\partial_i \omega_{i_1 i_2 \dots i_p} - \partial_{i_1}\omega_{i i_2 \dots i_p}$$

$$- \partial_{i_2}\omega_{i_1 i i_3 \dots i_p} - \dots - \partial_{i_p}\omega_{i_1 i_2 \dots i})dx^i \wedge dx^{i_1} \wedge \dots \wedge dx^{i_p}.$$

We see that $d^2\omega = d(d\omega) = 0$. The *codifferential* $\delta\omega$ of ω is the (p-1)-form with the local expression

$$\delta\omega = -\frac{1}{(p-1)!}(g^{ji}\nabla_j \omega_{i i_2 \dots i_p})dx^{i_2} \wedge \dots \wedge dx^{i_p}.$$

For a scalar function f, we put $\delta f = 0$. We see that $\delta^2\omega = \delta(\delta\omega) = 0$.

We can also define the codifferential δT of a more general tensor field T. For example, let T be a tensor field of type (0,3). Then T is the tensor field of type (0,2):

$$\delta T : -g^{tj}\nabla_t T_{jih} = -\nabla^j T_{jih},$$

where we have put $\nabla^j = g^{tj}\nabla_t$ and T_{jih} are local components of T.

A p-form ω, or a skew-symmetric tensor field of type (0,p), is said to be *harmonic* if $d\omega = \delta\omega = 0$. If we put

$$\Delta = -\delta d - d\delta,$$

then we have $\Delta\omega = 0$ for a harmonic p-form ω.

For a vector field X with local components X^h there is associated 1-form ξ given by

$$\xi = g_{ji}X^j dx^i = X_i dx^i.$$

The codifferential $\delta\xi$ of ξ is given by

$$\delta\xi = -\nabla_i X^i = -g^{ji}\nabla_j X_i.$$

We denote it by δX. The famous Green's theorem can now be stated as follows.

GREEN'S THEOREM. In a compact orientable Riemannian manifold M without boundary, we have

$$\int_M (\delta X)*1 = 0$$

for any vector field X on M, or

$$\int_M (\nabla_i X^i)*1 = 0 \quad \text{or} \quad \int_M (g^{ji}\nabla_j X_i)*1 = 0,$$

X^i and X_i being components of X.

On the other hand, the *divergence* of X, denoted by div X, is given by

$$(\text{div } X)_x = \text{trace of the endomorphism } V \longrightarrow \nabla_V X,\ V \in T_x(M).$$

Therefore, the integral equation in Green's Theorem can be stated as

$$\int_M (\text{div } X)*1 = 0.$$

If we have a function f, we can form the codifferential δdf of the differential df of f. This has the local expression

$$\Delta f = g^{ji}\nabla_j\nabla_i f = \nabla^i\nabla_i f.$$

The differential operator $\Delta = g^{ji}\nabla_j\nabla_i$ or $\nabla^i\nabla_i$ is sometimes called the *Laplacian.*

Applying Green's Theorem, we have

THEOREM 4.2. In a compact and orientable Riemannian manifold M without boundary, we have

$$\int_M \Delta f * 1 = 0$$

for any function f on M.

Using this theorem, we have

HOPF'S LEMMA. Let M be a compact orientable Riemannian manifold without boundary. If f is a function on M such that $f \geq 0$ everywhere (or $f \leq 0$ everywhere), then f is a constant function.

Let M be an n-dimensional Riemannian manifold. We denote by $\{e_i\}$ an orthonormal frame of M. For any vector field X of M, we have

$$\begin{aligned}&\operatorname{div}(\nabla_X X) - \operatorname{div}((\operatorname{div}X)X)\\ &\qquad = S(X,X) + \sum_{i,j} g(\nabla_{e_j}X,e_i)g(e_j,\nabla_{e_i}X) - (\operatorname{div}X)^2,\end{aligned}$$

where S is the Ricci tensor of M. On the other hand, we have

$$\begin{aligned}&\sum_{i,j}(g(\nabla_{e_j}X,e_i) + g(\nabla_{e_i}X,e_j))^2\\ &\qquad = 2\sum_{i,j}(g(\nabla_{e_j}X,e_i)g(\nabla_{e_i}X,e_j) + g(\nabla_{e_j}X,e_i)^2).\end{aligned}$$

Thus we have

$$\sum_{i,j} g(\nabla_{e_j}X,e_i)g(\nabla_{e_i}X,e_j) = -|\nabla X|^2 + \tfrac{1}{2}|L_X g|^2.$$

We have therefore (Yano [1])

THEOREM 4.3. For any vector field X of a Riemannian manifold M, we have

$$(4.3) \qquad \operatorname{div}(\nabla_X X) - \operatorname{div}((\operatorname{div}X)X) = S(X,X) + \tfrac{1}{2}|L_X g|^2 - |\nabla X|^2 - (\operatorname{div}X)^2.$$

In the following, we shall give applications of Theorem 4.3. Let $\xi = X_i dx^i$ be a 1-form associated to the vector field X. Then we have

$$|d\xi|^2 = \sum_{i,j} (g(\nabla_{e_j} X, e_i) - g(\nabla_{e_i} X, e_j))^2$$
$$= -2 \sum_{i,j} g(\nabla_{e_j} X, e_i) g(\nabla_{e_i} X, e_j) + 2|\delta\xi|^2.$$

Thus, (4.3) can be rewritten as

$$(4.4) \qquad \operatorname{div}(\nabla_X X) - \operatorname{div}((\operatorname{div}X)X) = S(X,X) - \tfrac{1}{2}|d\xi|^2 + |\nabla X|^2 - (\delta\xi)^2.$$

If M is compact orientable, by (4.4), we have

$$\int_M [S(X,X) + |\nabla X|^2] *1 = 0$$

for a harmonic vector field X. Therefore, we have (Bochner [1])

THEOREM 4.4. If the Ricci tensor S of a compact orientable Riemannian manifold M satisfies $S(X,X) \geq 0$, then a harmonic vector field X in M has a vanishing covariant derivative. If the Ricci tensor S is positive-definite, then a harmonic vector field other than zero does not exist in M.

The existence of harmonic forms in M is closely related to the topology of M. In fact we have (Hodge [1])

THEOREM OF HODGE. In a compact orientable Riemannian manifold, the number of linearly independent (with constant real coefficients) harmonic p-forms is equal to the p-th Betti number B_p of the manifold.

Combing this with Theorem 4.4, we obtain (Bochner [1], Myers [1])

THEOREM 4.5. In a compact orientable Riemannian manifold M with positive Ricci tensor, the first Betti number vanishes.

From (4.3) we also have (Bochner [1])

THEOREM 4.6. Let M be a compact and orientable Riemannian manifold. If the Ricci tensor S of M satisfies $S(X,X) \leq 0$, then a Killing vector field X in M has vanishing covariant derivative. If S is negative-definite, then a Killing vector field other than zero does not exist in M.

Let X be a conformal Killing vector field such that $L_X g = 2\rho g$, where ρ is given by $\rho = \frac{-1}{n}\delta X$. Then (4.3) implies

$$\int_M [S(X,X) - |\nabla X|^2 - \frac{n-2}{n}(\delta X)^2]*1 = 0,$$

from which we have (Yano [1])

THEOREM 4.7. Let M be a compact and orientable Riemannian manifold. If the Ricci tensor S of M satisfies $S(X,X) \leq 0$, a conformal Killing vector field X in M has vanishing covariant derivative, and if S is negative-definite, a conformal Killing vector field other than the zero vector field does not exist in M.

5. FIBRE BUNDLES AND COVERING SPACES

Let G be a differentiable manifold with a countable basis. If G is a group such that the group operation $(a,b) \in G \times G \longrightarrow ab^{-1} \in G$ is a differentiable mapping, then G is called a *Lie group*.

A Lie group is clearly a locally compact group with a countable basis. Let G_0 be the connected component of G containing the identity element of G. We see that G_0 is a closed normal subgroup of G_0.

Moreover, since G_0 is locally connected, G_0 is an open submanifold of G. G_0 is also a Lie group.

We denote by L_a (resp. R_a) the left (resp. right) translation of G by an element $a \in G$: $L_a x = ax$ (resp. $R_a x = xa$) for every $x \in G$. For $a \in G$, ad a is the inner automorphism of G defined by $(\mathrm{ad}\ a)x = axa^{-1}$ for every $x \in G$. A vector field X on G is called *left invariant* (resp. *right invariant*) if it is invariant by all L_a, i.e., $(L_a)_* X = X$ for all $a \in G$ (resp. R_a, i.e., $(R_a)_* X = X$ for all $a \in G$). Let $\mathfrak{g}$ be the set of all left invariant vector fields on G. We call $\mathfrak{g}$ the *Lie algebra* of the Lie group G. In fact, $\mathfrak{g}$ is closed for the usual addition, scalr multiplication and bracket operation. As a vector space, $\mathfrak{g}$ is isomorphic with the tangent space $T_e(G)$ at the identity, the isomorphism being given by the mapping which sends $X \in \mathfrak{g}$ into X_e, the value of X at e. Thus $\mathfrak{g}$ is a Lie subalgebra of dimension n (= dim G) of the Lie algebra of vector fields $\mathfrak{X}(G)$.

Every $A \in \mathfrak{g}$ generates a (global) 1-parameter group of transformations of G.

We say that a Lie group G is a *Lie transformation group* on a manifold M or that G acts on M if the following conditions are satisfied:

(1) Every element $a \in G$ induces a transformation of M, denoted by $x \longrightarrow xa$, where $x \in M$;

(2) $(a,x) \in G \times M \longrightarrow xa \in M$ is a differentiable mapping;

(3) $x(ab) = (xa)b$ for all $a,b \in G$ and $x \in M$.

We also write $R_a x$ for xa and say that G acts on M on the right. If we write ax and assume $(ab)x = a(bx)$ instead of (3), we say that G acts on M on the left and write $L_a x$ for ax also. Note that $R_{ab} = R_a \cdot R_b$ and $L_{ab} = L_a \cdot L_b$. From (3) and from the fact that each R_a or L_a is 1 : 1 on M, it follows that R_e and L_e are the identity transformation of M.

We say that G acts *effictively* (resp. *freely*) on M if $R_a x = x$ for all $x \in M$ (resp. for some $x \in M$) implies that $a = e$.

Definition. Let M be a manifold and G a Lie group. A (*differentiable*) *principal fibre bundle over* M *with group* G consists of a manifold P and an action of G on P satisfying the following conditions:

(1) G acts freely on P on the right: $(u,a) \in P \times G \longrightarrow ua = R_a u \in P$;

(2) M is the quotient space of P by the equivalence relation induced by G, $M = P/G$, and the canonical projection $\pi : P \longrightarrow M$ is differentiable;

(3) P is locally trivial, that is, every point x of M has a neighborhood U such that $\pi^{-1}(U)$ is isomorphic with $U \times G$ in the sense that there is a diffeomorphism $\psi : \pi^{-1}(U) \longrightarrow U \times G$ such that $\psi(u) = (\pi(u), \phi(u))$ where ϕ is a mapping of $\pi^{-1}(U)$ into G satisfying $\phi(ua) = (\phi(u))a$ for all $u \in \pi^{-1}(U)$ and $a \in G$.

A principal fibre bundle will be denoted by $P(M,G,\pi)$, $P(M,G)$ or simply P. We call P the *total space* or the *bundle space*, M the *base space*, G the *structure group* and π the *projection*. For each point x of M, $\pi^{-1}(x)$ is a closed submanifold of P, called the *fibre* over x. If u is a point of $\pi^{-1}(x)$, then $\pi^{-1}(x)$ is the set of points ua, $a \in G$, and is called the fibre through u. Every fibre is diffeomorphic to G.

Given a Lie group G and a manifold M, G acts freely on $P = M \times G$ on the right as follows. For each $b \in G$, R_b maps $(x,a) \in M \times G$ into $(x,ab) \in M \times G$. The principal fibre bundle $P(M,G)$ thus obtained is called *trivial*.

From local triviality of $P(M,G)$ we see that if W is a submanifold of M, then $\pi^{-1}(W)(W,G)$ is a principal fibre bundle. We call it the *restriction* of P to W and denote it by $P|W$.

Given a principal fibre bundle $P(M,G)$, the action of G on P induces a homomorphism σ of the Lie algebra $\mathfrak{g}$ of G into the Lie algebra $\mathfrak{X}(P)$ of vector fields on P. σ can be defined as follows: For every u, let σ_u be the mapping $a \in G \longrightarrow ua \in P$. Then $(\sigma_u)_* A_e = (\sigma A)_u$. For each $A \in$, $A^* = \sigma(A)$ is called the *fundamental vector field* corresponding to A. Since the action of G sends each fibre into itself, A^*_u is tangent to the fibre at each $u \in P$. As G acts freely on P, A^* never vanishes on P if $A \neq 0$. The dimension of each fibre being equal to that of $\mathfrak{g}$, the mapping $A \longrightarrow (A^*)_u$ of $\mathfrak{g}$ into $T_u(P)$ is a linear isomorphism of $\mathfrak{g}$ onto the tangent space at u of the fibre through. We also see that for each $a \in G$, $(R_a)_* A^*$ is the fundamental vector field corresponding to $(ad(a^{-1}))A \in \mathfrak{g}$.

We now give the concept of transitive functions. For a principal fibre bundle P(M,G), we can choose an open covering $\{U_i\}$ of M, each $\pi^{-1}(U_i)$ provided with a diffeomorphism $u \longrightarrow (\pi(u),\phi_i(u))$ of $\pi^{-1}(U_i)$ onto $U_i \times G$ such that $\phi_i(ua) = (\phi_i(u))a$. If $u \in \pi^{-1}(U_i \cap U_j)$, then $\phi_j(ua)(\phi_i(ua))^{-1} = \phi_j(u)(\phi_i(u))^{-1}$, which shows that $\phi_j(u)(\phi_i(u))^{-1}$ depends only on $\pi(u)$ not on u.

We define a mapping $\psi_{ji} : U_i \cap U_j \longrightarrow G$ by $\psi_{ji}(\phi(u)) = \phi_j(u)(\phi_i(u))^{-1}$. The family of mappings ψ_{ji} are called *transitive functions* of the bundle P(M,G) corresponding to the open covering $\{U_i\}$ of M. It is easy to verify that

$$\psi_{ki}(x) = \psi_{kj}(x)\cdot\psi_{ji}(x) \quad \text{for } x \in U_i \cap U_j \cap U_k. \tag{5.1}$$

Conversely, we have (cf. Kobayashi-Nomizu [1])

PROPOSITION 5.1. Let M be a manifold, $\{U_i\}$ an open covering of M and G a Lie group. Given a mapping $\psi_{ji} : U_i \cap U_j \longrightarrow G$ for every non-empty $U_i \cap U_j$, in such a way that the relation (5.1) are satisfied, we can construct a (differentiable) principal fibre bundle P(M,G) with transitive functions ψ_{ji}.

A *homomorphism* f of a principal fibre bundle P'(M',G') into another principal fibre bundle P(M,G) consists of a mapping $f' : P' \longrightarrow P$ and a homomorphism $f'' : G' \longrightarrow G$ such that $f'(u'a') = f'(u')f''(a)$ for all $u' \in P'$ and $a' \in G'$. For the sake of simplicity, we shall denote f' and f'' by the same letter f. Every homomorphism $f : P' \longrightarrow P$ maps each fibre of P' into fibre of P and hence induces a mapping of M' into M, which will be also denoted by f. A homomorphism $f : P'(M',G') \longrightarrow P(M,G)$ is called an *imbedding* or *injection* if the induced mapping $f : M' \longrightarrow M$ is an imbedding and if $f : G' \longrightarrow G$ is a monomorphism. By identifying P' with f(P'), G' with f(G') and M' with f(M'), we say that P'(M',G') is a *subbundle* of P(M,G). If, moreover, M' = M and the induced mapping $f : M' \longrightarrow M$ is the identity transformation of M, $f : P'(M',G') \longrightarrow P(M,G)$ is called a *reduction* of the structure group G of P(M,G) to G'. The subbundle P'(M,G') is called a *reduced bundle*. Given P(M,G) and a Lie subgroup G' of G, we say that G is reducible to

G' if there is a reduced bundle P'(M',G').

Let P(M,G) be a principal fibre bundle and F a manifold on which G acts on the left : $(a,\xi) \in G \times F \longrightarrow a\xi \in F$. We shall construct a fibre bundle E(M,F,G,P) associated with P with standard fibre F. On the product manifold $P \times F$, we let G act on the right as follows: $a \in G$ maps $(u,\xi) \in P \times F$ into $(ua, a^{-1}\xi) \in P \times F$. The quotient space of $P \times F$ by this group action is denoted by $E = P \times_G F$. The mapping $P \times F \longrightarrow M$ which maps (u,ξ) into $\pi(u)$ induces a mapping π_E, called the projection, of E onto M. For each $x \in M$, $\pi_E^{-1}(x)$ is called the fibre of E over x. Every point x of M has a neighborhood U such that $\pi^{-1}(U)$ is isomorphic to $U \times G$. Identifying $\pi^{-1}(u)$ with $U \times G$, we see that the action of G on $\pi^{-1}(U) \times F$ on the right is given by

$$(x,a,\xi) \longrightarrow (x,ab,b^{-1}\xi) \quad \text{for } (x,a,\xi) \in U \times G \times F \text{ and } b \in G.$$

It follows that the isomorphism $\pi^{-1}(U) \approx U \times G$ induces an isomorphism $\pi_E^{-1}(U) \approx U \times F$. We can therefore introduce a differentiable structure in E by the requirement that $\pi_E^{-1}(U)$ is an open submanifold of E which is diffeomorphic with $U \times F$ under the isomorphism $\pi_E^{-1}(U) \approx U \times F$. The projection π_E is then a differentiable mapping of E onto M. We call E(M,F,G,P) or simply E the *fibre bundle* over the base space M, with standard fibre F and structure group G, which is associated with the principal fibre bundle P.

We recall here some results on covering spaces. Given a connected, locally arcwise connected topological space M, a connected space E is called a *covering space* over M with projection $p : E \longrightarrow M$ if every point x of M has a connected open neighborhood U such that each connected component of $p^{-1}(U)$ is open in E and is mapped homeomorphically onto U by p.

Two covering spaces $p : E \longrightarrow M$ and $p' : E' \longrightarrow M$ are *isomorphic* if there exists a homeomorphism $f : E \longrightarrow E'$ such that $p' \cdot f = p$. A covering space $p : E \longrightarrow M$ is a *universal covering space* if E is simply connected. If M is a manifold, every covering space has a unique structure of manifold such that p is differentiable.

PROPOSITION 5.2. (1) Given a connected manifold M, there is a unique (unique up to an isomorphism) universal covering manifold, which will be denoted by $\hat{M}$.

(2) The universal covering manifold $\hat{M}$ is a principal fibre bundle over M with group $\pi_1(M)$ and the projection $p : \hat{M} \longrightarrow M$, where $\pi_1(M)$ is the first homotopy group of M.

For the proof, see Steenrod [1;pp.67-71].

PROPOSITION 5.3. Let M be a Riemannian manifold with metric g. Let $p : E \longrightarrow M$ be a covering manifold of M. Then p^*g is a Riemannian metric on E. Moreover, E is complete if and only if M is complete.

Example 5.1. Bundle of linear frames: Let M be an n-dimensional manifold. A linear frame u at a point x of M is an ordered basis $X_1,\ldots,X_n$ of $T_x(M)$. Let L(M) be the set of all linear frames u at all points of M and let π be the mapping of L(M) onto M which maps a linear frame u at x into x. The general linear group GL(n;R) acts on L(M) on the right as follows. Let $a = (a^i_j) \in GL(n;R)$ and $u = (X_1,\ldots,X_n)$ be a linear frame at x. Then ua is the frame $(Y_1,\ldots,Y_n)$ at x defined by $Y_i = \Sigma_j a^j_i X_j$. It is clear that GL(n;R) acts freely on L(M) and $\pi(u) = \pi(v)$ if and only if $v = ua$ for some $a \in GL(n;R)$. Let $(x^1,\ldots,x^n)$ be a local coordinate system in a coordinate neighborhood U in M. Every frame u at $x \in U$ can be expressed uniquely in the form $u = (X_1,\ldots,X_n)$ with $X_i = \Sigma_k X^k_i(\partial/\partial x^k)$, where (X^k_i) is a non-singular matrix. This shows that $\pi^{-1}(U)$ is in 1 : 1 correspondence with $U \times GL(n;R)$. We can make L(M) into a differentiable manifold by taking (x^j) and (X^k_i) as a local coordinate system in $\pi^{-1}(U)$. It is easy to verify that L(M)(M,GL(n;R)) is a principal fibre bundle. We call L(M) the bundle of linear frames over M. A linear frame u at $x \in M$ can be defined as a non-singular linear mapping of R^n onto $T_x(M)$. The two definitions are related to each other as follows. Let $e_1,\ldots,e_n$ be the natural basis for R^n: $e_1 = (1,0,\ldots,0),\ldots,e_n = (0,\ldots,0,1)$. A linear frame $u = (X_1,\ldots,X_n)$ at x can be given as a linear mapping $u : R^n \longrightarrow T_x(M)$ such that $ue_i = X_i$ for $i = 1,\ldots,n$. The action of GL(n;R) on L(M) can be accordingly interpreted as follows. Consider $a = (a^i_j) \in GL(n;R)$ as a linear transformation of R^n which maps e_j into $\Sigma_i a^i_j e_i$. Then $ua : R^n \longrightarrow T_x(M)$ is the composite of the following two mappings:

$$R^n \xrightarrow{\ a\ } R^r \xrightarrow{\ u\ } T_x(M).$$

Example 5.2. Tangnet bundle: Let $GL(n;R)$ act on R^n as above. The tangent bundle $T(M)$ over M is the bundle associated with $L(M)$ with standard fibre R^n. The fibre of $T(M)$ over $x \in M$ may be considered as $T_x(M)$.

Example 5.3. Tensor bundle: Let T^r_s be the tensor space of type (r,s) over the vector space R^n. The group $GL(n;R)$ can be regarded as a group of linear transformations of T^r_s. With this standard fibre T^r_s, we obtain the tensor bundle $T^r_s(M)$ of type (r,s) over M which is associated with $L(M)$. It is easy to see that the fibre of $T^r_s(M)$ over $x \in M$ may be considered as the tensor space over $T_x(M)$ of type (r,s).

Example 5.4. Vector bundle: Let F be either the real number field R or the complex number field C, F^n the vector space of all n-tuples of elements of F and $GL(n;F)$ the group of all (n,n)-non-singular matrices with entries from F. $GL(n;F)$ acts on F^n on the left in a natural manner; if $a = (a^i_j) \in GL(n;F)$ and $\xi = (\xi^1,\ldots,\xi^n) \in F^n$, then $a\xi = (\Sigma_j a^1_j \xi^j,\ldots,\Sigma_j a^n_j \xi^j) \in F^n$. Let $P(M,G)$ be a principal fibre bundle and ρ a representation of G into $GL(n;F)$. Let $E(M,F^n,G,P)$ be the associated bundle with standard fibre F^n on which G acts through ρ. We call E a real or complex vector bundle over M according as $F = R$ or $F = C$. Each fibre $\pi_E^{-1}(x)$, $x \in M$, of E has the structure of a vector space such that every $u \in P$ with $\pi(u) = x$, considered as a mapping of F^n onto $\pi_E^{-1}(x)$, is a linear isomorphism of F^n onto $\pi_E^{-1}(x)$.

We give examples of universal covering manifolds.

Example 5.5. Let $e_1,\ldots,e_n$ be any basis of R^n, and let G be the subgroups of R^n generated by $e_1,\ldots,e_n$: $G = \{\Sigma m_i e_i : m_i \text{ integers}\}$. The action of G on R^n is properly discontinuous and R^n is the universal covering manifold of R^n/G. The quotient manifold R^n/G is called an n-dimensional torus.

Example 5.6. Let $S^n = \{(x^1,\ldots,x^{n+1}) \in R^{n+1} : \Sigma(x^i)^2 = 1\}$ and G be the group consisting of the identity transformation of S^n and the transformation of S^n which maps (x^i) into $(-x^i)$. Then S^n $(n \geq 2)$ is the universal covering manifold of S^n/G. The quotient manifold S^n/G is called the n-dimensional real projective space.

EXERCISES

A. CONCIRCULAR CURVATURE TENSOR: Let M be an n-dimensional Riemannian manifold with metric tensor field g. We denote by R, Q and r the Riemannian curvature tensor, the Ricci operator and the scalar curvature of M respectively. We put

$$Z(X,Y)W = R(X,Y)W - \frac{r}{n(n-1)}(g(Y,W)X - g(X,W)Y),$$

$$GX = QX - \frac{r}{n}X$$

for any vector fields X, Y and W on M. We call Z the *concircular curvature tensor* of M, which represents deviation of the manifold from constant curvature. We also see that $G = 0$ if and only if M is an Einstein manifold. We have

$$|Z|^2 = |R|^2 - \frac{2}{n(n-1)}r^2,$$

$$|G|^2 = |Q|^2 - \frac{1}{n}r^2.$$

We put

$$P(X,Y)W = R(X,Y)W - \frac{1}{n-1}(g(Y,W)QX - g(X,W)QY).$$

We call P the Weyl's *projective curvature tensor*. The vanishing of P is equivalent to the fact that M is of constant curvature. We have

$$|P|^2 = |R|^2 - \frac{2}{n-1}|Q|^2.$$

Let C be the conformal curvature tensor of M. Then we have

$$|C|^2 = |R|^2 - \frac{4}{n-2}|Q|^2 + \frac{2}{(n-1)(n-2)}r^2.$$

B. CONFORMALLY FLAT RIEMANNIAN MANIFOLDS: The conditions for a conformally flat Riemannian manifold to be a space of constant curvature are given by the following (Goldberg-Okumura [1])

THEOREM 1. Let M be an n-dimensional compact conformally flat Riemannian manifold with constant scalar curvature r. If the length of the Ricci tensor is less than $r/(n-1)^{1/2}$, $n \geq 3$, then M is of constant curvature.

THEOREM 2. In an n-dimensional compact conformally flat Riemannian manifold M, if the length of the Ricci tensor is constant and less than $r/(n-1)^{1/2}$, then M is of constant curvature.

Theorem 1 is a generalization of the theorem of Goldberg [4].

C. NOMIZU'S PROBLEM: If a Riemannian manifold is locally symmetric, then its curvature tensor R satisfies

(*) $R(X,Y)R = 0$ for all tangent vectors X and Y,

where the endomorphism $R(X,Y)$ operates on R as a derivation of the tensor algebra at each point of M. Conversely, does this algebraic condition (*) on the curvature tensor field R imply that M is locally symmetric?.

H. Takagi [1] showed that, in a 4-dimensional Euclidean space R^4, there exists an irreducible and complete hypersurface M which satisfies the condition (*) but is not locally symmetric.

Let (x,y,z,w) be a Cartesian coordinate system in R^4. We consider the hypersurface M defined by

$$w = (x^2z - y^2z - 2xy)/2(z^2 + 1)$$

or

$$2z^2w - x^2z + y^2z + 2w + 2xy = 0,$$

which satisfies the non-linear partial differential equation

$$w_x^2 - w_y^2 + 2w_z = 0.$$

Then M is a desired manifold.

For Riemannian manifolds satisfying the condition (*) see Sekigawa [ ].

D. EQUATION OF MMAURER-CARTAN: A differential form ω on a Lie group G is called left invariant if $(L_a)^*\omega = \omega$ for every $a \varepsilon G$. The vector space $\mathfrak{g}^*$ formed by all left invariant 1-forms is the dual space of the Lie algebra $\mathfrak{g}$: if $A \varepsilon \mathfrak{g}$ and $\omega \varepsilon \mathfrak{g}^*$, then the function $\omega(A)$ is constant on G. If ω is a left invariant form, then so is $d\omega$. We have the equation of Mourer-Cartan:

$$d\omega(A,B) = -\tfrac{1}{2}\omega([A,B]) \quad \text{for } \omega \varepsilon \mathfrak{g}^*,\ A,B \varepsilon \mathfrak{g}.$$

The canonical 1-form θ on G is the left invariant $\mathfrak{g}$-valued 1-form uniquely determined by $\theta(A) = A$ for $A \varepsilon \mathfrak{g}$.

E. CONNECTION: We define a connection in a principal fibre bundle P(M,G) (see Kobayashi-Nomizu [1]). For each $u \varepsilon P$, let $T_u(P)$ be the tangent space of P at u and G_u the subspace of $T_u(P)$ consisting of vectors tangent to the fibre through u. A *connection* Γ in P is an assignment of a subspace Q_u of $T_u(P)$ to each $u \varepsilon P$ such that

(a) $T_u(P) = G_u + Q_u$ (direct sum);
(b) $Q_{ua} = (R_a)_* Q_u$ for every $u \varepsilon P$ and $a \varepsilon G$;
(c) Q_u depends differentiably on u.

Condition (b) means that the distribution $u \longrightarrow Q_u$ is invariant by G. We call G_u the *vertical subspace* and Q_u *the horizontal subspace* of $T_u(P)$. A vector $X \varepsilon T_u(P)$ is called *vertical* (resp. *horizontal*) if it lies in G_u (resp. Q_u). By (a), any vector X of P at u can be uniquely written as

$$X = Y + Z, \quad Y \varepsilon G_u,\ Z \varepsilon Q_u.$$

We call Y (resp. Z) the vertical (resp. horizontal) component of X.

Given a connection Γ in P, we define a 1-form ω on P with values in the Lie algebra $\mathfrak{g}$ of G as follows. For each $X \in T_u(P)$, we define $\omega(X)$ to be the unique $A \in \mathfrak{g}$ such that $(A^*)_u$ is equal to the vertical component of X, where A^* is the fundamental vector field corresponding to A. It is clear that $\omega(X) = 0$ if and only if X is horizontal. The form ω is called the *connection form* of the given connection Γ. The projection $\pi : P \longrightarrow M$ induces a linear mapping $\pi : T_u(P) \longrightarrow T_x(M)$ for each $u \in P$, where $x = \pi(u)$. When a connection is given, π maps Q_u isomorphically onto $T_x(M)$. The *horizontal lift* or simply, *lift* of a vector field X on M is a unique vector field X^* on P which is horizontal and $\pi(X^*_u) = X_{\pi(u)}$ for every $u \in P$.

THEOREM 1. Given a connection in P and a vector field X on M, there is a unique horizontal lift X^* of X. The lift X^* is invariant by R_a for every $a \in G$. Conversely, every horizontal vector field X^* on P invariant by G is the lift of a vector field X on M.

Let P(M,G) be principal fibre bundle and A a subset of M. We say that a connection is defined over A if, at every point $u \in P$ with $\pi(u) \in A$, a subspace Q_u of $T_u(P)$ is given in such a way that conditions (a) and (b) for connections are satisfied and Q_u depends differentiably on u in the following way. For every point $x \in A$, there exist an open neighborhood U and a connection in $P|U = \pi^{-1}(U)$ such that the horizontal subspace at every point $u \in \pi^{-1}(A)$ is the given space Q_u. We have

THEOREM 2. Let P(M,G) be a principal fibre bundle and A a closed subset of M (A may be empty). If M is paracompact, every connection defined over A can be extended to a connection in P. In particular, P admits a connection if M is paracompact.

F. CANONICAL FLAT CONNECTION: Let $P = M \times G$ be a trivial principal fibre bundle. The *canonical flat connection* in P is defined by taking the tangent space to $M \times \{a\}$ at $u = (x,a) \in M \times G$ as the horizontal subspace at u. Let θ be the canonical 1-form on G and $f : M \times G \longrightarrow G$ be the natural projection. Then $\omega = f^*\theta$ is the connection form of the canonical flat connection in P. From the Maurer-Cartan equation of θ we see that the canonical flat connection has zero curvature. A connection in any principal fibre bundle P(M,G) is called *flat* if every point x of M has a neighborhood U such that the induced connection in $P|U = \pi^{-1}(U)$ is isomorphic with the canonical flat connection in $U \times G$. More precisely, there is an isomorphism $f : \pi^{-1}(U) \longrightarrow U \times G$ which maps the horizontal subspace at each $u \in \pi^{-1}(U)$ upon the horizontal subspace at f(u) of the canonical flat connection in $U \times G$.

THEOREM 1. A connection in P(M,G) is flat if and only if the curvature form vanishes identically.

THEOREM 2. Let Γ be a connection in P(M,G) such that the curvature vanishes identically. If M is paracompact and simply connected, then P is isomorphic with the trivial bundle $M \times G$ and Γ is isomorphic with the canonical flat connection in $M \times G$.

G. STRUCTURE EQUATION: Let P(M,G) be a principal fibre bundle with a connection Γ. Let $h : T_u(P) \longrightarrow Q_u$ be the projection and ω be the connection form of Γ. We put $D\omega = (d\omega)h$. $D\omega$ is called the curvature form of ω, which will be denoted by Ω. Then

THEOREM (Structure equation). Let ω be a connection form and Ω its curvature form. Then

$$d\omega(X,Y) = -\tfrac{1}{2}[\omega(X),\omega(Y)] + \Omega(X,Y), \quad X,Y \in T_u(P).$$

(See Ambrose-Singer [1].)

H. LINEAR CONNECTION: Let L(M) be the bundle of linear frames over a manifold M and GL(n;R) the general linear group, n = dim M. The *caninical form* θ of L(M) is the R^n-valued 1-form on L(M) defined by

$$\theta(X) = u^{-1}(\pi(X)) \quad \text{for } X \in T_u(P),$$

where u is considered as a linear mapping of R^n onto $T_{\pi(u)}(M)$

(cf. Example 5.1). A connection in L(M) over M is called a *linear connection* of M. Let U be a coordinate neighborhood in M with local coordinate system (x^i). We put $X_i = \partial/\partial x^i$, $i = 1,\dots,n$, in U. Every linear frame at a point x of U can be uniquely expressed by

$$(\sum_i X_1^i (X_i)_x, \dots, \sum_i X_h^i (X_i)_x), \qquad \det(X_i^j) \neq 0.$$

We take (x^i, X_k^j) as a local coordinate system in $\pi^{-1}(U) \subset L(M)$. Let (Y_k^j) be the inverse matrix of (X_k^j) so that $\Sigma_j X_i^j Y_j^k = \Sigma_j Y_i^j X_j^k = \delta_i^k$. Let $e_1,\dots,e_n$ be the natural basis for R^n and set

$$\theta = \sum_i \theta^i e_i,$$

which is the canonical form. Then θ is expressed by

$$\theta^i = \sum_j Y_j^i dx^j.$$

Let ω be the connection form of a linear connection Γ of M. With respect to the basis $\{E_i^j\}$ of the Lie algebra $\mathfrak{gl}(n;R)$, we put

$$\omega = \sum_{i,j} \omega_j^i E_i^j.$$

Let σ be the cross section of L(M) over U which assigns to each $x \in U$ the linear frame $((X_1)_x,\dots,(X_n)_x)$. We set

$$\omega_U = \sigma^* \omega.$$

Then ω_U is a $\mathfrak{gl}(n;R)$-valued 1-form defined on U. We define

$$\omega_U = \sum_{i,j,k} (\Gamma_{jk}^i dx^j) E_i^k.$$

The functions Γ_{jk}^i are called the components (or Christoffel's symbols) of the linear connection Γ with respect to the local coordinate system (x^i). We obtain

$$\nabla_{X_j} X_i = \sum_k \Gamma^k_{ji} X_k$$

(see the definition of affine connection in §2).

Traditionally, the words "linear connection" and "affine connection" have been used interchangeably. But, in the book of Kobayashi-Nomizu [1], these words are made a logical distinction.

CHAPTER II

SUBMANIFOLDS OF RIEMANNIAN MANIFOLDS

In this chapter we give the fundamental results and some theorems concerning geometry of submanifolds which will be needed for the later treatment of submanifolds.

In §1, we discuss the induced connection on submanifolds and second fundamental forms of immersions. We give the basis formulas for submanifolds which are called the Gauss and Weingarten formulas. In §2, we prepare equations of Gauss, Codazzi and Ricci which give the relationship between submanifolds and ambient manifolds. In §3, we compute the Laplacian of the second fundamental forms of immersions and give the Simons' type formula on submanifolds. §4 is devoted to the study of submanifolds of space forms. In the last §5, we discuss minimal submanifolds of space forms. Moreover, in §§4 and 5, we also discuss submanifolds with flat normal connections or with parallel second fundamental forms.

For the general theory of submanifolds we refer to Chen [1] and Kobayashi-Nomizu [2].

1. INDUCED CONNECTION AND SECOND FUNDAMENTAL FORM

Let M be an n-dimensional manifold isometrically immersed in an m-dimensional Riemannian manifold $\bar{M}$. We put $m = n+p$, $p > 0$. Since the discussion is local, we may assume, if we want, that M is imbedded in $\bar{M}$. If the manifold $\bar{M}$ is covered by a system of coordinate neighborhoods $\{V;u^A\}$ and M is covered by a system of coordinate neighborhoods $\{U;x^h\}$, where, here and in the sequel, the indices A, B, C,... run over the range 1,2,...,m and i, j, k,... run over the range 1,2,...,n. Then the submanifold M can be represented locally by

$$u^A = u^A(x^h).$$

In the following, we shall identify vector fields in M and their images under the differential mapping, that is, if i denotes the immersion of M into $\bar{M}$ and X is a vector field in M, we identify X and $i_*(X)$. Thus, if X is a vector field in M with local expression $X = X^h\partial_h$ ($\partial_h = \partial/\partial x^h$), then X also has the local expression $X = B_h{}^A X^h \partial_A$ in $\bar{M}$ ($\partial_A = \partial/\partial u^A$), and $B_h{}^A = \partial u^A/\partial x^h$, where we used the Einstein convention, that is, repeated indices, with one upper index and one lower index, denote summation over its range.

We denote by g the Riemannian metric tensor field of $\bar{M}$. Then the submanifold M is also a Riemannian manifold with the Riemannian metric h given by h(X,Y) = g(X,Y) for any vector fields X and Y in M. The Riemannian metric h on M is called the *induced metric* on M. We see that $g_{ji} = g_{BA}B_j{}^B B_i{}^A$ with $h = g_{ji}dx^j dx^i$ and $g = g_{BA}du^B du^A$.

Throughout this chapter, the induced metric h will be denoted by the same g as that of the ambient manifold $\bar{M}$ to simplyfy the notation because it may cause no confusion.

If a vector V of $\bar{M}$ at a point x of M satisfies g(X,V) = 0 for any vector X of M at x, then V is called a normal vector of M in $\bar{M}$ at x. A unit normal vector field of M in $\bar{M}$ is sometimes called a *normal section* on M.

Let $T(M)^\perp$ denote the vector bundle of all normal vectors of M in $\bar{M}$. Then the tangent bundle of $\bar{M}$, restricted to M, is the direct sum of the tangent bundle T(M) of M and the normal bundle $T(M)^\perp$ of M in $\bar{M}$. We denote by $\bar{\nabla}$ the operator of covariant differentiation on $\bar{M}$.

LEMMA 1.1. Let X and Y be vector fields on M and let $\bar{X}$ and $\bar{Y}$ be extensions of X and Y, respectively. Then $[\bar{X},\bar{Y}]|_M$ is independent of the extensions and $[\bar{X},\bar{Y}]|_M = [X,Y]$, and $(\bar{\nabla}_{\bar{X}}\bar{Y})|_M$ is also independent of the extensions.

Proof. Let $\bar{X}^A(u)$ and $\bar{Y}^A(u)$ be extensions of $B_h{}^A X^h(u)$ and $B_h{}^A Y^h(U)$, respectively. Then

$$\bar{X}^A(u(x)) = B_h{}^A X^h(u), \qquad \bar{Y}^A(u(x)) = B_h{}^A Y^h(u).$$

Thus we have

$$\begin{aligned}
[\bar{X},\bar{Y}]^A|_{u=u(x)} &= (\bar{X}^B \partial_B \bar{Y}^A - \bar{Y}^B \partial_B \bar{X}^A)|_{u=u(x)} \\
&= (B_h{}^B X^h \partial_B \bar{Y}^A - B_h{}^B Y^h \partial_B \bar{X}^A)|_{u=u(x)} \\
&= X^h \partial_h (B_i{}^A Y^i) - Y^h \partial_h (B_i{}^A X^i) \\
&= B_i{}^A (X^h \partial_h Y^i - Y^h \partial_h X^i) \\
&= B_i{}^A [X,Y]^i.
\end{aligned}$$

This shows that $[\bar{X},\bar{Y}]|_M$ does not depend on the extensions $\bar{X}$ and $\bar{Y}$ of X and Y and is equal to [X,Y].

In the next place, the components of $\bar{\nabla}_{\bar{X}}\bar{Y}$ are given by

$$\bar{X}^B(\partial_B \bar{Y}^C + \Gamma^C_{BA}\bar{Y}^A),$$

from which

$$\begin{aligned}
&\bar{X}^B(\partial_B \bar{Y}^C + \Gamma^C_{BA}\bar{Y}^A)|_{u=u(x)} \\
&\qquad = B_i{}^B X^i(\partial_B \bar{Y}^C + \Gamma^C_{BA} B_j{}^A Y^j)|_{u=u(x)} \\
&\qquad = X^i(\partial_i (B_j{}^C Y^j) + \Gamma^C_{BA} B_i{}^B B_j{}^A Y^j).
\end{aligned}$$

Therefore $\bar{\nabla}_{\bar{X}}\bar{Y}$ does not depend on the extensions. QED.

We denote $(\bar{\nabla}_{\bar{X}}\bar{Y})|_M$ by $\bar{\nabla}_X Y$. We put

$$(1.1) \qquad \bar{\nabla}_X Y = \nabla_X Y + B(X,Y),$$

where $\nabla_X Y$ is the tangential component of $\bar{\nabla}_X Y$ and B(X,Y) the normal component of $\bar{\nabla}_X Y$. Then ∇ is the operator of covariant differentiation with respect to the induced metric on M. We shall prove this fact in the following:

Let a and b be functions on M. Then

$$\bar{\nabla}_{aX} bY = a\bar{\nabla}_X bY = a[(Xb)Y + b\bar{\nabla}_X Y]$$
$$= [a(Xb)Y + ab\nabla_X Y] + abB(X,Y),$$

from which

$$\nabla_{aX} bY = a(Xb)Y + ab\nabla_X Y$$

and

$$B(aX,bY) = abB(X,Y).$$

The first equation shows that ∇ defines an affine connection on M and the second equation shows that B is bilinear in X and Y, since additivity is trivial. We next prove ∇ has no torsion and $\nabla g = 0$. Since $\bar{\nabla}$ has no torsion, we have

$$0 = \bar{\nabla}_X Y - \bar{\nabla}_Y X - [X,Y]$$
$$= \nabla_X Y + B(X,Y) - \nabla_Y X - B(Y,X) - [X,Y].$$

Comparing the tangential and normal parts of the equation above, we find

$$\nabla_X Y - \nabla_Y X - [X,Y] = 0$$

and

$$B(X,Y) = B(Y,X).$$

These equations show that ∇ has no torsion and B is symmetric. Moreover, since $\bar{\nabla} g = 0$, we see that

$$\begin{aligned}\nabla_X(g(Y,Z)) &= \bar{\nabla}_X(g(Y,Z)) \\ &= g(\bar{\nabla}_X Y,Z) + g(Y,\bar{\nabla}_X Z) \\ &= g(\nabla_X Y,Z) + g(Y,\nabla_X Z),\end{aligned}$$

which means that $\nabla g = 0$.

We call the Riemannian connection ∇ the *induced connection* and B the *second fundamental form* of M (or of the immersion i). For each poin x of M, B(X,Y) at x depends only X_x and Y_x.

Next, let V be a normal vector field on M and X be a vector field on M. We put

$$(1.2) \qquad \bar{\nabla}_X V = -A_V X + D_X V,$$

where $-A_V X$ and $D_X V$ are, respectively, the tangential component and the normal component of $\bar{\nabla}_X V$. It is easily verified that the vector fields $A_V X$ and $D_X V$ are differentiable on M.

For any functions a and b on M we have

$$\begin{aligned}\bar{\nabla}_{aX} bV &= a\bar{\nabla}_X bV = a[(Xb)V + b\bar{\nabla}_X V] \\ &= a(Xb)V - abA_V X + abD_X V,\end{aligned}$$

from which

$$A_{bV}(aX) = abA_V X$$

and

$$D_{aX}(bV) = a(Xb)V + abD_X V.$$

Since the additivity is trivial, $A_V X$ is bilinear in V and X and $A_V X$ at x depends only on V_x and X_x. We give a relation between B and A. Let X, Y be vector fields on M and V be a vector field normal to M. Then

$$0 = g(\bar{\nabla}_X Y, V) + g(Y, \bar{\nabla}_X V) = g(B(X,Y),V) - g(Y, A_V X),$$

from which

$$g(B(X,Y),V) = g(A_V X, Y).$$

Consequently, A_V is a symmetric linear transformation of $T_x(M)$, that is, A is an element of $\mathrm{Hom}(T(M)^{\perp}, S(M))$, where S(M) is the bundle whose fibre at each point is a symmetric linear transformation of $T_x(M)$.

We call A the *associated second fundamental form* to B or simply the second fundamental form of M.

LEMMA 1.2. D is a metric connection in the normal bundle $T(M)^{\perp}$ of M in $\bar{M}$ with respect to the induced metric on $T(M)^{\perp}$.

Proof. We easily see that D defines an affine connection in the normal bundle $T(M)^{\perp}$. Moreover, for any vector fields V and U in $T(M)^{\perp}$, we have

$$\begin{aligned} g(D_X V, U) + g(V, D_X U) &= g(\bar{\nabla}_X V, U) + g(V, \bar{\nabla}_X U) \\ &= \bar{\nabla}_X g(V,U) = D_X g(V,U), \end{aligned}$$

which shows that D is metric for the fibre metric in $T(M)^{\perp}$, namely, the restriction of g to the normal spaces. QED.

We have the first set of basic formulas for submanifolds, namely,

$$\bar{\nabla}_X Y = \nabla_X Y + B(X,Y) \quad \text{and} \quad \bar{\nabla}_X V = -A_V X + D_X V.$$

The first formula is called the *Gauss formula* and the second formula is called the *Weingarten formula*.

A normal vector field V on M is said to be *parallel* in the normal bundle, or simply parallel, if $D_X V = 0$ for all vector fields X tangent to M. A submanifold M is said to be *totally geodesic* if its second fundamental form vanishes identically, that is, $B = 0$ or equivalently $A = 0$. For a normal section V on M, if A_V is everywhere proportional to the identity transformation I, that is, $A_V = aI$ for some function a, then V is called an *umbilical section* on M, or M is said to be umbilical with respect to V. If the submanifold M is umbilical with respect to every local normal section of M, then M is said to be *totally umbilical*.

Let $e_1,\dots,e_n$ be an orthonormal basis in $T_x(M)$. The *mean curvature vector* μ of M is defined to be $\mu = (1/n)(\mathrm{Tr}B)$, where $\mathrm{Tr}B = \Sigma_i B(e_i,e_i)$, which is independent of the choice of a basis. We obtain $\mu = \Sigma_a[(\mathrm{Tr}A_a)/n]e_a$, where $e_{n+1},\dots,e_m$ is an orthonormal basis in $T_x(M)^{\perp}$, and we denote A_{e_a} by A_a for simplicity, $a = n+1,\dots,m$. If $\mu = 0$, then M is said to be *miminal*.

We notice that any submanifold M which is minimal and totally umbilical is totally geodesic.

2. EQUATIONS OF GAUSS, CODAZZI AND RICCI

For the second fundamental form B we define its covariant derivative $\nabla_X B$ by

$$(\nabla_X B)(Y,Z) = D_X B(Y,Z) - B(\nabla_X Y,Z) - B(Y,\nabla_X Z)$$

for any vector fields X, Y and Z tangent to M, which is defined equivalently by putting

$$(\nabla_X A)_V Y = \nabla_X(A_V Y) - A_{D_X V}Y - A_V \nabla_X Y$$

for any vector fields X, Y tangent to M and any vector field V normal to M. If $\nabla_X B = 0$ for all X, then the second fundamental form of M is said to be *parallel*, which is equivalent to $\nabla_X A = 0$ for all X.

Let $\bar{R}$ and R be the Riemannian curvature tensor fields of $\bar{M}$ and M respectively. Then the Gauss and Weingarten formulas imply

$$\begin{aligned}
\bar{R}(X,Y)Z &= \bar{\nabla}_X\bar{\nabla}_Y Z - \bar{\nabla}_Y\bar{\nabla}_X Z - \bar{\nabla}_{[X,Y]}Z \\
&= \bar{\nabla}_X(\nabla_Y Z + B(Y,Z)) - \bar{\nabla}_Y(\nabla_X Z + B(X,Z)) \\
&\qquad - (\nabla_{[X,Y]}Z + B([X,Y],Z)) \\
&= \nabla_X\nabla_Y Z + B(X,\nabla_Y Z) + (\nabla_X B)(Y,Z) + B(\nabla_X Y,Z) \\
&\quad + B(Y,\nabla_X Z) - \nabla_Y\nabla_X Z - B(Y,\nabla_X Z) - (\nabla_Y B)(X,Z) \\
&\quad - B(\nabla_Y X,Z) - B(X,\nabla_Y Z) - \nabla_{[X,Y]}Z - B([X,Y],Z) \\
&\quad - A_{B(Y,Z)}X + A_{B(X,Z)}Y \\
&= R(X,Y)Z - A_{B(Y,Z)}X + A_{B(X,Z)}Y \\
&\qquad + (\nabla_X B)(Y,Z) - (\nabla_Y B)(X,Z),
\end{aligned}$$

from which

$$\bar{R}(X,Y)Z = R(X,Y)Z - A_{B(Y,Z)}X + A_{B(X,Z)}Y + (\nabla_X B)(Y,Z) - (\nabla_Y B)(X,Z) \tag{2.1}$$

for any vector fields X, Y and Z tangent to M. For any vector field W tangent to M, (2.1) gives *equation of Gauss*

$$g(\bar{R}(X,Y)Z,W) = g(R(X,Y)Z,W) - g(B(X,W),B(Y,Z)) + g(B(Y,W),B(X,Z)). \tag{2.2}$$

Taking the normal component of (2.1), we obtain *equation of Codazzi*

$$(\bar{R}(X,Y)Z)^{\perp} = (\nabla_X B)(Y,Z) - (\nabla_Y B)(X,Z). \tag{2.3}$$

We now define the *curvature tensor* $R^{\perp}$ *of the normal bundle* of M by

$$R^{\perp}(X,Y)V = D_X D_Y V - D_Y D_X V - D_{[X,Y]}V. \tag{2.4}$$

For any vector fields X, Y tangent to M and any vector field V normal to M, from the Gauss and Weingarten formulas, we have

$$\begin{aligned}
\bar{R}(X,Y)V &= \bar{\nabla}_X\bar{\nabla}_Y V - \bar{\nabla}_Y\bar{\nabla}_X V - \bar{\nabla}_{[X,Y]}V \\
&= \bar{\nabla}_X(-A_V Y + D_Y V) - \bar{\nabla}_Y(-A_V X + D_X V) - (-A_V[X,Y] + D_{[X,Y]}V) \\
&= -\nabla_X(A_V Y) - B(X,A_V Y) - A_{D_Y V}X + D_X D_Y V \\
&\quad +\nabla_Y(A_V X) + B(Y,A_V X) + A_{D_X V}Y - D_Y D_X V \\
&\quad + A_V[X,Y] - D_{[X,Y]}V \\
&= R^{\perp}(X,Y)V - B(X,A_V Y) + B(Y,A_V X) \\
&\quad - (\nabla_X A)_V Y + (\nabla_Y A)_V X,
\end{aligned}$$

that is,

$$\bar{R}(X,Y)V = R^{\perp}(X,Y)V - B(X,A_V Y) + B(Y,A_V X) - (\nabla_X A)_V Y + (\nabla_Y A)_V X. \tag{2.5}$$

Let U be a vector field normal to M. Then we have

$$\begin{aligned}
&-g(B(X,A_V Y),U) + g(B(Y,A_V X),U) \\
&\qquad = -g(A_U X,A_V Y) + g(A_U Y,A_V X) = g([A_U,A_V]X,Y),
\end{aligned}$$

where $[A_U,A_V] = A_U A_V - A_V A_U$. Thus (2.5) implies *equation of Ricci*

$$g(\bar{R}(X,Y)V,U) = g(R^{\perp}(X,Y)V,U) + g([A_U,A_V]X,Y). \tag{2.6}$$

If $R^{\perp}$ vanishes identically, then the normal connection of M is said to be *flat* (or *trivial*). When $(\bar{R}(X,Y)V)^{\perp} = 0$, the normal connection of M is flat if and only if the second fundamental form of M is commutative, that is, $[A_V, A_U] = 0$ for all U, V. Particularly, if $\bar{M}$ is of constant curvature, $(\bar{R}(X,Y)V)^{\perp} = 0$, and hence the normal connection of M is flat if and only if the second fundamental form of M is commutative. If $\bar{R}(X,Y)Z$ is tangent to M, equation of Codazzi (2.3) reduces to

$$(\nabla_X B)(Y,Z) = (\nabla_Y B)(X,Z), \tag{2.7}$$

which is equivalent to

$$(\nabla_X A)_V Y = (\nabla_Y A)_V X. \tag{2.8}$$

Particularly, if $\bar{M}$ is of constant curvature, $\bar{R}(X,Y)Z$ is tangent to M. If $\bar{M}$ is of constant curvature c, then equation of Gauss reduces to

$$\begin{aligned} g(R(X,Y)Z,W) &= c[g(Y,Z)g(X,W) - g(X,Z)g(Y,W)] \\ &\quad + g(B(Y,Z),B(X,W)) - g(B(X,Z),B(Y,W)). \end{aligned} \tag{2.9}$$

On the other hand, we have

$$\begin{aligned} &g(B(Y,Z),B(X,W)) - g(B(X,Z),B(Y,W)) \\ &\quad = \sum_a [g(B(Y,Z),e_a)g(e_a,B(X,W)) - g(B(X,Z),e_a)g(e_a,B(Y,W))] \\ &\quad = \sum_a [g(A_aY,Z)g(A_aX,W) - g(A_aX,Z)g(A_aY,W)]. \end{aligned}$$

Thus (2.9) can be rewritten as

$$\begin{aligned} g(R(X,Y)Z,W) &= c[g(Y,Z)g(X,W) - g(X,Z)g(Y,W)] \\ &\quad + \sum_a [g(A_aY,Z)g(A_aX,W) - g(A_aX,Z)g(A_aY,W)]. \end{aligned} \tag{2.10}$$

We can give a similar equation to (2.10) for (2.2) by using A_a.

Let S be the Ricci tensor of M. Then (2.10) gives

$$S(X,Y) = (n-1)cg(X,Y) + \sum_a \mathrm{Tr}A_a g(A_a X,Y) - \sum_a g(A_a X, A_a Y). \tag{2.11}$$

Therefore, the scalar curvature r of M is given by

$$r = n(n-1)c + \sum_a (\mathrm{Tr}A_a)^2 - \sum_a \mathrm{Tr}A_a^2, \tag{2.12}$$

$\Sigma_a \mathrm{Tr}A_a^2$ is the square of the length of the second fundamental form of M, which will be denoted by $|A|^2$. We also have

$$|B|^2 = \sum_{i,j} g(B(e_i,e_j),B(e_i,e_j)) = |A|^2.$$

We shall give the structure equations of an n-dimensional submanifold M of an m-dimensional Riemannian manifold $\bar{M}$. We choose a local field of orthonormal frames $e_1,\dots,e_m$ in $\bar{M}$ in such a way that, restricted to M, the vectors $e_1,\dots,e_n$ are tangent to M and hence $e_{n+1},\dots,e_m$ are normal to M. With respect to this frame field of $\bar{M}$, let $\omega^1,\dots,\omega^m$ be the field of dual frames. We shall make use the following convention on the ranges of indices:

$$A,B,C,\dots = 1,\dots,m; \qquad i,j,k,\dots = 1,\dots,n;$$
$$a,b,c,\dots = n+1,\dots,m.$$

Then the structure equations of $\bar{M}$ are given by

$$d\omega^A = -\omega^A_B \wedge \omega^B, \qquad \omega^A_B + \omega^B_A = 0,$$
$$d\omega^A_B = -\omega^A_C \wedge \omega^C_B + \Phi^A_B, \qquad \Phi^A_B = \tfrac{1}{2}\bar{R}^A_{BCD}\omega^C \wedge \omega^D.$$

We restrict these forms to M. Then $\omega^a = 0$. Since $0 = d\omega^a = -\omega^a_i \wedge \omega^i$, by Cartan's lemma, we obtain

$$\omega^a_i = h^a_{ij}\omega^j, \qquad h^a_{ij} = h^a_{ji}. \tag{2.13}$$

We can see that $h^a_{ij} = g(A_a e_i, e_j)$. Thus h^a_{ij} are components of the second fundamental form A_a with respect to e_a. Therefore A_a can be considered as a symmetric (n,n)-matrix $A_a = (h^a_{ij})$. Moreover, we have the following equations:

$$(2.14) \qquad d\omega^i = -\omega^i_k \wedge \omega^k, \qquad \omega^i_k + \omega^k_i = 0,$$

$$(2.15) \qquad d\omega^i_j = -\omega^i_k \wedge \omega^k_j + \Omega^i_j, \qquad \Omega^i_j = \tfrac{1}{2} R^i_{jkl}\omega^k \wedge \omega^l,$$

$$(2.16) \qquad R^i_{jkl} = \bar{R}^i_{jkl} + \sum_a (h^a_{ik}h^a_{jl} - h^a_{il}h^a_{jk}),$$

$$(2.17) \qquad d\omega^a_b = -\omega^a_c \wedge \omega^c_b + \Omega^a_b, \qquad \Omega^a_b = \tfrac{1}{2} R^a_{bkl}\omega^k \wedge \omega^l,$$

$$(2.18) \qquad R^a_{bkl} = \bar{R}^a_{bkl} + \sum_i (h^a_{ik}h^a_{il} - h^a_{il}h^a_{ik}).$$

Equation (2.16) is the local expression of the Gauss equation (2.1) and equation (2.18) is the local expression of the Ricci equation (2.6). The forms (ω^i_j) define the Riemannian connection of M and the forms (ω^a_b) define the connection induced in the normal bundle of M. The second fundamental form of M is represented by $h^a_{ij}\omega^i\omega^j e_a$ and is sometimes denoted by its components h^a_{ij}. The second fundamental form of M is commutative if and only if $\Sigma_j(h^a_{ij}h^b_{jk} - h^a_{jk}h^b_{ij}) = 0$ for all a, b, i and k. If $R^a_{bkl} = 0$ for all a, b, k and l, then the normal connection of M is flat. The mean curvature vector μ of M is given by $(1/n)(\Sigma_k h^a_{kk}e_a)$. If $h^a_{ij} = (1/n)(\Sigma_k h^a_{kk})\delta_{ij}$ for all a, then M is totally umbilical, and if $\Sigma_k h^a_{kk} = 0$ for all a, then M is minimal.

The covariant derivative h^a_{ijk} of h^a_{ij} is given by

$$(2.19) \qquad h^a_{ijk}\omega^k = dh^a_{ij} - h^a_{il}\omega^l_j - h^a_{lj}\omega^l_i + h^b_{ij}\omega^a_b$$

and we put

$$(2.20) \qquad h^a_{ijkl}\omega^l = dh^a_{ijk} - h^a_{ljk}\omega^l_i - h^a_{ilk}\omega^l_j - h^a_{ijl}\omega^l_k + h^b_{ijk}\omega^a_b.$$

If $h^a_{ijk} = 0$ for all indices, then the second fundamental form of M is parallel.

3. LAPLACIAN OF THE SECOND FUNDAMENTAL FORM

Let M be an n-dimensional submanifold of an m-dimensional Riemannian manifold $\bar{M}$. Let $e_1,\ldots,e_n$ denote an orthonormal basis of $T_x(M)$. We denote by the same letters local, orthonormal vector fields on M which extend $e_1,\ldots,e_n$, and which are covariant constant with respect to ∇. Let $e_{n+1},\ldots,e_m$ be an orthonormal basis in $T_x(M)^\perp$.

We now compute the Laplacian of the second fundamental form of M. First of all, from equation of Codazzi (2.3), we have

$$(3.1)\qquad \begin{aligned}(\nabla^2 B)(X,Y) &= \sum_i(\nabla_{e_i}\nabla_{e_i}B)(X,Y)\\ &= \sum_i[(R(e_i,X)B)(e_i,Y) + (\bar{\nabla}_X(\bar{R}(e_i,Y)e_i)^\perp)^\perp\\ &\qquad + (\bar{\nabla}_{e_i}(\bar{R}(e_i,X)Y)^\perp)^\perp] + D_XD_Y(\mathrm{Tr}B),\end{aligned}$$

where $\mathrm{Tr}B = \Sigma_i B(e_i,e_i)$. On the other hand, we obtain

$$(3.2)\qquad \sum_i(R(e_i,X)B)(e_i,Y) = \sum_i[R^\perp(e_i,X)B(e_i,Y) - B(R(e_i,X)e_i,Y) - B(e_i,R(e_i,X)Y)],$$

$$(3.3)\qquad \begin{aligned}\sum_i(\bar{\nabla}_X(\bar{R}(e_i,Y)e_i)^\perp)^\perp &= \sum_i[(\bar{\nabla}_X\bar{R})(e_i,Y)e_i + \bar{R}(B(X,e_i),Y)e_i\\ &\quad + \bar{R}(e_i,B(X,Y))e_i + \bar{R}(e_i,Y)B(X,e_i)]^\perp - \sum_i B(X,\bar{R}(e_i,Y)e_i)^T),\end{aligned}$$

$$(3.4)\qquad \begin{aligned}\sum_i(\bar{\nabla}_{e_i}(\bar{R}(e_i,X)Y)^\perp)^\perp &= \sum_i[(\bar{\nabla}_{e_i}\bar{R})(e_i,X)Y + \bar{R}(B(e_i,e_i),X)Y\\ &\quad + \bar{R}(e_i,B(e_i,X))Y + \bar{R}(e_i,X)B(e_i,Y)]^\perp - \sum_i B(e_i,(\bar{R}(e_i,X)Y)^T).\end{aligned}$$

For any vector V normal to M, (3.1), (3.2), (3.3) and (3.4) imply

$$(3.5)\quad g((\nabla^2B)(X,Y),V) = g(D_XD_Y(\mathrm{Tr}B),V) + \sum_i[g((\bar\nabla_X\bar R)(e_i,Y)e_i,V)$$
$$- g((\bar\nabla_{e_i}\bar R)(e_i,X)Y,V)] + \sum_i[g(R^\perp(e_i,X)B(e_i,Y),V)$$
$$+ 2g(\bar R(e_i,Y)B(X,e_i),V) + g(\bar R(e_i,B(X,Y))e_i,V)$$
$$+ g(\bar R(B(e_i,e_i),X)Y,V) + g(\bar R(e_i,X)B(e_i,Y),V)$$
$$- g(A_VX,\bar R(e_i,Y)e_i) - g(A_Ve_i,\bar R(e_i,X)Y)]$$
$$- \sum_i[g(R(e_i,X)e_i,A_VY) + g(R(e_i,X)Y,A_Ve_i)],$$

where we used the first Bianchi identity. Moreover, we have

$$(3.6)\quad - \sum_i[g(R(e_i,X)e_i,A_VY) + g(R(e_i,X)Y,A_Ve_i)]$$
$$= - \sum_i[g(\bar R(e_i,X)e_i,A_VY) + g(\bar R(e_i,X)Y,A_Ve_i)]$$
$$- \sum_a[g(A_VA_a^2X,Y) - \mathrm{Tr}A_a g(A_VA_aX,Y) + \mathrm{Tr}A_aA_V g(A_aX,Y)$$
$$- g(A_aA_VA_aX,Y)],$$

$$(3.7)\quad \sum_i g(R^\perp(e_i,X)B(e_i,Y),V) = \sum_i g(\bar R(e_i,X)B(e_i,Y),V)$$
$$+ \sum_a[g(A_aA_VA_aX,Y) - g(A_a^2A_VX,Y)].$$

Substituting (3.6) and (3.7) into (3.5), we get

$$(3.8)\quad g((\nabla^2B)(X,Y),V) = g(D_XD_Y(\mathrm{Tr}B),V) + \sum_i[g((\bar\nabla_X\bar R)(e_i,Y)e_i,V)$$
$$- g((\bar\nabla_{e_i}\bar R)(e_i,X)Y,V)] + \sum_i[2g(\bar R(e_i,Y)B(X,e_i),V)$$
$$+ 2g(\bar R(e_i,X)B(e_i,Y),V) - g(A_VX,\bar R(e_i,Y)e_i) - g(A_VY,\bar R(e_i,X)e_i)$$
$$+ g(\bar R(e_i,B(X,Y))e_i,V) + g(\bar R(B(e_i,e_i),X)Y,V)$$
$$- 2g(A_Ve_i,\bar R(e_i,X)Y)] + \sum_a[\mathrm{Tr}A_a g(A_VA_aX,Y) - \mathrm{Tr}A_aA_V g(A_aX,Y)$$
$$+ 2g(A_aA_VA_aX,Y) - g(A_a^2A_VX,Y) - g(A_VA_a^2X,Y)].$$

PROPOSITION 3.1. Let M be a submanifold of a locally symmetric Riemannian manifold $\bar{M}$. If the mean curvature vector of M is parallel, then

$$(3.9)\quad g((\nabla^2 B)(X,Y),V) = \sum_i [2g(\bar{R}(e_i,Y)B(X,e_i),V) + 2g(\bar{R}(e_i,X)B(Y,e_i),V)$$
$$- g(A_V X,\bar{R}(e_i,Y)e_i) - g(A_V Y,\bar{R}(e_i,X)e_i + g(\bar{R}(e_i,B(X,Y))e_i,V)$$
$$+ g(\bar{R}(B(e_i,e_i),X)Y,V) - 2g(A_V e_i,\bar{R}(e_i,X)Y)]$$
$$+ \sum_a [\mathrm{Tr}A_a g(A_V A_a X,Y) - \mathrm{Tr}A_a A_V g(A_a X,Y) + 2g(A_a A_V A_a X,Y)$$
$$- g(A_a^2 A_V X,Y) - g(A_V A_a^2 X,Y)].$$

If M is minimal, then

$$(3.10)\quad g((\nabla^2 B)(X,Y),V) = \sum_i [2g(\bar{R}(e_i,Y)B(X,e_i),V)$$
$$+ 2g(\bar{R}(e_i,X)B(Y,e_i),V) - g(A_V X,\bar{R}(e_i,Y)e_i)$$
$$- g(A_V Y,\bar{R}(e_i,X)e_i) + g(\bar{R}(e_i,B(X,Y))e_i,V)$$
$$- 2g(A_V e_i,\bar{R}(e_i,X)Y)] + \sum_a [2g(A_a A_V A_a X,Y) - g(A_a^2 A_V X,Y)$$
$$- g(A_V A_a^2 X,Y)].$$

We now assume that $\bar{R}(X,Y)Z$ is tangent to M for any vector fields X, Y and Z tangent to M. Then (3.1) reduces to

$$(3.11)\quad (\nabla^2 B)(X,Y) = \sum_i (R(e_i,X)B)(e_i,Y) + D_X D_Y(\mathrm{Tr}B).$$

From (3.2) and (3.11) we have

PROPOSITION 3.2. Let M be a submanifold of a Riemannian manifold $\bar{M}$. If $\bar{R}(X,Y)Z$ is tangent to M for any vector fields X, Y and Z tangent to M, then

$$(3.12) \qquad (\nabla^2 B)(X,Y) = \sum_i [R^\perp(e_i,X)B(e_i,Y) - B(R(e_i,X)e_i,Y) - B(e_i,R(e_i,X)Y)] + D_X D_Y(\mathrm{Tr}B).$$

We now give a necessary and sufficient condition for a submanifold to have a flat normal connection (cf. Chen [1]).

PROPOSITION 3.3. Let M be an n-dimensional submanifold of an m-dimensional Riemannian manifold $\bar{M}$. Then the normal connection D of M in $\bar{M}$ is flat if and only if there exist locally m-n mutually orthogonal unit normal vector fields e_a such that each of the e_a is parallel.

Proof. Suppose that there exist locally m-n mutually orthogonal unit normal vector fields e_a such that $De_a = 0$. Then (2.4) gives $R^\perp(X,Y)e_a = 0$, which shows that the normal connection of M is flat.

Conversely, suppose that the normal connection of M is flat. Then we have

$$D_X D_Y e_a - D_Y D_X e_a - D_{[X,Y]} e_a = 0$$

for any m-n mutually orthogonal unit vector fields e_a normal to M. If we put

$$D_X e_a = \omega_a^b(X) e_b,$$

then we see that $\omega_a^b = -\omega_b^a$ and

$$0 = [X\omega_a^b(Y) - Y\omega_a^b(X) - \omega_a^b([X,Y])]e_b + [\omega_a^c(Y)\omega_c^b(X) - \omega_a^c(X)\omega_c^b(Y)]e_b,$$

that is,

$$d\omega_a^b = -\omega_a^c \wedge \omega_c^b, \qquad \omega_a^b = -\omega_b^a.$$

Thus we know that there exists an (m-n,m-n)-matrix $A = (\lambda_a^b)$ of functions satisfying

$$dA = -A\Omega, \qquad {}^tA = A^{-1},$$

where $\Omega = (\omega_a^b)$. This equation has the local expression

$$d\lambda_a^c = \sum_b \lambda_a^b \omega_c^b = -\sum_b \lambda_a^b \omega_b^c.$$

We put $e'_a = \lambda_a^b e_b$. Then e'_a are also m-n mutually orthogonal unit normal vector fields of M and we have

$$\omega'^b_a \lambda_b^c = d\lambda_a^c + \lambda_a^b \omega_b^c = 0,$$

where ω'^b_a are defined by $De'_a = \omega'^b_a e'_b$. This shows that each of e'_a is parallel in the normal bundle. This proves our assertion. QED.

Let M be an n-dimensional submanifold of an m-dimensional Riemannian manifold $\bar{M}$. For $x \in M$, the *first normal space* $N_1(x)$ is the orthogonal complement in $T_x(M)^\perp$ of the set

$$N_0(x) = \{V \in T_x(M) : A_V = 0\}.$$

If, for any vector field V with $V_x \in N_1(x)$, we have $D_X V \in N_1(x)$ for any vector field X of M at x, then the first normal space $N_1(x)$ is said to be *parallel* with respect to the normal connection. We have the following reduction theorem of codimension (cf. Erbacher [1]).

THEOREM 3.1. Let M be an n-dimensional submanifold of an m-dimensional complete simply connected space form $\bar{M}^m(c)$. Suppose the first normal space $N_1(x)$ has constant dimension k, and is parallel with respect to the normal connection. Then there is a totally geodesic (n+k)-dimensional submanifold $M^{n+k}(c)$ of $\bar{M}^m(c)$ which contains M.

4. SUBMANIFOLDS OF SPACE FORMS

In this section we shall consider submanifolds of space forms and give some examples.

First of all, we prove

PROPOSITION 4.1. A totally umbilical submanifold M of a Riemannian manifold $\bar{M}$ of constant curvature c is also of constant curvature.

Proof. Since M is totally umbilical, we have $B(X,Y) = g(X,Y)\mu$ for any vector fields X and Y tangent to M, where μ is the mean curvature vector of M. Thus (2.9) implies

$$g(R(X,Y)Z,W) = (c+|\mu|^2)[g(Y,Z)g(X,W) - g(X,Z)g(Y,W)].$$

This shows that M is of constant curvature $c+|\mu|^2$ for dim M > 2. If dim M = 2, $|\mu|^2$ is a constant by equation of Codazzi (2.7). Consequently, M is of constant curvature. QED.

We now give applications of the expression for the Laplacian of the second fundamental form.

PROPOSITION 4.2. Let M be an n-dimensional submanifold of a space form $\bar{M}^m(c)$. Suppose that the normal connection of M is flat and the mean curvature vector of M is parallel. Then

$$\tfrac{1}{2}\Delta|A|^2 = |\nabla A|^2 + \tfrac{1}{2}\sum_{a,i,j} K_{ij}(\lambda_i^a - \lambda_j^a)^2, \tag{4.1}$$

where K_{ij} denotes the sectional curvature of M for the section spanned by e_i and e_j and λ_i^a denote the eigenvalues of A_a with respect to the basis $\{e_i\}$.

Proof. From the assumption and (3.12) we have

$$(\nabla^2 B)(X,Y) = -\sum_i [B(R(e_i,X)e_i,Y) + B(e_i,R(e_i,X)Y].$$

On the other hand, we have

$$g((\nabla^2 B)(X,Y),V) = g((\nabla^2 A)_V X,Y)$$

for any vector field V normal to M. Thus we get

$$g((\nabla^2 A)_V X,Y) = -\sum_i [g(A_V Y,R(e_i,X)e_i) + g(A_V e_i,R(e_i,X)Y)],$$

from which

$$(4.2) \qquad g(\nabla^2 A,A) = \sum_{a,i} g((\nabla^2 A)_a e_i,A_a e_i)$$
$$= -\sum_{a,i,j} [g(A_a^2 e_j,R(e_i,e_j)e_i)+g(A_a e_i,R(e_i,e_j)A_a e_j)].$$

Since the normal connection of M is flat, we see that the second fundamental form of M is commutative. Thus we can choose $e_1,\ldots,e_n$ such that $A_a e_i = \lambda_i^a e_i$ for all a. Therefore, (4.2) becomes

$$g(\nabla^2 A,A) = \sum_{a,i,j} K_{ij}(\lambda_j^a \lambda_j^a - \lambda_i^a \lambda_j^a) = \tfrac{1}{2}\sum_{a,i,j} K_{ij}(\lambda_i^a - \lambda_j^a)^2.$$

Since we have

$$g(\nabla^2 A,A) = -|\nabla A|^2 + \tfrac{1}{2}\Delta|A|^2,$$

we obtain (4.1) from the equation above. QED.

As applications of Proposition 4.2 we have the following propositions.

PROPOSITION 4.3. Let M be an n-dimensional submanifold of $\bar{M}^m(c)$ with flat normal connection and parallel mean curvature vector. If M is compact and M has non-negative sectional curvature, then the second fundamental form of M is parallel.

PROPOSITION 4.4. Let M be an n-dimensional submanifold of $\bar{M}^m(c)$ with flat normal connection and parallel mean curvature vector. If $|A|^2$ is constant on M, and M has non-negative sectional curvature, then the second fundamental form of M is parallel.

We now give a few examples of n-dimensional submanifolds discussed above in an m-dimensional Euclidean space R^m with usual inner product (x,y) (cf. Yano-Ishihara [3]).

Example 4.1. For integers $p_1,\ldots,p_N \geq 1$, $p_1+\ldots+p_N = n$, consider R^m given

$$R^m = R^{p_1+1} \times \ldots \times R^{p_N+1}, \qquad N = m-n.$$

We put

$$S^{p_i}(r_i) = \{x_i \in R^{p_i+1} : (x_i,x_i) = r_i^2\}, \quad i = 1,\ldots,N.$$

Then the pythagorean product

$$\Pi S^{p_i}(r_i) = S^{p_1}(r_1) \times \ldots \times S^{p_N}(r_N)$$
$$= \{(x_1,\ldots,x_N) \in R^m : x_i \in S^{p_i}(r_i),\ i = 1,\ldots,N\}$$

is an n-dimensional submanifold M^n of essential codimension m-n in R^m. The mean curvature vector μ of M^n is given by

$$\mu = \frac{1}{n}(p_1 r_1^{-2} x_1 + \ldots + p_N r_N^{-2} x_N)$$

at $(x_1,\ldots,x_N) \in M^n$, which is parallel in the normal bundle of M^n and the function $|A|^2$ is given by

$$|A|^2 = 1/r_1^2 + \ldots + 1/r_N^2,$$

which is constant on M^n. Moreover, the normal connection of M^n is flat.

For integers $p_1,\ldots,p_N,p$ such that $p_1,\ldots,p_N,p \geq 1$, $p_1 + \ldots + p_N + p = n$, consider R^m given

$$R^m = R^{p_1+1} \times \ldots \times R^{p_N+1} \times R^p, \quad N = m-n.$$

Then the pythagorean product

$$S^{p_1}(r_1) \times \ldots \times S^{p_N}(r_N) \times R^p$$
$$= \{(x_1,\ldots,x_N,x) \in R^m : x_i \in S^{p_i}(r_i),\ i = 1,\ldots,N,\ x \in R^p\}$$

is an n-dimensional submanifold M^n of essential codimension $N = m-n$ in R^m. M^n has parallel mean curvature vector and flat normal connection. Moreover $|A|^2$ is constant.

Example 4.2. In R^{m+1} we consider an m-dimensional sphere of radius $a > 0$:

$$S^m(a) = \{x \in R^{m+1} : (x,x) = a^2\}.$$

For mutually orthogonal unit vectors $b_1,\ldots,b_{m-n}$ in R^{m+1}, a submanifold $\Sigma^n(r)$ in $S^m(a)$ defined by

$$\Sigma^n(r) = \{x \in S^m(a) : (x,b_t) = d_t,\ t = 1,\ldots,m-n\}$$

is called an n-dimensional *small sphere* of $S^m(a)$ with radius r if $(d_1,\ldots,d_{m-n}) \neq (0,\ldots,0)$, where $r^2 = a^2 - d_1^2 - \ldots - d_{m-n}^2 > 0$ and $1 < n < m$. $\Sigma^n(r)$ is called an n-dimensional *great sphere* of $S^m(a)$ if $(d_1,\ldots,d_{m-n}) = (0,\ldots,0)$, that is, if $r = a$. If $r \neq a$, then $\Sigma^n(r)$ is a totally umbilical submanifold of essential codimension m-n in $S^m(a)$, and the mean curvature of $\Sigma^n(r)$ is given by

$$|\mu| = d/(a(a^2-d^2)^{1/2}), \quad d^2 = d_1^2 + \ldots + d_{m-n}^2 \ (d > 0).$$

A great sphere $\Sigma^n(a)$ is totally geodesic in $S^m(a)$.

Example 4.3. For integers $p_1,\ldots,p_N$ such that $p_1,\ldots,p_N \geq 1$, $p_1 + \ldots + p_N = n$, consider

$$R^{m+1} = R^{p_1+1} \times \ldots \times R^{p_N+1}, \quad n = m-n+1.$$

$$(4.3) \quad S^{p_1}(r_1) \times \ldots \times S^{p_N}(r_N)$$
$$= \{(x_1,\ldots,x_N) \in R^{m+1} : x_i \in S^{p_i}(r_i),\ i = 1,\ldots,N\},$$

where $S^{p_i}(r_i) \subset R^{p_i+1}$ $(i = 1,\ldots,N)$, is an n-dimensional submanifold M^n of essential codimension m-n imbedded in $S^m(a)$ if $r_i^2 = a^2$. The mean curvature vector μ of M^n relative to $S^m(a)$ is given by

$$\mu = \frac{1}{n}(p_1r_1^{-2}x_1 + \ldots + p_Nr_N^{-2}x_N) - a^{-2}(x_1 + \ldots + x_N)$$

at $(x_1,\ldots,x_N) \in M^n$, which is parallel in the normal bundle of M^n relative to $S^m(a)$. We also have

$$|A|^2 = r_1^2(r_1^{-2} - a^{-2})^2 + \ldots + r_N^2(r_N^{-2} - a^{-2})^2 + N(N-1)a^{-2},$$

which is constant in M^n. It is easily verified that the connection of the normal bundle of M^n is flat.

Let $\Sigma^{m-1}(r)$ be an (m-1)-dimensional small sphere of $S^m(a)$ $(0 < r < a)$. For integers $p_1,\ldots,p_{N'}$ such that $p_1,\ldots,p_{N'} \geq 1$, $p_1 + \ldots + p_{N'} = n$, $N' = m-n$, in $\Sigma^{m-1}(r)$ consider an n-dimensional submanifold $\tilde{M}^n$ of the form

$$(4.4) \quad \Sigma^{p_1}(r_1) \times \ldots \times \Sigma^{p_{N'}}(r_{N'}) \subset \Sigma^{m-1}(r),$$

where $r_1^2 + \ldots + r_{N'}^2 = r^2 < a^2$, $\Sigma^{p_i}(r_i)$ $(i = 1,\ldots,N')$ is a p_i-dimensional sphere with radius r_i and $\tilde{M}^n$ is constructed in $\Sigma^{m-1}(r)$ in the same way as that used in constructing in $S^m(a)$ a submanifold M^n of the form (4.3). Then $\tilde{M}^n$ is an n-dimensional submanifold of essential codimension m-n-1 in $\Sigma^{m-1}(r)$ and therefore m-n in $S^m(a)$. The mean curvature vector of $\tilde{M}^n$ relative to $S^m(a)$ is parallel, $|A|^2$ relative to $S^m(a)$ is constant and the normal connection of $\tilde{M}^n$ relative to $S^m(a)$ is flat.

We now prove the following lemma (Erbacher [2], Yano-Ishihara [3]).

LEMMA 4.1. Let M be an n-dimensional submanifold of $\bar{M}^m(c)$ with flat normal connection and parallel mean curvature vector. Suppose that M has non-negative sectional curvature, and M has constant scalar curvature or M is compact. Then for every point x of M, there exist orthonormal vector fields $v_1,\dots,v_p$ $(p = m-n)$ defined in a neighborhood U of x such that

(1) v_a are all parallel in the normal bundle on U;

(2)

$$A_a = \begin{bmatrix} 0 & 0 & 0 \\ 0 & \lambda_a I_k & 0 \\ 0 & 0 & 0 \end{bmatrix},$$

where I_k is the (p_a,p_a) identity matrix, the zero matrix in the upper left-hand corner is of degree $p_1+\dots+p_{a-1}$ and A_a's are expressed with respect to their common orthonormal eigenvectors $e_1,\dots,e_n$. Note that $A_a = 0$ if $p_1+\dots+p_{a-1} = n$ and we may assume that $A_a = 0$ implies $A_b = 0$ for $b > a$;

(3) Each λ_a is constant on U.

Proof. First of all we notice that $|A|^2$ is constant because of the scalar curvature of M is constant and the mean curvature vector of M is parallel.

Since the normal connection of M is flat, there exist orthonormal vector fields $v_1,\dots,v_p$ $(p = m-n)$ such that $Dv_a = 0$ in U. Then we see that $(\nabla_X A)_a = \nabla_X(A_a) = 0$ by Propositions 4.3 and 4.4. Thus the eigenvalues of A_a are constant. If $v'_b = \Sigma_a O_{ab} v_a$, (O_{ab}) an orthogonal matrix with constant entries, then $Dv'_b = 0$ and $A'_b = \Sigma_a O_{ab} A_a$, where A'_b is the second fundamental tensor corresponding to v'_b. In what follows we will begin with any v_a such that $Dv_a = 0$ in U and show that there exists an orthogonal matrix (O_{ab}) with constant entries such that A'_b have the desired property (2). The claim is clearly true if all the A_a's = 0 at x (and therefore by constancy of the eigenvalues $A_a = 0$ in U). If this is not the case we distinguish three cases:

(i) all sectional curvatures of $M > 0$ at x,

(ii) all sectional curvatures of $M = 0$ at x,

(iii) at least one non-zero sectional curvature at x and at least one sectional curvature that is zero at x.

Suppose that v_a and U have been chosen such that (1) is satisfied. Thus each A_a has constant eigenvalues in U.

Case (i): From Proposition 4.2 we see that $A_a = \lambda_a I$. We may assume that $\lambda_1 \neq 0$. Let

$$v_1' = (\sum_a \lambda_a v_a)/(\sum_a \lambda_a^2)^{1/2}$$

and

$$\bar{v}_b = (\lambda_1 v_b - \lambda_b v_1)/(\lambda_1^2 + \lambda_b^2)^{1/2}$$

for $b > 1$. Then $A_1' = \lambda I$, $\lambda \neq 0$ and $\bar{A}_b = 0$ for $b > 1$, $\bar{v}_b \perp v_1'$. Use the Gram-Schmit orthogonalization process on $\bar{v}_2, \ldots, \bar{v}_p$ to obtain $v_2', \ldots, v_p'$. Then $A_b' = 0$ for $b > 1$.

Case (ii): Let

$$A_a = \begin{pmatrix} \lambda_{a1} & & & & \\ & \cdot & & & \\ & & \cdot & & \\ & & & \cdot & \\ & & & & \lambda_{an} \end{pmatrix}$$

expressed with respect to the common eigenvectors $e_1, \ldots, e_n$ of the A_a's. We may assume that $\lambda_{11} \neq 0$. Let

$$v_1' = (\sum_a \lambda_{a1} v_a)/(\sum_a \lambda_{a1}^2)^{1/2},$$

$$\bar{v}_a = (\lambda_{11} v_a - \lambda_{a1} v_1)/(\lambda_{a1}^2 + \lambda_{a1}^2)^{1/2} \quad \text{for } a > 1.$$

Again, $\bar{v}_2 \perp v'_1$. Use the Gram-Schmit orthogonalization process on $\bar{v}_2,\ldots,\bar{v}_p$ to obtain $v'_2,\ldots,v'_p$ Then, for $a \geq 2$,

$$A'_a = \begin{pmatrix} 0 & 0 \ldots 0 \\ 0 & * \\ \cdot & \cdot \\ \cdot & \quad \cdot \\ \cdot & \quad\quad \cdot \\ 0 & \quad\quad\quad * \end{pmatrix}$$

and $\lambda'_{11} \neq 0$. Thus, we may assume that $\lambda_{a1} = 0$ for $a > 1$, $\lambda_{11} \neq 0$. Since $0 = K_{1j} = \Sigma_a \lambda_{a1}\lambda_{aj} = \lambda_{11}\lambda_{1j}$ for $j > 1$, we have $\lambda_{1j} = 0$ for $j > 1$. If one of the A'_as, for $a \geq 2$, is not zero, we may assume that it is A_2 and apply the above argument to $v_2,\ldots,v_p$ and $A_2,\ldots,A_p$ restricted to the span $\{e_2,\ldots,e_n\}$. We obtain $\lambda_{2j} = 0$ for $j > 2$ and $\lambda_{a2} = 0$ for $a > 2$. It is now clear that an induction argument will work.

Case (iii): Order $e_1,\ldots,e_n$ so that $K_{1t} > 0$ for $2 \leq t \leq p_1$, and $K_{1t} = 0$ for $t > p_1$. Then (4.1) implies that $\lambda_{a1} = \lambda_{at}$ for $1 \leq t \leq p_1$. Define v'_a as in case (ii). We see that we may assume that $\lambda_{at} = 0$ for $1 \leq t \leq p_1$, $2 \leq a \leq p$. Then $K_{1t} = \lambda_{11}\lambda_{1t} = 0$ for $t > p_1$ and thus $\lambda_{1t} = 0$ for $t > p_1$. If $K_{ij} \neq 0$ for some $i,j > p_1$, we repeat the above argument applied to $v_2,\ldots,v_p$ and $A_2,\ldots,A_p$ restricted to the span $\{e_{p_1+1},\ldots,e_n\}$. If $K_{ij} = 0$ for all $i,j > p_1$, we apply the argument of case (ii) to $v_2,\ldots,v_p$ and $A_2,\ldots,A_p$ restricted to the span $\{e_{p_1+1},\ldots,e_n\}$. In either case we obtain the desired form for A_1 and A_2. It is clear that an induction argument will work. QED.

We shall prove the following theorems (Erbacher [2], Yano-Ishihara [3]).

THEOREM 4.1. Let M be an n-dimensional complete submanifold of R^m with non-negative sectional curvature. Suppose that the normal connection of M is flat and the mean curvature vector of M is parallel. If the scalar curvature of M is constant, then M is a sphere $S^m(r)$, n-dimensional plane R^n, a pythagorean product of the form

$$(4.5) \qquad S^{p_1}(r_1) \times \ldots \times S^{p_N}(r_N), \qquad \Sigma p_i = n,\ 1 < N \leq m-n,$$

or a pythagorean product of the form

$$S^{p_1}(r_1) \times \ldots \times S^{p_N}(r_N) \times R^p, \quad \Sigma p_i + p = n, \ 1 < N \leq m-n. \tag{4.6}$$

Proof. Let v_a be chosen as in Lemma 4.1. We may assume that $\lambda_b \neq 0$ for $1 \leq b \leq c-1$ and $\lambda_b = 0$ for $b \geq c$ (if all $\lambda_a = 0$, then M is totally geodesic).

Define distributions $T_1, \ldots, T_k$ by

$$T_b(x) = \{X \in T_x(M) : A_b X = \lambda_b X\} \quad \text{for } b < c,$$

$$T_c(x) = \{X \in T_x(M) : A_a X = 0, \ 1 \leq a \leq m-n\}.$$

Let $p_a = \dim T_a$ (p_a may be zero). Assume that M is simply connected and complete. Then each T_a is globally defined (for $v \in T_x(M)^{\perp}$, the parallel translate of v with respect to the normal connection is independent of path if the normal connection is flat and M is simply connected). Each T_a has constant dimension and is differentiable (the eigenspaces of A_a have constant dimension, and thus we may find differentiable orthonormal eigenvectors). The T_a's are orthogonal to each other and

$$T_x(M) = T_1(x) \oplus \ldots \oplus T_c(x) \quad \text{(orthogonal direct sum).}$$

Let Y be in T_a. For any vector field X tangent to M we have

$$A_a \nabla_X Y = \nabla_X (A_a Y) - (\nabla_X A)_a Y = \lambda_a \nabla_X Y.$$

Thus we have $\nabla_X Y \in T_a$. Therefore, each T_a is parallel. For each point x of M, let M^{p_a} be the maximal integral submanifold of T_a through x. Then M is the Riemannian product

$$M^{p_1} \times \ldots \times M^{p_c}.$$

If $p_a = 1$, then the image of M^{p_a} is a circle since we have assumed that M is simply connected and complete. If $p_a > 1$, then the curvature tensor of M^{p_a} is the restriction of the curvature tensor of M, since M^{p_a} is totally geodesic in M. Hence, the sectional curvature of M^{p_a} is constant and equal to λ_a^2. On the other hand, we see that $M^{p_c} = R^{p_c}$. Thus, M is a product of small spheres and possibly a great sphere. Clearly, the corresponding local result is true if we do not assume completeness since we only used completeness to obtain M^{p_a} as the entire sphere.

The second fundamental forms and the normal connection forms of our submanifold with respect to v_a, chosen as in Lemma 4.1, are the same as Example 4.1. Thus, by the fundamental theorem of submanifolds (cf. Kobayashi-Nomizu [;p.45] for the case p = 1), we see that the submanifold M is of the form as in the theorem up to a rigid motion.

If M is not simply connected, let $\hat{M}$ be its simply connected Riemannian covering manifold and let π be the covering map. Then the composition mapping ψ of $\hat{M}$ into R^m under π and the immersion of M satisfies the assumptions of the theorem and, by the above, there exists an isometry Φ of R^m such that $\psi = \Phi \circ i$, i being the immersion of Example 4.1. Thus, M is immersed as the submanifold of the form as in the theorem. This completes the proof of the theorem. QED.

Remark. If M is a pythagorean product of the form (4.5) or (4.6), then M is of essential codimension N.

THEOREM 4.2. Let M be an n-dimensional compact submanifold of R^m with non-negative sectional curvature, and suppose that the normal connection of M is flat. If the mean curvature vector of M is parallel, then M is a sphere $S^n(r)$ or a pythagorean product of the form (4.5), which is of essential codimension N.

If the ambient manifold $\bar{M}$ is a sphere $S^m(a)$, we just imbedded $S^m(a)$ in R^{m+1} as Example 4.2. Then M regarded as a submanifold of R^{m+1} having the properties in Theorem 4.1 by the following lemma, which will be proved easily.

LEMMA 4.2. Let M be an n-dimensional submanifold of $S^m(a)$ such that

(1) $|A|^2$ is constant on M;

(2) The mean curvature vector μ of M is parallel;

(3) The normal connection of M is flat.

Then, if we consider $S^m(a)$ as isometrically immersed in R^{m+1}, conditions (1), (2) and (3) are also satisfied. (of course, $|A|^2$, μ and the normal connection are now taken with respect to M in R^{m+1}.)

From Theorem 4.1 and Lemma 4.2 we have (Erbacher [2], Yano-Ishihara [3])

THEOREM 4.3. Let M be an n-dimensional complete submanifold of $S^m(a)$ with non-negative sectional curvature. Suppose that the mean curvature vector of M is parallel and the normal connection of M is flat. If the scalar curvature of M is constant, then M is a small sphere $\Sigma^n(r)$, a great sphere $\Sigma^n(a)$ or a pythagorean product of a certain number of spheres. Moreover, if M is of essential codimension m-n, then M is a pythagorean product of the form (4.3) with $r_1^2 + \dots + r_N^2 = a^2$; $N = m-n+1$, or of the form (4.4) with $r_1^2 + \dots + r_{N'}^2 = r^2 < a^2$, $N' = m-n$. If M is a pythagorean product of the form (4.4) with $r_1^2 + \dots + r_N^2 = r^2 < a^2$; $N = m-n$, then M is contained in a small sphere $\Sigma^{m-1}(r)$ of $S^m(a)$.

THEOREM 4.4. Let M be an n-dimensional compact submanifold of $S^m(a)$ with non-negative sectional curvature. Suppose that the normal connection of M is flat and the mean curvature vector of M is parallel. If M is of essential codimension m-n, then we have the same conclusion as that in Theorem 4.3.

Remark. If M^n in $S^m(a)$ satisfies the condition of Theorem 4.3 or 4.4 and if M^n is of essential codimension s less than m-n, then M^n is contained in a great sphere $S^{n+s}(a)$ of $S^m(a)$ (see Theorem 3.1).

5. MINIMAL SUBMANIFOLDS

Let M be an n-dimensional submanifold of $S^m(a)$ and satisfy the conditions in Theorem 4.3 or 4.4. Then the mean curvature vector of M is given by

$$\mu = \frac{1}{n}(p_1\mu_1 + \dots + p_N\mu_N),$$

where $\mu_1,\dots,\mu_N$ are distinct vectors of eigenvalues, and $p_1,\dots,p_N$ the multiplicities of $\mu_1,\dots,\mu_N$ respectively. Thus M is minimal if and only if

$$\mu = p_1\mu_1 + \dots + p_N\mu_N = 0. \tag{5.1}$$

Therefore, the vectors $\mu_1,\dots,\mu_N$ are linearly dependent. Hence we see that M is of essential codimension N-1 if M is a pythagorean product of the form (4.3). Since we have

$$0 = \frac{1}{n}(p_1 r_1^{-2} x_1 + \dots + p_N r_N^{-2} x_N) - a^{-2}(x_1 + \dots + x_N), \tag{5.2}$$

where $r_i = a(p_i/n)^{1/2}$ $(i = 1,\dots,N)$. Therefore, Theorems 4.3 and 4.4 imply the following theorems (Yano-Ishihara [3]).

THEOREM 5.1. Let M be an n-dimensional complete minimal submanifold of $S^m(a)$ with non-negative sectional curvature, and suppose that the normal connection of M is flat. If the scalar curvature of M is constant, then M is a great sphere of $S^m(a)$ or a pythagorean product of the form

$$S^{p_1}(r_1) \times \dots \times S^{p_N}(r_N), \qquad \Sigma p_i = n,\ 1 < N \le m-n+1 \tag{5.3}$$

with essential codimension N-1, where $r_i = a(p_i/n)^{1/2}$ $(i = 1,\dots,N)$.

THEOREM 5.2. Let M be an n-dimensional compact minimal submanifold of $S^m(a)$. If M has non-negative sectional curvature and the normal connection of M is flat, then we have the same conclusion as that in Theorem 5.1.

Example 5.1. Let n and p be two positive integers such that $n > p$. We put

$$M_{p,n-p} = S^p((p/n)^{1/2}) \times S^{n-p}(((n-p)/n)^{1/2}).$$

We imbed $M_{p,n-p}$ into $S^{n+1} = S^{n+1}(1)$ as follows. Let (x_1,x_2) be a point of $M_{p,n-p}$ where x_1 (resp. x_2) is a vector in R^{p+1} (resp. R^{n-p+1}) of length $(p/n)^{1/2}$ (resp. $((n-p)/n)^{1/2}$). We consider (x_1,x_2) as a unit vector in $R^{n+2} = R^{p+1} \times R^{n-p+1}$. Then $M_{p,n-p}$ is a minimal hypersurface of S^{n+1}. $M_{p,n-p}$ is called a *Clifford minimal hypersurface*. In particular, if $n = 2$ and $p = 1$, $M_{1,1}$ is a flat minimal surface of S^3. We call this minimal surface the *Clifford torus*.

Example 5.2. Let (x,y,z) be the standard coordinate system in R^3 and (u^1,u^2,u^3,u^4,u^5) be the standard coordinate system in R^5. We consider the mapping defined by

$$u^1 = (1/3)^{1/2}yz, \quad u^2 = (1/3)^{1/2}zx, \quad u^3 = (1/3)^{1/2}xy,$$

$$u^4 = (1/12)^{1/2}(x^2 - y^2), \quad u^5 = (1/6)(x^2 + y^2 - 2z^2).$$

This defines an isometric immersion of $S^2(\sqrt{3})$ into $S^4 = S^4(1)$. Two points (x,y,z) and $(-x,-y,-z)$ of $S^2(\sqrt{3})$ are mapped into the same point of S^4, and this mapping defines an imbedding of the real projective plane into S^4. This real projective plane imbedded in S^4 is called the *Veronese surface*. It is minimal surface of S^4.

Let M be an n-dimensional minimal submanifold of $\bar{M}^m(c)$ $(m-n = p)$. Then (3.10) implies

$$(5.4) \quad \tfrac{1}{2}\Delta|A|^2 = nc|A|^2 - \sum_{a,b}(\mathrm{Tr}A_aA_b)^2 + \sum_{a,b}\mathrm{Tr}[A_a,A_b]^2 + |\nabla A|^2.$$

We need the following lemma (Chern-do Carmo-Kobayashi [1]).

LEMMA 5.1. Let A and B be symmetric (n,n)-matrices. Then

$$-\mathrm{Tr}(AB - BA)^2 \leq 2\mathrm{Tr}A^2\mathrm{Tr}B^2,$$

and the equality holds for non-zero matrices A and B if and only if A and B can be transformed simultaneousely by an orthogonal matrix into scalar multiples of $\bar{A}$ and $\bar{B}$ respectively, where

$$\bar{A} = \left(\begin{array}{cc|c} 0 & 1 & \\ 1 & 0 & 0 \\ \hline \multicolumn{2}{c|}{0} & 0 \end{array}\right), \qquad \bar{B} = \left(\begin{array}{cc|c} 1 & 0 & \\ 0 & -1 & 0 \\ \hline \multicolumn{2}{c|}{0} & 0 \end{array}\right).$$

Moreover, if A_1, A_2 and A_3 are (n,n)-symmetric matrices and if

$$-\mathrm{Tr}(A_iA_j - A_jA_i)^2 = 2\mathrm{Tr}A_i^2\mathrm{Tr}A_j^2, \quad 1 \leq i,j \leq 3,$$

then at least one of the matrices A_i must be zero.

Proof. We may assume that B is diagonal and we denote by $b_1,\ldots,b_n$ the diagonal entries in B. Then we have

$$-\mathrm{Tr}(AB - BA)^2 = \sum_{i\neq k} a_{ik}^2(b_i - b_k)^2,$$

where $A = (a_{ij})$. Since $(b_i - b_k)^2 \leq 2(b_i^2 + b_k^2)$, we obtain

$$\begin{aligned} -\mathrm{Tr}(AB - BA)^2 &= \sum_{i\neq k} a_{ik}^2(b_i - b_k)^2 \leq 2\sum_{i\neq k} a_{ik}^2(b_i^2 + b_k^2) \\ &\leq 2(\sum_{i,k} a_{ik}^2)(\sum_i b_i^2) = 2\mathrm{Tr}A^2\mathrm{Tr}B^2. \end{aligned}$$

Now, assume that A and B are nonzero matrices and that the equality holds. From the second equality in the inequalities above, we have $a_{11} = \ldots = a_{nn} = 0$ and $b_i + b_k = 0$ if $a_{i,k} \neq 0$. Without loss of generality, we may assume that $a_{12} \neq 0$. Then $b_1 = -b_2$. From the third equality, we now obtain $b_3 = \dots = b_n = 0$. Since $B \neq 0$, we must have $b_1 = -b_2 \neq 0$ and we conclude that $a_{ik} = 0$ for $(i,k) \neq (1,2)$. To prove the last statement, let A_1, A_2, A_3 be all nonzero symmetric matrices. From the second statement we have just proved, we see that one of these matrices can be transformed to B a scalar multiple of $\bar{A}$ as well as to a scalar multiple of $\bar{B}$ by orthogonal matrices. But this is impossible since $\bar{A}$ and $\bar{B}$ are not orthogonally equivalent. QED.

From Lemma 5.1 we obtain

$$
\begin{aligned}
(5.5) \qquad \sum_{a,b} (\mathrm{Tr} A_a A_b)^2 - \sum_{a,b} \mathrm{Tr}[A_a, A_b]^2 &\leq \sum_a (\mathrm{Tr} A_a^2)^2 + 2 \sum_{a \neq b} \mathrm{Tr} A_a^2 \mathrm{Tr} A_b^2 \\
&= 2(\sum_a \mathrm{Tr} A_a^2)^2 - \sum_a (\mathrm{Tr} A_a^2)^2 \\
&= (2 - 1/p)(\sum_a \mathrm{Tr} A_a^2)^2 - \frac{1}{p} \sum_{a>b} (\mathrm{Tr} A_a^2 - \mathrm{Tr} A_b^2)^2 \\
&= (2 - 1/p)|A|^4,
\end{aligned}
$$

where $p = m-n$ and we used a frame $\{e_a\}$ of $T_x(M)^\perp$ for which $\mathrm{Tr} A_a A_b = 0$ if $a \neq b$, that is, the symmetric (p,p)-matrix $(\mathrm{Tr} A_a A_b)$ can be diagonalized. From (5.4) and (5.5) we have

$$
(5.6) \qquad |\nabla A|^2 - \tfrac{1}{2}\Delta |A|^2 \leq [(2 - 1/p)|A|^2 - nc]|A|^2.
$$

Consequently, we have (Simons [1])

THEOREM 5.3. Let M be an n-dimensional compact minimal submanifold of $\bar{M}^m(c)$. Then

$$
0 \leq \int_M |\nabla A|^2 * 1 \leq \int_M [(2 - 1/p)|A|^2 - nc]|A|^2 * 1,
$$

where $p = m-n$.

THEOREM 5.4. Let M be an n-dimensional compact minimal submanifold of $\bar{M}^m(c)$ $(c > 0)$. Then either M is totally geodesic, or $|A|^2 = nc/q$, or at some point of M, $|A|^2 \;\; nc/q$, where $q = 2 - 1/(m-n)$.

Proof. Suppose that $|A|^2 \leq nc/q$ everywhere on M. Then Theorem 5.4 implies that the second fundamental form of M is parallel and hence $|A|^2$ is constant. Thus $|A|^2 = 0$ and M is totally geodesic or $|A|^2 = nc/q$. Except for these possibilities, $|A|^2 > nc/q$ at some point of M.

QED.

We shall consider the case that $|A|^2 = nc/q$. In this case the second fundamental form of M is parallel. Therefore, we first study the submanifolds with parallel second fundamental form (Yano-Kon[11]).

THEOREM 5.5. Let M be an n-dimensional submanifold of $\bar{M}^m(c)$ $(c \geq 0)$ with flat normal connection. If the second fundamental form of M is parallel, then the sectional curvature of M is non-negative.

Proof. Since the normal connection of M is flat, the A_a's are simultaneously diagonalizabe at each poin of M. Let λ_i^a $(1 \leq i \leq n,$ $n+1 \leq a \leq m)$ be the eigenvalues of A_a corresponding to eigenvectors $e_1,\ldots,e_n$. We can choose a local field of orthonormal frames $e_{n+1},\ldots,e_m$ in the normal bundle such that $De_a = 0$ $(a = n+1,\ldots,m)$. Since the second fundamental form of M is parallel, we have

$$A_a R(e_i,e_j)e_i = R(e_i,e_j)A_a e_i = \lambda_i^a R(e_i,e_j)e_i.$$

Thus the eigenspace T_i^a corresponding to λ_i^a has the property that

$$R(e_i,e_j)T_i^a \subset T_i^a \quad \text{for all } a = n+1,\ldots,m.$$

If $\lambda_i^a \neq \lambda_j^a$ for some a, then we have

$$g(R(e_i,e_j)e_i,e_j) = 0,$$

which means that the sectional curvature K_{ij} of M with respect to the section spanned by e_i, e_j vanishes. On the other hand, from equation of Gauss (2.9), the sectional curvature K_{ij} of M is given by

$$K_{ij} = c + \sum_a \lambda_i^a \lambda_j^a.$$

If $\lambda_i^a \neq \lambda_j^a$ for some a, then $K_{ij} = 0$, and if $\lambda_i^a = \lambda_j^a$ for all a, then $K_{ij} \geq 0$. Thus the sectional curvature K_{ij} with respect to the section spanned by e_i, e_j is non-negative.

In the following, we prove that the sectional curvature $g(R(X,Y)Y,X)$ for any orthonormal vectors X and Y is non-negative. From (2.10) we have

$$g(R(X,Y)Y,X) = c + \sum_a g(A_aX,X)g(A_aY,Y) - \sum_a g(A_aX,Y)^2.$$

We now put $X = \Sigma_i\alpha_ie_i$ and $Y = \Sigma_i\beta_ie_i$. Then we see that

$$\begin{aligned} g(R(X,Y)Y,X) &= c + \sum_{a,i,j} \alpha_i^2\beta_j^2\lambda_i^a\lambda_j^a - \sum_{a,i,j} \alpha_i\alpha_j\beta_i\beta_j\lambda_i^a\lambda_j^a \\ &= c + \sum_{i,j} (K_{ij} - c)\alpha_i^2\beta_j^2 - \sum_{i,j} (K_{ij} - c)\alpha_i\alpha_j\beta_i\beta_j. \end{aligned}$$

Since we have $\Sigma_{i,j}\alpha_i^2\beta_j^2 = 1$ and $\Sigma_i\alpha_i\beta_i = 0$, the equation above becomes

$$\begin{aligned} g(R(X,Y)Y,X) &= \sum_{i,j} K_{ij}\alpha_i^2\beta_j^2 - \sum_{i,j} K_{ij}\alpha_i\alpha_j\beta_i\beta_j \\ &= \sum_{i>j} K_{ij}(\alpha_i\beta_j - \alpha_j\beta_i)^2. \end{aligned}$$

We have already seen that $K_{ij} \geq 0$, and hence $g(R(X,Y)Y,X) \geq 0$. Therefore, the sectional curvature of M is non-negative. QED.

PROPOSITION 5.1. Let M be an n-dimensional minimal submanifold of $\bar{M}^m(c)$ with parallel second fundamental form.

(1) If $c \leq 0$, then M is totally geodesic;

(2) If $C > 0$ and $|A|^2 \geq n(m-n)c$, then $R^\perp = 0$.

Proof. Since the second fundamental form of M is parallel, we see that $|A|^2$is constant. Thus (5.4) implies

$$0 \leq - \sum_{a,b} \mathrm{Tr}[A_a,A_b]^2 = nc|A|^2 - \sum_{a,b} (\mathrm{Tr}A_aA_b)^2.$$

We consider the symmetric (m-n,m-n)-matrix $(\mathrm{Tr}A_aA_b)$. Then we can choose a suitable basis $e_{n+1},\ldots,e_m$ for which the matrix $(\mathrm{Tr}A_aA_b)$ can be assumed to be diagonal. Thus we have

$$(5.7) \qquad 0 \leq -\sum_{a,b} \mathrm{Tr}[A_a, A_b]^2 = nc|A|^2 - \frac{1}{m-n}(\sum_a \mathrm{Tr}A_a^2)^2$$

$$\frac{1}{m-n}\sum_{a>b}(\mathrm{Tr}A_a^2 - \mathrm{Tr}A_b^2)^2$$

$$\leq \frac{1}{m-n}[n(m-n)c - |A|^2]|A|^2.$$

If $c \leq 0$, (5.7) implies that $|A|^2 = 0$ and hence M is totally geodesic. If $c > 0$ and $|A|^2 \geq n(m-n)c$, then (5.7) shows that $[A_a, A_b] = 0$ for all a and b, which means that the normal connection of M is flat, that is, $R^{\perp} = 0$. QED.

We prove the following theorem (Yano-Kon [11]).

THEOREM 5.6. Let M be an n-dimensional complete minimal submanifold of S^m with parallel second fundamental form. If $|A|^2 \geq n(m-n)$, then M is a pythagorean product of the form

$$S^{p_1}(r_1) \times \ldots \times S^{p_N}(r_N), \quad r_t = (p_t/n)^{1/2} \ (t = 1,\ldots,N),$$

where $1 \leq p_1,\ldots,p_N < n$, $p_1 + \ldots + p_N = n$, $m-n = N-1$.

Proof. From Theorem 5.5 and Proposition 5.1, M has non-negative sectional curvature and flat normal connection. Thus Theorem 5.1 implies that the essential codimension of N is n-1. On the other hand, by (5.7), we have $\mathrm{Tr}A_a^2 = \mathrm{Tr}A_b^2$ for all a, b. Since $|A|^2 = n(m-n)$, M is not totally geodesic and hence $\mathrm{Tr}A_a^2 = |A|^2/(m-n) \neq 0$. Thus we must have $m-n = N-1$. The other statements are trivial consequences of Theorem 5.1. QED.

From Theorem 4.3 and Theorem 5.5 we have the following theorem (Yano-Kon [11]).

THEOREM 5.7. Let M be a complete n-dimensional submanifold of S^m with flat normal connection. If the second fundamental form of M is parallel, then M is a small sphere, a great sphere or a pythagorean product of a certain numbers of spheres. Moreover, if M is of essential codimension m-n, then M is a pythagorean product of the form

$$S^{p_1}(r_1) \times \ldots \times S^{p_N}(r_N), \quad \Sigma r_i^2 = 1, \; m\text{-}n = N\text{-}1,$$

or a pythagorean product of the form

$$S^{p_1}(r_1) \times \ldots \times S^{p_{N'}}(r_{N'}) \subset S^{m-1} \subset S^m, \quad \Sigma r_i^2 = r^2 \;\; 1, \; m\text{-}n = N'.$$

Remark. The local version of Theorem 5.6 and Theorem 5.7 is also true (see the proof of Theorem 4.1). See also Waerden [1].

Let M be an n-dimensional minimal submanifold of S^m with $|A|^2 = n/(2 - 1/p)$, $p = m\text{-}n$. Then the second fundamental form of M is parallel. Moreover, (5.5) implies

$$\text{(5.8)} \qquad \mathrm{Tr}A_a^2 = \mathrm{Tr}A_b^2 \quad \text{for all } a, b.$$

From Lemma 5.1 we may assume that

$$\text{(5.9)} \qquad A_a = 0 \quad \text{for } a = n+3,\ldots,m.$$

Since M is not totally geodesic, (5.8) and (5.9) imply that $p = m\text{-}n \leq 2$.

Let $p = 1$. Then M is a hypersurface of S^m. Then the normal connection of M is flat and $|A|^2 = n$. Thus, from Theorem 5.6, M is locally a Clifford minimal hypersurface

$$M_{t,n-t} = S^t((t/n)^{1/2}) \times S^{n-t}(((n-t)/n)^{1/2}).$$

We next consider the case that $p = 2$. From Lemma 5.1 we have

$$A_{n+1} = \lambda\bar{A}, \quad A_{n+2} = \mu\bar{B}, \quad \lambda, \mu \neq 0,$$

where $\bar{A}$ and $\bar{B}$ are defined in Lemma 5.1. In other words,

$$\omega_1^{n+1} = \lambda\omega^2, \qquad \omega_2^{n+1} = \lambda\omega^1, \qquad \omega_i^{n+1} = 0 \text{ for } i = 3,\ldots,n,$$

$$\omega_1^{n+2} = \mu\omega^1, \qquad \omega_2^{n+2} = -\mu\omega^2, \qquad \omega_i^{n+2} = 0 \text{ for } i = 3,\ldots,n.$$

Since the second fundamental form of M is parallel, (2.19) implies

$$(5.10) \qquad dh^a_{ij} = h^a_{ik}\omega^k_j + h^a_{kj}\omega^k_i - h^b_{ij}\omega^a_b.$$

Setting $a = n+1$, $i = 1$ and $j = 2$, we see that $d\lambda = dh^{n+1}_{12} = 0$, that is, λ is constant. Setting $a = n+1$, $i = 1$ and $j \geq 3$, we see that

$$\omega^2_j = 0 \quad \text{for } j \geq 3.$$

Setting $a = n+1$, $i = 2$ and $j \geq 3$, we see that

$$\omega^1_j = 0 \quad \text{for } j \geq 3.$$

Similarly, setting $a = n+2$, and $i = j = 1$, we see that μ is a constant. Thus, if $j \geq 3$, we obtain

$$0 = d\omega^1_j = -\omega^1_k \wedge \omega^k_j + \omega^1 \wedge \omega^j = \omega^1 \wedge \omega^j.$$

Since $\omega^1,\ldots,\omega^n$ are orthonormal, $\omega^1 \wedge \omega^j = 0$ implies $\omega^j = 0$ for $j \geq 3$. This shows that dim M = 2 and hence dim $\bar{M} = m = 4$. On the other hand, from (5.8), it follows that $\lambda^2 = \mu^2$. Since $4/3 = |A|^2 = 4\lambda^2$, we have $\lambda^2 = \mu^2 = 1/3$. Thus we may assume that $-\lambda = \mu = (1/3)^{1/2}$. Setting $a = 3$ and $i = j = 1$, we obtain

$$\omega^3_4 = (2\lambda/\mu)\omega^2_1 = -2\omega^2_1.$$

The curvature of M is given by

$$\Omega^1_2 = \omega^1 \wedge \omega^2 + \omega^3_1 \wedge \omega^3_2 + \omega^4_1 \wedge \omega^4_2 = (1-\lambda^2-\mu^2)\omega^1 \wedge \omega^2 = \frac{1}{3}\omega^1 \wedge \omega^2.$$

From these considerations, the connection form (ω_B^A) of $\bar{M}$, restricted to M, is given by

$$\begin{pmatrix} 0 & \omega_2^1 & \mu\omega^2 & -\mu\omega^1 \\ \omega_1^2 & 0 & \mu\omega^1 & \mu\omega^2 \\ \lambda\omega^2 & \lambda\omega^1 & 0 & 2\omega_2^1 \\ -\lambda\omega^1 & \lambda\omega^2 & 2\omega_1^2 & 0 \end{pmatrix} \qquad -\lambda = \mu = (1/3)^{1/2}.$$

This coincides the form of the Veronese surface in Example 5.2 (see Chern-do Carmo-Kobayashi [1]). Therefore, M coincides locally with the Veronese surface. Consequently, we obtain the following theorem (Chern-do Carmo-Kobayashi [1]).

THEOREM 5.8. The Veronese surface in S^4 and the Clifford hypersurface $M_{t,n-t}$ in S^{n+1} are the only compact minimal submanifold of dimension n in S^{n+p} satisfying $|A|^2 = n/(2 - 1/p)$. The corresponding local result also holds.

If the normal connection of a minimal submanifold M of S^m is flat, then we have

$$\tfrac{1}{2}\Delta|A|^2 = n|A|^2 - \sum_{a,b}(\mathrm{Tr}A_aA_b)^2 + |\nabla A|^2.$$

If M is compact, then

$$\int_M [|A|^2 - n]|A|^2 {*1} \geq 0.$$

Using this, Kenmotsu [4] proved

THEOREM 5.8. Let M be an n-dimensional compact minimal submanifold of S^m with flat normal connection. If $|A|^2 = n$, then there exists an (n+1)-dimensional sphere S^{n+1} containing M as a Clifford minimal hypersurface.

EXERCISES

A. TOTALLY UMBILICAL SUBMANIFOLDS: Let $\bar{M}^n(k)$ be a simply connected complete space form of constant curvature k (see §3 of Chapter I). We have (cf. Chen [1])

THEOREM. A totally umbilical submanifold M^m in $\bar{M}^n(k)$ is either totally geodesic in $\bar{M}^n(k)$ or contained in a hypersphere of an $(m+1)$-dimensional totally geodesic submanifold of $\bar{M}^n(k)$.

B. SUBMANIFOLD WITH $R(X,Y)R = 0$: Nomizu [2] studied the effect of the condition

(*) $R(X,Y)R = 0$ for any tangent vectors X and Y

for hypersurfaces M of the Euclidean space R^{n+1}, where R denotes the Riemannian curvature tensor and $R(X,Y)$ operates on the tensor algebra at each point as a derivation, and prove the following

THEOREM 1. Let M be a connected and complete hypersurface of R^{n+1} so that the type number is greater than 2 at least at one point. If M satisfies the condition (*), then it is of the form $M = S^k \times R^{n-k}$, where S^k is a hypersphere in R^{k+1} of R^{n+1} and R^{n-k} is a Euclidean subspace orthogonal to R^{k+1}.

Ryan [1] treated the same condition for hypersurfaces of spaces of non-zero constant curvature. Moreover, Tanno [3] discussed the effect of the condition

(**) $R(X,Y)Q = 0$ for any tangent vectors X and Y

for hypersurfaces of the Euclidean space, where Q denotes the Ricci operator. Obviously, the condition (*) implies the condition (**). If the ambient manifold is of non-zero constant curvature, the condition (*) and (**) are equivalent. Tanno-Takahashi [1] proved

THEOREM 2. Let M be a connected and complete hypersurface of a sphere $S^{n+1}(\bar{c})$ $(n \geq 4)$ of curvature $\bar{c}$. Then M satisfies the condition (**) if and only if M is one of the following:

(1) $M = S^n(\bar{c})$, great sphere;

(2) $M = S^n(c)$, small sphere, where $c > \bar{c}$;

(3) $M = S^p(c_1) \times S^{n-p}(c_2)$, where $p, n-p \geq 2$ and $c_1 > \bar{c}$, $c_2 > \bar{c}$ such that $c_1^{-1} + c_2^{-1} = \bar{c}^{-1}$.

C. EINSTEIN HYPERSURFACES: Fialkow [1] proved a local classification theorem for Einstein hypersurfaces of space forms.

THEOREM. Let M be a hypersurface of a space form $\bar{M}^{n+1}(c)$ $(n > 2)$. Let M be Einstein: $S = \rho g$. If $\rho > (n-1)c$, then M is umbilical and M is of constant curvature $\rho/(n-1)$. If $\rho = (n-1)c$, then M is of constant curvature c. If $\rho < (n-1)c$, then $c > 0$, $\rho = (n-2)c$ and M is locally $M^p(\frac{n-2}{p-1}c) \times M^{n-p}(\frac{n-2}{n-p-1}c)$, where $1 < p < n-1$.

(See also Thomas [1], Cartan (see Thomas [2]), Ryan [1].)

D. MINIMAL IMMERSIONS: Let M be a Riemannian manifold and $x : M \longrightarrow R^{n+p}$ be an isometric immersion of M in a Euclidean (n+p)-space R^{n+p}. Then we have (Tsunero Takahashi [1])

THEOREM 1. An isometric immersion $x : M \longrightarrow R^{n+p}$ is minimal if and only if $\Delta x = 0$.

THEOREM 2. If an isometric immersion $x : M \longrightarrow R^{n+p}$ satisfies $\Delta x = \lambda x$ for some constant $\lambda \neq 0$, then λ is necessarily positive and x realize a minimal immersion in a sphere S^{n+p-1} of radius $(n/\lambda)^{1/2}$ in R^{n+p}. Conversely, if x realizes a minimal immersion in a sphere of radius a in R^{n+p}, then x satisfies $\Delta x = \lambda x$ up to a parallel displacement in R^{n+p} and $\lambda = n/a^2$.

For minimal submanifolds of space forms, the Simons' type formula and its applications were studied by many authors. Yau [1] proved the following

THEOREM 3. Let M be an n-dimensional minimal submanifold of a space form $\bar{M}^{n+p}(1)$. If M has non-negative curvature and $|A|^2 = np$, then M is an open piece of the product $\Pi S^{m_i}((m_i/n)^{1/2})$, $\sum_{i=1}^{p} m_i = n$.

Matsuyama [3] proved the following

THEOREM 4. Let M be an n-dimensional $(n \geq 4)$ minimal submanifold of S^m which has at most two principal curvatures (If exactly two are distinct, then we assume these multiplicities ≥ 2) in the direction of any normal. Then the second fundamental form is parallel and $|A|^2$ holds $|A|^2 = 0$ or $n \leq |A|^2 \leq n^2/4$.

ISOPERIMETRIC SECTION: Let M be an n-dimensional submanifold of an m-dimensional Riemannian manifold $\bar{M}$. A parallel section ξ $(\neq 0)$ of the normal bundle of M is called an *isoperimetric section* if TrA_ξ = constant $\neq 0$ and a *minimal section* if $TrA_\xi = 0$. If M admits an isoperimetric section, then we can consider the Laplacian for $|A_\xi|^2$ and Smyth [3] proved

THEOREM 1. Let M be an n-dimensional compact irreducible submanifold of a space form $\bar{M}^m(c)$ with non-negative sectional curvature. If the normal bundle of M admits an isoperimetric (resp. minimal) section then M lies in a small (resp. great) hypersphere orthogonal to this section.

THEOREM 2. Let M be an n-dimensional compact irreducible submanifold of a space form $\bar{M}^{n+p}(c)$ with non-negative sectional curvature. If the mean curvature vector of M is non-zero parallel, then M must lie minimally in a hypersphere S^{n+p-1} of $\bar{M}^{n+p}(c)$ of constant positive curvature. When p = 2, M is imbedded as a great hypersphere of S^{n+p-1}.

F. PRODUCT IMMERSION: Let M_i be a compact connected Riemannian manifold of dimension $n_i \geq 2$ $(1 \leq i \leq p)$, and M the Riemannian product manifold $M_1 \times M_2 \times \ldots \times M_p$. For isometric immersions of M into a Euclidean m-space R^m, Moore [1] proved

THEOREM 1. For $1 \leq i \leq p$, let M_i be a connected non-flat Riemannian manifold of dimension n_i, and let M_0 be a connected flat Riemannian manifold of dimension n_0. Let $M_0 \times M_1 \times \ldots \times M_p$ be the Riemannian product manifold, and R^m a Euclidean space of dimension $m = n_0 + \Sigma n_i + p$. Then any isometric immersion

$$f : M_0 \times M_1 \times \ldots \times M_p \longrightarrow R^m$$

is a product immersion, that is, there exist isometric immersions $f_i : M_i \longrightarrow R^{n_i+1}$ $(1 \leq i \leq p)$ and a decomposition of R^m into a Riemannian product $R^m = R^{n_0} \times R^{n_1+1} \times \ldots \times R^{n_p+1}$ so that $f(m_0,m_1,\ldots,m_p) = (f_0(m_0),f_1(m_1),\ldots,f_p(m_p))$, $m_i \in M_i$ $(0 \leq i \leq p)$, $f_0 : M_0 \longrightarrow R^{n_0}$.

THEOREM 2. For $1 \leq i \leq p$, let M_i be a complete connected Riemannian manifold of dimension $n_i \geq 2$, $M = M_1 \times \ldots \times M_p$ the Riemannian product, and R^m a Euclidean space of dimension $m = \Sigma n_i + p$. Then any isometric immersion $f : M \longrightarrow R^m$ satisfies at least one of the following conditions: (a) It is a product of hypersurface immersions; (b) It carries a complete geodesic onto a straight line in R^m.

Alexander-Maltz [1] consider the following condition:

(*) No M_i contains an open submanifold which is isometric to the Riemannian product $R^{n_i-1} \times (-\varepsilon,\varepsilon)$.

Then

THEOREM 3. Let $M_1,\ldots,M_k$ be connected complete non-flat Riemannian manifold satisfying condition (*). Then any k-codimensional isometric immersion of the Riemannian product $M = M_1 \times \ldots \times M_k$ in Euclidean space is a product of hypersurface immersions.

G. PLANER GEODESIC IMMERSIONS: Let M be an n-dimensional connected complete Riemannian manifold and $\bar{M}$ an m-dimensional connected complete Riemannian manifold. An isometric immersion of M into $\bar{M}$ is called a *planer geodesic immersion* if every geodesic in M is mapped locally into a 2-dimensional totally geodesic submanifold of $\bar{M}$. Then the immersion is isotropic in the sense of O'Neill [1], that is, the second fundamental form B of the immersion satisfies $|B(X,X)|^2 = \lambda^2$ for all unit vector X where λ is a function, and the second fundamental form is parallel. Hong [1] studied planer geodesic immersions into Euclidean space and proved

THEOREM 1. Let $f : M^n \longrightarrow R^{n+p}$ be a planer geodesic immersion. Then the sectional curvature of M is 1/4-pinched except for the totally geodesic case and moreover if M has constant positive sectional curvature, then f(M) is an n-dimensional sphere or a Veronese manifold.

Sakamoto [2] studied the planer geodesic immersions into any space forms and constructed models of planer geodesic immersions. Moreover, the classification theorem of planer geodesic immersions was obtained. In [2] Sakamoto showed that

THEOREM 2. Let $f : M^n \longrightarrow \bar{M}^{n+p}(c)$ be a planer geodesic immersion. Then M^n is isometric to a symmetric space of rank one or an Euclidean space and the immersion is rigid.

H. ISOTROPIC IMMERSIONS: An isotropic immersion is an isometric immersion such that $|B(X,X)|^2 = \lambda^2$ for all unit vector X where λ is a function. If the second fundamental form B satisfies $(\nabla_X B)(Y,Z) = (\nabla_Y B)(X,Z)$, then λ is constant.

Itoh-Ogiue [1] proved the following theorems.

THEOREM 1. Let M be an n-dimensional space form of constant curvature c, and $\bar{M}$ be an $(n+\frac{1}{2}n(n+1)-1)$-dimensional space form of constant curvature $\bar{c}$. If $c < \bar{c}$, and M is an isotropic submanifold of $\bar{M}$ with parallel second fundamental form, then $c = (n/2(n+1))\bar{c}$, and the immersion is rigid.

THEOREM 2. Let M be an n-dimensional space form of constant curvature c, and $\bar{M}$ be an $(n+\frac{1}{2}n(n+1)-1)$-dimensional space form of constant curvature $\bar{c}$. If $c < \bar{c}$, and M is an isotropic submanifold of $\bar{M}$, then $c = (n/2(n+1))\bar{c}$, and the immersion is rigid provided that $n \leq 4$.

CHAPTER III

COMPLEX MANIFOLDS

In this chapter, we study the various almost complex manifolds and complex manifolds. In §1, we first prepare some algebraic results on real and complex vector space, and we give the fundamental properties of almost complex manifolds and complex manifolds. In §2, we give some examples of almost complex manifolds and complex manifolds. §3 is devoted to the study of Hermitian manifolds. We give the definitions of Kaehlerian manifolds and nearly Kaehlerian manifolds. In §4, we discuss Kaehlerian manifolds. We define the holomorphic sectional curvature of a Kaehlerian manifold and give the typical examples of complex space forms, that is, Kaehlerian manifolds of constant holomorphic sectional curvature. In §5, we consider nearly Kaehlerian manifolds. We give the condition for a nearly Kaehlerian manifold to be Einstein. The main theorem of this section is that there does not exist non-Kaehlerian nearly Kaehlerian manifold of constant holomorphic sectional curvature when the dimension of the manifold is not 6. In the last §6, following S. Ishihara [3], we study quaternion manifolds by using tensor calculas.

In this chapter, we refered to Goldberg [1]. Kobayashi-Nomizu [2], Matsushima [1] and Yano [5].

1. ALMOST COMPLEX MANIFOLDS AND COMPLEX MANIFOLDS

First of all, we prepare some linear algebraic results on real and complex vector spaces, which will be applied to tangent spaces of manifolds.

Let V be a vector space over R. The *complexification* V^c of V is the set of all symbols $X + iY$ ($i = \sqrt{-1}$), $X, Y \in V$ satisfying the following conditions: We suppose that $X + iY = X' + iY'$ if and only if $X = X'$ and $Y = Y'$. The sum of $X + iY$ and $X' + iY'$ is defined to be

$$(X + iY) + (X' + iY') = (X + X') + i(Y + Y').$$

The product of $a + ib \in C$ and $X + iY$ is defined by

$$(a + ib)(X + iY) = (aX - bY) + i(bX + aY).$$

Then V^c becomes a vector space over C. By identifying $X + i0$ of V^c with X of V we may consider $V \subset V^c$. For $Z = X + iY \in V^c$ we put $\bar{Z} = X - iY$, which will be called the *conjugate* of Z with respect to V. The *complex conjugation* in V^c is a conjugate linear map from V^c to V^c defined by $Z \longrightarrow \bar{Z}$, i.e., $\overline{Z + W} = \bar{Z} + \bar{W}$ and $\overline{\lambda Z} = \bar{\lambda}\bar{Z}$ ($\lambda \in C$).

We assume that V is of n-dimensional. Let $\{e_1, \ldots, e_n\}$ be a basis of V over R. For any vectors X, Y of V we put $X = \Sigma a^j e_j$, $Y = \Sigma b^j e_j$. Then

$$X + iY = \sum_j (a^j e_j + ib^j e_j) = \sum_j \lambda^j e_j,$$

where we have put $\lambda^j = a^j + ib^j$, $j = 1, \ldots, n$. If we consider $e_1, \ldots, e_n$ as elements of V^c, they are linearly independent over C. In fact, if $\Sigma\lambda^j e_j = 0$, then $\Sigma a^j e_j = 0$ and $\Sigma b^j e_j = 0$. Thus we have $a^j = b^j = 0$ for all j, and hence $\lambda^j = 0$ for all j. Therefore $e_1, \ldots, e_n$ are linearly independent over C. From this we see that $\{e_1, \ldots, e_n\}$ is a basis of V^c. A linear endomorphism J of real vector space V satisfying $J^2 = -I$ is called a *complex structure* on V, where I stands for the identity transformation of V.

Let V be a real vector space with a complex structure J. We can define the product λX of a complex number $\lambda = a + ib$ and an element X of V by

$$\lambda X = (a + ib)X = aX + bJX.$$

Then we can consider V as a vector space over C. Clearly, the real dimension of V must be even.

Conversely, given a complex vector space V of complex dimension n, let J be the linear endomorphism of V defined by $JX = iX$ for all $X \in V$. If we consider V as a real vector space of real dimension 2n, then J is a complex structure of V.

Now let V be a real vector space with a complex structure J. Then, we can extend J to a complex linear endomorphism of V^c, denoted also by J, by setting

$$J(X + iY) = JX + iJY.$$

Clearly, $J^2 = -I$.

In a 2n-dimensional real vector space V with complex structure J, there exist elements $X_1,\ldots,X_n$ of V such that $\{X_1,\ldots,X_n,JX_1,\ldots,JX_n\}$ forms a basis of V. We set

$$Z_k = \tfrac{1}{2}(X_k - iJX_k), \quad \bar{Z}_k = \tfrac{1}{2}(X_k + iJX_k), \quad k = 1,\ldots,n.$$

Then $\{Z_1,\ldots,Z_k,\bar{Z}_1,\ldots,\bar{Z}_n\}$ forms a basis of V^c, and we have

$$JZ_k = iZ_k, \qquad J\bar{Z}_k = -i\bar{Z}_k, \quad k = 1,\ldots,n.$$

Hence if we set

$$V^{1,0} = \{Z \in V^c : JZ = iZ\}, \qquad V^{0,1} = \{Z \in V^c : JZ = -iZ\},$$

then we have the complex vector space direct sum:

$$V^c = V^{1,0} + V^{0,1}.$$

We notice that $\bar{V}^{1,0} = V^{0,1}$. For any $Z \in V^c$ we have

$$Z = \tfrac{1}{2}(Z - iJZ) + \tfrac{1}{2}(Z + iJZ).$$

The first term of the right hand side of this equation belongs to $V^{1,0}$ and the second belongs to $V^{0,1}$. Therefore, we easily see that

$$V^{1,0} = \{X - iJX : X \varepsilon V\}, \quad V^{0,1} = \{X + iJX : X \varepsilon V\}.$$

We denote by V^* the dual space of a real vector space V. Then we can construct the complexification V^{*c} of V^*. Naturally, we can identify V^{*c} and the dual space V^{c*} of the complexification V^c of V. A complex structure J on V induces a complex structure on V^*, denoted also by J, by setting

$$\langle JX, X^*\rangle = \langle X, JX^*\rangle, \quad X \varepsilon V, \ X^* \varepsilon V^*.$$

Then we have the following decomposition

$$V^{*c} = V_{1,0} + V_{0,1},$$

where

$$V_{1,0} = \{X^* \varepsilon V^{*c} : \langle X, X^*\rangle = 0 \text{ for all } X \varepsilon V^{0,1}\},$$

$$V_{0,1} = \{X^* \varepsilon V^{*c} : \langle X, X^*\rangle = 0 \text{ for all } X \varepsilon V^{1,0}\}.$$

Let M be a real differentiable manifold. A tensor field J on M is called an *almost complex structure* on M if, at every point x of M, J is an endomorphism of the tangent space $T_x(M)$ such that $J^2 = -I$. A manifold M with a fixed almost complex structure J is called an *almost complex manifold*. Every almost complex manifold is of even dimensional.

We shall prove that every complex manifold M carries a natural almost complex structure. Let $(z^1,\dots,z^n)$ be a complex local coordinate system on a neighborhood U of a point p of M. We put $z^j = x^j + iy^j$, $j = 1,\dots,n$. We define an endomorphism J of $T_p(M)$ by

(1.1) $\quad J(\partial/\partial x^j) = \partial/\partial y^j, \quad J(\partial/\partial y^j) = -(\partial/\partial x^j), \quad j = 1,\dots,n.$

We prove that the definition of J does not depend on the choice of the complex local coordinate system. Let $T^c_p(M)$ be the complexification of $T_p(M)$. We extend J to $T^c_p(M)$ and we have

$$(1.2)\qquad J(\partial/\partial z^j) = i(\partial/\partial z^j),\quad J(\partial/\partial \bar{z}^j) = -i(\partial/\partial \bar{z}^j),\quad j = 1,\ldots,n,$$

where

$$(1.3)\qquad (\partial/\partial z^j) = \tfrac{1}{2}\{(\partial/\partial x^j)-i(\partial/\partial y^j)\},\quad (\partial/\partial \bar{z}^j) = \tfrac{1}{2}\{(\partial/\partial x^j)+i(\partial/\partial y^j)\}.$$

Hence if an element Z of $T^c_p(M)$ is a linear combination of $(\partial/\partial z^j)$ ($j = 1,\ldots,n$) only, then $JZ = iZ$, and if Z is a linear combination of $(\partial/\partial \bar{z}^j)$ ($j = 1,\ldots,n$) only, then $JZ = -iZ$.

Now if $(w^1,\ldots,w^n)$ is another complex local coordinate system on U at p, and if $w^k = u^k + iv^k$, $k = 1,\ldots,n$, then define an endomorphism J' of $T_p(M)$ by

$$J'(\partial/\partial u^k) = \partial/\partial v^k,\qquad J'(\partial/\partial v^k) = -(\partial/\partial u^k),\quad k = 1,\ldots,n.$$

Extending J' to $T^c_p(M)$, we obtain

$$J'(\partial/\partial w^k) = i(\partial/\partial w^k),\quad J'(\partial/\partial \bar{w}^k) = -i(\partial/\partial \bar{w}^k),\quad k = 1,\ldots,n.$$

On the other hand, at $p \in M$,

$$\frac{\partial}{\partial w^k} = \sum_j \frac{\partial F^j}{\partial w^k}(p)\frac{\partial}{\partial z^j},\qquad \frac{\partial}{\partial \bar{w}^k} = \sum_j \overline{\frac{\partial F^j}{\partial w^k}}(p)\frac{\partial}{\partial \bar{z}^j},\quad k = 1,\ldots,n.$$

Hence $\partial/\partial w^k$ and $\partial/\partial \bar{w}^k$ are linear combinations of $\partial/\partial z^j$ and $\partial/\partial \bar{z}^j$ respectively. Thus we have

$$J(\partial/\partial w^k) = i(\partial/\partial w^k),\qquad J(\partial/\partial \bar{w}^k) = -i(\partial/\partial \bar{w}^k).$$

Consequently, we see that J and J' coincide at $p \in M$, and hence J does not depend on the choice of the complex local coordinate system in the neighborhood of p. It is clear that $J^2 = -I$. Thus J is an almost complex structure on M.

Let M and M' be almost complex manifolds with almost complex structures J and J' respectively. A mapping $f: M \longrightarrow M'$ is said to be *almost complex* if $J' \cdot f_* = f_* \cdot J$.

PROPOSITION 1.1. Let M and M' be complex manifolds. A mapping $f: M \longrightarrow M'$ is holomorphic if and only if f is almost complex with respect to the complex structures of M and M'.

Proof. Let J and J' be almost complex structures of M and M', respectively. Let $(z^1,\ldots,z^n)$ and $(w^1,\ldots,w^m)$ be complex local coordinate systems on neighborhoods of $p \in M$ and $f(p) \in M'$, respectively, and set $z^k = x^k + iy^k$, $w^j = u^j + iv^j$. If we put

$$f^*u^j = \alpha^j(x^1,\ldots,x^n,y^1,\ldots,y^n), \quad f^*v^j = \beta^j(x^1,\ldots,x^n,y^1,\ldots,y^n),$$

then

$$f_*(\partial/\partial x^k) = \sum_j\{(\partial\alpha^j/\partial x^k)(p)(\partial/\partial u^j)+(\partial\beta^j/\partial y^k)(p)(\partial/\partial v^j)\},$$

$$f_*(\partial/\partial y^k) = \sum_j\{(\partial\alpha^j/\partial y^k)(p)(\partial/\partial u^j)+(\partial\beta^j/\partial y^k)(p)(\partial/\partial v^j)\}.$$

Comparing $f_*(J(\partial/\partial x^k))$ with $J'(f_*(\partial/\partial x^k))$, and $f_*(J(\partial/\partial y^k))$ with $J'(f_*(\partial/\partial y^k))$, we see that f is almost complex if and only if

$$(\partial\alpha^j/\partial x^k)(p) = (\partial\beta^j/\partial y^k)(p), \quad (\partial\alpha^j/\partial y^k)(p) = -(\partial\beta^j/\partial x^k)(p)$$

for all j, k. But this is the Cauchy-Riemann equation of $f^*w^j = f^*u^j + if^*v^j = \alpha^j + i\beta^j$. Thus f is almost complex if and only if f is holomorphic. QED.

We call $T_x^c(M)$ the *complex tangent space* of M at x. An element of $T_x^c(M)$ is called a *complex tangent vector* at x. Let M be an almost complex manifold with almost complex structure J. Then

$$T_x^c(M) = T_x^{1,0}(M) + T_x^{0,1}(M),$$

where $T_x^{1,0}(M)$ and $T_x^{0,1}(M)$ are the eigenspaces of J corresponding to the eigenvalues i and -i respectively. A complex tangent vector (field) is said to be of type (1,0) (resp. (0,1)) if it belongs to $T_x^{1,0}(M)$ (resp. $T_x^{0,1}(M)$). We see that a complex tangent vector Z is of type (1,0) (resp. (0,1)) if and only if $Z = X - iJX$ (resp. $Z = X + iJX$) for some $X \in T_x(M)$.

Let M be a real 2n-dimensional complex manifold with the natural almost complex structure J. Let $(z^1,\dots,z^n)$ be a complex local coordinate system. Then, from (1.2), we see that $\{\partial/\partial z^1,\dots,\partial/\partial z^n\}$ is a basis of $T_x^{1,0}(M)$ and $\{\partial/\partial \bar{z}^1,\dots,\partial/\partial \bar{z}^n\}$ is a basis of $T_x^{0,1}(M)$. Moreover, $\{\partial/\partial z^1,\dots,\partial/\partial z^n,\partial/\partial \bar{z}^1,\dots,\partial/\partial \bar{z}^n\}$ is a basis of $T_x^c(M)$. We denote by $T_x^{*c}(M)$ the complexification of the dual space $T_x^*(M)$ of $T_x(M)$. We put $z^j = x^j + iy^j$, $j = 1,\dots,n$, and set

$$(1.4) \qquad dz^j = dx^j + idy^j, \quad d\bar{z}^j = dx^j - idy^j, \quad j = 1,\dots,n.$$

Then $\{dz^1,\dots,dz^n,d\bar{z}^1,\dots,d\bar{z}^n\}$ forms a basis of $T_x^{*c}(M)$. We also have

$$dz^j(\partial/\partial z^k) = d\bar{z}^j(\partial/\partial \bar{z}^k) = \delta_k^j,$$

$$dz^j(\partial/\partial \bar{z}^k) = d\bar{z}^j(\partial/\partial z^k) = 0,$$

$j,k = 1,\dots,n$.

Let $D^r(M)$ be the space of r-forms on a manifold M. The complexification $C^r(M)$ of $D^r(M)$ is the set of all $\omega = \omega_1 + i\omega_2$, where ω_1 and ω_2 are real r-forms. We call $\omega \in C^r(M)$ a *complex* r-*form* on M. A complex r-form ω on M gives an element of $\Lambda^r T_x^{*c}(M)$ at each $x \in M$, that is, a skew-symmetric r-linear mappings $T_x^c(M) \times \dots \times T_x^c(M) \longrightarrow C$ at each $x \in M$. More generally, we can define the space of complex tensor fields on M as the complexification of the space of real tensor fields. Such operations as contractions, brackets, exterior differentiations, Lie differentiations, etc. can be extended by linearlity to complex tensor fields or complex forms.

Let V be a vector space over R. Then we have seen that $V^{*c} = V_{1,0} + V_{0,1}$. The exterior algebras $\Lambda V_{1,0}$ and $\Lambda V_{0,1}$ can be considered as subalgebras of ΛV^{*c}. Let $\Lambda^{p,q}V^{*c}$ be the subspace of ΛV^{*c} spanned by $\alpha \wedge \beta$, where $\alpha \in \Lambda^p V_{1,0}$ and $\beta \in \Lambda^q V_{0,1}$. Then we have the decomposition

$$\Lambda V^{*c} = \sum_{r=0}^{n} \Lambda^r V^{*c} \quad \text{with} \quad \Lambda^r V^{*c} = \sum_{p+q=r} \Lambda^{p,q} V^{*c}.$$

The complex conjugation in V^{*c} gives a real linear isomorphism between $\Lambda^{p,q}V^{*c}$ and $\Lambda^{q,p}V^{*c}$. Applying these results to $T_x^{*c}(M)$, we have the decomposition

$$C(M) = \sum_{r=0}^{n} C^r(M) = \sum_{p,q=0}^{n} C^{p,q}(M), \quad C^r(M) = \sum_{p+q=r} C^{p,q}(M),$$

where C(M) is the space of complex differential forms on M. An element of $C^{p,q}(M)$ is called a (*complex*) *form of degree* (p,q). A complex 1-form ω is of degree (1,0) if and only if $\omega(Z) = 0$ for all complex vector fields Z of type (0,1), and ω is of degree (0,1) if and only if $\omega(Z) = 0$ for all Z of type (1,0).

Let $\{\omega^1,\ldots,\omega^n\}$ be a local basis of $C^{1,0}(M)$. Then its complex conjugate $\{\bar{\omega}^1,\ldots,\bar{\omega}^n\}$ is a local basis of $C^{0,1}(M)$. It follows that the set of forms $\omega^{j_1} \wedge \ldots \wedge \omega^{j_p} \wedge \bar{\omega}^{k_1} \wedge \ldots \wedge \bar{\omega}^{k_q}$, $1 \le j_1 < \ldots < j_p \le n$ and $1 \le k_1 < \ldots < k_q \le n$, is a local basis of $C^{p,q}(M)$. Since C(M) is locally generated by $C^{0,0}(M)$, $C^{1,0}(M)$ and $C^{0,1}(M)$, we have

$$dC^{0,0}(M) \subset C^{1,0}(M) + C^{0,1}(M), \quad dC^{1,0}(M) \subset C^{2,0}(M)+C^{1,1}(M)+C^{0,2}(M),$$
$$dC^{0,1}(M) \subset C^{2,0}(M) + C^{1,1}(M) + C^{0,2}(M).$$

From these inclusions we obtain

$$dC^{p,q}(M) \subset C^{p+2,q-1}(M) + C^{p+1,q}(M) + C^{p,q+1}(M) + C^{p-1,q+2}(M).$$

Here we define the torsion tensor field N of type (1,2) of an almost complex structure J by

$$(1.5) \qquad N(X,Y) = [JX,JY] - J[X,JY] - J[JX,Y] - [X,Y]$$

for any vector fields X and Y on M.

THEOREM 1.1. For an almost complex manifold M, the following conditions are equivalent:

(a) If Z and W are complex vector fields of type (1,0), so is [Z,W];

(b) If Z and W are complex vector fields of type (0,1), so is [Z,W];

(c) $dC^{1,0}(M) \subset C^{2,0}(M)+C^{1,1}(M)$, $dC^{0,1}(M) \subset C^{1,1}(M)+C^{0,2}(M)$;

(d) $dC^{p,q}(M) \subset C^{p+1,q}(M) + C^{p,q+1}(M)$ for $p,q = 0,1,\ldots,n$;

(e) The almost complex structure has no torsion.

Proof. For any complex vector fields $Z = X + iY$ and $W = X' + iY'$, the bracket [Z,W] is given by

$$[Z,W] = ([X,X'] - [Y,Y']) + i([X,Y'] + [Y,X']).$$

Thus we have $\overline{[Z,W]} = [\bar{Z},\bar{W}]$, which gives the equivalence of (a) and (b).

Let $\omega \in C^{1,0}(M)$. If Z and W are complex vector fields of type (0,1), then (b) implies

$$2d\omega(Z,W) = Z(\omega(W)) - W(\omega(Z)) - \omega([Z,W]) = 0.$$

Hence $d\omega$ does not have the component of degree (0,2). Similarly, we see that if $\omega \in C^{0,1}(M)$, then $d\omega$ does not have the component of degree (2,0). To prove the converse (c) $\longrightarrow$ (a), let Z and W be vector fields of type (1,0). Then $\omega([Z,W]) = 0$ for all forms of ω of degree (0,1). Hence, [Z,W] is of type (1,0). The proof of (c) $\longrightarrow$ (b) is similar. Since C(M) is locally generated by $C^{0,0}(M)$, $C^{1,0}(M)$ and $C^{0,1}(M)$, (c) implies (d). The converse (d) $\longrightarrow$ (c) is trivial.

We next prove the equivalence of (a) and (e). Let X and Y be real vector fields and put $Z = [X-iJX,Y-iJY]$. Then (a) holds if and only if

Z is of type (1,0) for all X and Y. On the other hand, we have

$$Z + iJZ = -N(X,Y) - iJN(X,Y).$$

Since Z + iJZ = 0 if and only if Z is of type (1,0), (a) holds if and only if N(X,Y) = 0 for all X and Y. QED.

If the condition (a) holds, we say that the almost complex structure J is *integrable*. That is, J is integrable if and only if N vanishes identically.

We now give the conditions that an almost complex manifold M to be a complex manifold.

THEOREM 1.2. Let M be an almost complex manifold with almost complex structure J. Then J is a complex structure if and only if J has no torsion.

Proof. Let $(x^1,\ldots,x^{2n})$ be a local coordinate system in M. With respect to this the components N^i_{jk} of N is given by

$$N^i_{jk} = \sum_{h=1}^{2n} (J^h_j \partial_h J^i_k - J^h_k \partial_h J^i_j - J^i_h \partial_j J^h_k + J^i_h \partial_k J^h_j),$$

where J^i_j denote the components of J. If J is a complex structure on M, then, by (1.1), we see that the components of J are constant. Thus we have $N^i_{jk} = 0$.

The converse is beyond the scope of this book. For the complete proof of the converse is given by Newlander-Nierenberg [1]. (See also Kobayashi-Nomizu [2], Matsushima [1].) QED.

THEOREM 1.3. Let M be a real 2n-dimensional almost complex manifold with almost complex structure J. Suppose that there exists an open covering {U} of M satisfying the following conditiond: There is a local coordinate system $(x^1,\ldots,x^n,y^1,\ldots,y^n)$ on each U, such that for each point of U,

$$J(\partial/\partial x^j) = \partial/\partial y^j, \quad J(\partial/\partial y^j) = -(\partial/\partial x^j), \quad j = 1,\ldots,n.$$

Then M is a complex manifold.

Proof. Let (x^j, y^j) and (u^j, v^j) be local coordinate systems on U and V respectively satisfying the condition. On $U \cap V \neq \emptyset$ we set

$$u^j = \alpha^j(x^k, y^k), \qquad v^j = \beta^j(x^k, y^k).$$

Then we have

$$\partial/\partial x^j = \sum_k \{(\partial\alpha^k/\partial x^j)(\partial/\partial u^k) + (\partial\beta^k/\partial x^j)(\partial/\partial v^k)\},$$

$$\partial/\partial y^j = \sum_k \{(\partial\alpha^k/\partial y^j)(\partial/\partial u^k) + (\partial\beta^k/\partial y^j)(\partial/\partial v^k)\}.$$

Applying J on the both sides of the equations above, we obtain

$$\partial/\partial y^j = \sum_k \{(\partial\alpha^k/\partial x^j)(\partial/\partial v^k) - (\partial\beta^k/\partial x^j)(\partial/\partial u^k)\},$$

$$\partial/\partial x^j = -\sum_k \{(\partial\alpha^k/\partial y^j)(\partial/\partial v^k) + (\partial\beta^k/\partial y^j)(\partial/\partial u^k)\}.$$

From these equations we have

$$\partial\alpha^k/\partial x^j = \partial\beta^k/\partial y^j, \quad \partial\alpha^k/\partial y^j = -(\partial\beta^k/\partial x^j), \quad j,k = 1,\ldots,n.$$

We put $z^j = x^j + iy^j$, $w^j = u^j + iv^j$, then $(z^1,\ldots,z^n)$ and $(w^1,\ldots,w^n)$ are complex local coordinate systems of U and V respectively. We have

$$w^k = f^k(z^1,\ldots,z^n), \quad f^k = \alpha^k + i\beta^k, \quad k = 1,\ldots,n.$$

Since f^k is holomorphic, M is a complex manifold. QED.

If a vector field X on M satisfies $L_X J = 0$, L_X being the Lie differentiation with respect to X, then X is called an *infinitesimal automorphism* (*analytic vector field*) of an almost complex structure J. For any vector fields X and Y on M we have

$$(L_X J)Y = L_X JY - JL_X Y = [X,JY] - J[X,Y].$$

Thus we have

PROPOSITION 1.2. A vector field X on M is an infinitesimal automorphism of an almost complex structure J on M if and only if $[X,JY] = J[X,Y]$ for all vector fields Y on M.

THEOREM 1.4. Let ϕ_t be a 1-parameter group on a complex manifold M and X be the infinitesimal transformation of ϕ_t. Then X is an infinitesimal automorphism of an almost complex structure J on M if and only if ϕ_t is a holomorphic isomorphism of M for each t.

Proof. For any vector field Y on M we have

$$[X,JY]_x - J_x[X,Y]_x = \lim_{t\to 0}\frac{1}{t}[(JY)_x - (\phi_t)_*(JY)_{\phi_t^{-1}(x)} - J_x(Y_x - (\phi_t)_*Y_{\phi_t^{-1}(x)})]$$

$$= \lim_{t\to 0}\frac{1}{t}[J_x((\phi_t)_*Y_{\phi_t^{-1}(x)} - (\phi_t)_*(JY)_{\phi_t^{-1}(x)}]$$

at each point x of M. Thus, from Proposition 1.1, if ϕ_t is a holomorphic isomorphism, then $[X,JY] = J[X,Y]$. Hence X is an infinitesimal automorphism.

Conversely, we suppose that X is an infinitesimal automorphism. We put

$$\Phi(t;Y) = J(\phi_t)_*Y - (\phi_t)_*JY$$

for any vector field Y on M. Then we have $[d\Phi(t;Y)_x/dt]_{t=0} = 0$. On the other hand, we see that $\Phi(t+s;Y) = \Phi(s;(\phi_t)_*Y) + (\phi_s)_*\Phi(t;Y)$. Hence we obtain

$$\lim_{s\to 0}\frac{1}{s}[\Phi(t+s;Y)_x - \Phi(t;Y)_x] = -[X,\Phi(t;Y)]_x.$$

As a function of t, $\Phi(t;Y)_x$ is a solusion of the ordinary differential equation

$$\frac{d\Phi(t;Y)_x}{dt} = -[X,\Phi(t;Y)]_x.$$

We notice that $\Phi(0;Y) = 0$. Since the vector field which is constantly zero is a solusion of the equation above, the uniqueness of solusions implies that $\Phi(t;Y)_x = 0$. Therefore, ϕ_t is a holomorphic isomorphism. QED.

PROPOSITION 1.3. Let X be an infinitesimal automorphism of an almost complex structure J on M. Then JX is also an infinitesimal automorphism of J if and only if $N(X,Y) = 0$ for all vector fields Y on M.

Proof. From Proposition 1.2 we obtain

$$N(X,Y) = [JX,JY] - J[JX,Y]$$

for any vector field Y on M. From this we have our assertion. QED.

Let M be a complex manifold and $(z^1,\ldots,z^n)$, $z^j = x^j + iy^j$, be a complex local coordinate system of M. We define

$$d' : C^{p,q}(M) \longrightarrow C^{p+1,q}(M), \quad d'' : C^{p,q}(M) \longrightarrow C^{p,q+1}(M)$$

by

$$d = d'\omega + d''\omega \quad \text{for } \omega \in C^{p,q}(M).$$

We obtain

$$(d')^2 = 0, \ (d'')^2 = 0, \quad d'd'' + d''d' = 0.$$

A form ω of degree $(p,0)$ is said to be *holomorphic* if $d''\omega = 0$. We now express ω in terms of the local coordinate system $(z^1,\ldots,z^n)$:

$$\omega = \sum f_{j_1\ldots j_p} dz^{j_1} \wedge \ldots \wedge dz^{j_p}.$$

Then $d''\omega = 0$ if and only if $d''f_{j_1\cdots j_p} = 0$. For any function f on M we obtain

$$df = \sum(\partial f/\partial z^j)dz^j + \sum(\partial f/\partial \bar{z}^j)d\bar{z}^j,$$

from which

$$d'f = \sum(\partial f/\partial z^j)dz^j, \qquad d''f = \sum(\partial f/\partial \bar{z}^j)d\bar{z}^j.$$

Therefore, we see that $d''\omega = 0$ if and only if $\partial f_{j_1\cdots j_p}/\partial \bar{z}^j = 0$ for $j = 1,\ldots,n$. Hence ω is holomorphic if and only if the coefficient $f_{j_1\cdots j_p}$ of ω are all holomorphic.

A complex vector field Z of type (1,0) on a complex manifold M is said to be *holomorphic* if Zf is holomorphic for every locally defined holomorphic function f. We put

$$Z = \sum f^j(\partial/\partial z^j).$$

Then Z is holomorphic if and only if f^j are all holomorphic functions.

PROPOSITION 1.4. Let M be a complex manifold with almost complex structure J. If X is an infinitesimal automorphism of J on M, then X - iJX is a holomorphic vector field.

Proof. We can set

$$\tfrac{1}{2}(X - iJX) = \sum f^j(\partial/\partial z^j), \qquad \tfrac{1}{2}(Y - iJY) = \sum g^j(\partial/\partial z^j).$$

We easily see that $[X,JY] = J[X,Y]$ if and only if $\Sigma \bar{g}^k(\partial f^j/\partial \bar{z}^k) = 0$ for $j = 1,\ldots,n$. Since Y is an arbitrary vector field, we see that $\partial f^j/\partial \bar{z}^k = 0$ for all j, k. Thus X - iJX is holomorphic. QED.

Let Z be a holomorphic vector field on M. We have

$$Z = \sum f^j(\partial/\partial z^j) = \tfrac{1}{2}\sum_j\{\alpha^j(\partial/\partial x^j)+\beta^j(\partial/\partial y^j)\}+\tfrac{1}{2}i\sum_j\{\beta^j(\partial/\partial x^j)-\alpha^j(\partial/\partial y^j)\},$$

where $f^j = \alpha^j + i\beta^j$. If we put $X = \Sigma\{\alpha^j(\partial/\partial x^j) + \beta^j(\partial/\partial y^j)\}$, then $Z = \frac{1}{2}(X - iJX)$. X is independent of the choice of coordinate neighborhoods. Since $\partial f^j/\partial \bar{z}^k = 0$ for all j, k, we have $[X,JY] = J[X,Y]$ for all Y. Thus X is an infinitesimal automorphism. We take a mapping $\theta : X \longrightarrow \frac{1}{2}(X - iJX)$ form the Lie algebra of infinitesimal automorphisms of J to the Lie algebra of holomorphic vector fields. For any infinitesimal automorphisms X and Y of J we have

$$\theta([X,Y]) = [\theta X,\theta Y]$$

by using $J[X,Y] = [JX,Y] = [X,JY]$ and $[JX,JY] = -[X,Y]$. Claerly, θ is one-to-one. Therefore, θ is an isomorphism.

2. EXAMPLES OF COMPLEX MANIFOLDS AND ALMOST COMPLEX MANIFOLDS

In this section we give some examples of complex manifolds and almost complex manifolds. First of all, we give examples of complex structures on vector spaces.

Example 2.1. Let C^n be the complex vector space of all n-tuples of complex numbers $(z^1,\ldots,z^n)$. If we set $z^k = x^k + iy^k$, x^k, $y^k \in R$, $k = 1,\ldots,n$, then C^n can be identified with the real vector space R^{2n}, the identification will be given by $(z^1,\ldots,z^n) \longrightarrow (x^1,\ldots,x^n,y^1,\ldots,y^n)$. The complex structure of R^{2n} defined by $(x^1,\ldots,x^n,y^1,\ldots,y^n) \longrightarrow (y^1,\ldots,y^n,-x^1,\ldots,-x^n)$ is called the *canonical complex structure* of R^{2n}. In terms of the natural basis of R^{2n}, it is given by the matrix

$$J_0 = \begin{pmatrix} 0 & I_n \\ -I_n & 0 \end{pmatrix}.$$

Example 2.2. A Lie algebra $\mathfrak{g}$ over C is called a *complex Lie algebra*. Considering $\mathfrak{g}$ as a real vector space, we define a complex structure J on $\mathfrak{g}$ by $JX = iX$. The complex structure J satisfies $[JX,Y] = J[X,Y] = [X,JY]$ for all $X,Y \in \mathfrak{g}$, that is, $\mathrm{ad}(X)\cdot J = J\cdot\mathrm{ad}(X)$ for all X.

Conversely, let $\mathfrak{g}$ be a Lie algebra over R with a complex structure J such that $ad(X) \cdot J = J \cdot ad(X)$ for all $X \in \mathfrak{g}$. Then, defining $(a + ib)X = aX + bJX$, we have

$$[(a+ib)X,Y] = a[X,Y] + b[JX,Y] = a[X,Y] + bJ[X,Y] = (a+ib)[X,Y].$$

Thus we see that $\mathfrak{g}$ is a complex Lie algebra.

We next give some examples of complex manifolds.

Example 2.3. A *complex Lie group* G is a group which is at the same time a complex manifold such that the group operations $(a,b) \in G \times G \longrightarrow ab \in G$ and $a \in G \longrightarrow a^{-1} \in G$ are both holomorphic. We give some examples of complex Lie groups:

(a) G(n,C) is a complex Lie group;

(b) C^n is a complex Lie group with respect to addition;

(c) Direct product of two complex Lie groups is a complex Lie group;

(d) Every even-dimensional commutative Lie group is a complex Lie group.

If the Lie group G is a complex Lie group, then the Lie algebra $\mathfrak{g}$ of G is a complex Lie algebra. Conversely, if the Lie algebra $\mathfrak{g}$ of a real Lie group G is a complex Lie algebra, then G has the structure of a complex Lie group (see Matsushima [ ]).

Example 2.4. Let M be a complex manifold. We consider a covering space $\tilde{M}$ over M with projection $p : \tilde{M} \longrightarrow M$. Let $\{U_i\}$ be an open covering of $\tilde{M}$ such that for every i U_i is mapped holomorphically onto $p(U_i)$ by p. Such a covering of $\tilde{M}$ aleays exists. We denote by J a complex structure of M. We denote by p_i the restriction of p to U_i. We put

$$J_i = (p_i^{-1})^* \cdot J \cdot (p_i)^*.$$

On the intersection $U_i \cap U_j$, J_i and J_j coincide since $p_j^{-1} \cdot p_i$ is the identity mapping on $U_i \cap U_j$. Thus, the operator on $\tilde{M}$ having the J_i as its restrictions defines a complex structure on $\tilde{M}$. From the construction we see that the projection p is holomorphic with respect to this complex structure.

Example 2.5. Consider C^n as a real 2n-dimensional vector space. Let $\{a_1,\ldots,a_{2n}\}$ be a basis of C^n over R. We put

$$\Gamma = \sum_{j=1}^{2n}\{m_j a_j : m_j \text{ integers}\}.$$

We consider C^n as an abelian group with respect to the addition. Then Γ is a subgroup of C^n. We denote by T the quotient group C^n/Γ of C^n by Γ. Let $\pi : C^n \longrightarrow T$ be the natural mapping. Let U be the subset of T. Defining U to be an open set of T if $\pi^{-1}(U)$ is open in C^n, we introduce a topology on T. Then each point $\pi(a)$ of T $(a \in C^n)$ has a neighborhood homeomorphic to a neighborhood of $a \in C^n$. Thus we can see that T is a complex manifold of complex dimension n. We call T a *complex torus*.

Example 2.6. In the set $C^{n+1} - \{0\}$ we define an equivalence relation as follows: two points $z = (z^k)$ and $w = (w^k)$ of $C^{n+1} - \{0\}$ are equivalent if there is a nonzero complex number λ such that $Z = \lambda w$, that is, $z^k = \lambda w^k$ $(k = 0,1,\ldots,n)$. The set of equivalence class obtained by the equivalence relation defined above is called an n-dimensional *complex projective space*, which will be denoted by CP^n. The topology of CP^n is defined by the natural quotient topology. For each j, $j = 0,1,\ldots,n$, we put

$$U_j^* = \{z : z^j \neq 0\} \subset (C^{n+1} - \{0\}).$$

Let U_j be the image of U_j^* under the natural projection $C^{n+1} - \{0\} \longrightarrow CP^n$. We define a mapping ϕ_j from U_j to C^n which maps an equivalence class of z with $z^j \neq 0$ onto $(z^0/z^j,\ldots,z^{j-1}/z^j,z^{j+1}/z^j,\ldots,z^n/z^j)$ of C^n. Then $\{(U_j,\phi_j)\}$, $0 \leq j \leq n$, is a complex coordinate system of CP^n. The local coordinate system $(z^0/z^j,\ldots,z^{j-1}/z^j,z^{j+1}/z^j,\ldots,z^n/z^j)$ is called the *inhomogeneous coordinate system* of CP^n and the coordinate system $(z^0,z^1,\ldots,z^n)$ is called the *homogeneous coordinate system* of CP^n.

We denote by C^* the multiplicative group of nonzero complex numbers. C^* acts freely on $C^{n+1} - \{0\}$ by

$$(\lambda, z) \in C^* \times (C^{n+1} - \{0\}) \longrightarrow \lambda z \in C^{n+1} - \{0\}.$$

Then $CP^n = (C^{n+1} - \{0\})/C^*$. $C^{n+1} - \{0\}$ is a principal fibre bundle over CP^n with group C^*. We denote by π the projection $C^{n+1} - \{0\} \longrightarrow CP^n$. Then the local triviality $\psi_j : \pi^{-1}(U_j) \approx U_j \times C^*$ is given by

$$\psi_j(z) = (\pi(z), z^j) \in U_j \times C^*, \quad z = (z^j) \in C^{n+1} - \{0\}.$$

We define the trnsition functions $\psi_{kj} : U_j \cap U_k \longrightarrow C^*$ by

$$\psi_{kj} = z^k/z^j$$

for a homogeneous coordinate system $z^0, \ldots, z^n$.

Let S^{2n+1} be the unit sphere in C^{n+1}, that is,

$$S^{2n+1} = \{(z^0, \ldots, z^n) \in C^{n+1} : \sum_{k=0}^{n} |z^k|^2 = 1\}.$$

The unit circle $S^1 = \{z \in C : |z| = 1\}$ may be considered as a multiplicative group and identified with the additive group R/Z (the real numbers modulo 1) by

$$\lambda = e^{2\pi i\theta} \in S^1 \longrightarrow \frac{1}{2\pi i} \log\lambda = \theta \in R/Z.$$

We see that S^{2n+1} is a principal fibre bundle over CP^n with group S^1 which is a subbundle of $C^{n+1} - \{0\}$. We denote by π the projection $S^{2n+1} \longrightarrow CP^n$. Then the local trivialities $\psi_j : \pi^{-1}(U_j) \approx U_j \times S^1$ and the transition functions $\psi_{kj} : U_j \cap U_k \longrightarrow S^1$ are given respectively by

$$\psi_j(z) = (\pi(z), z^j/|z^j|) \in U_j \times S^1, \quad z \in S^{2n+1},$$

$$\psi_{kj} = (z^k|z^j|)/(z^j|z^k|).$$

If we consider S^{2n+1} as a principal fibre bundle over CP^n with group R/Z, then the transition functions $\psi_{kj} : U_j \cap U_k \longrightarrow R/Z$ are given by

$$\psi'_{kj} = \frac{1}{2\pi i} \log \psi_{kj} = \frac{1}{2\pi i}(\log z^k/z^j - \log|z^k|/|z^j|).$$

Example 2.7. Let S^{2p+1} and S^{2q+1} be two unit spheres. Then the product manifold $S^{2p+1} \times S^{2q+1}$ $(p,q \geq 0)$ admits a complex structure (Calabi-Eckmann [1]). Especially, the fact that $S^{2p+1} \times S^1$ admits a complex structure was discovered by Hopf [1].

Example 2.8. Let C denote the algebra of Cayley numbers with a basis $\{I, e_0, e_1, \ldots, e_6\}$, I being the unit element of C. The multiplication table is given by the following:

$$e_i^2 = -I, \quad e_i \cdot e_j = -e_j \cdot e_i \ (i \neq j), \ i,j = 0,1,\ldots,6,$$

$$e_0 \cdot e_1 = e_2, \quad e_0 \cdot e_3 = e_4, \quad e_0 \cdot e_5 = e_6,$$

$$e_1 \cdot e_4 = e_5, \quad e_1 \cdot e_3 = e_6, \quad e_2 \cdot e_3 = e_5, \quad e_2 \cdot e_4 = -e_6,$$

the other $e_i \cdot e_j$ being given by permuting the indices cyclically. The algebra C is non-associative.

Any element of C may be written as

$$xI + X, \quad x \in R,$$

where

$$X = \sum_{i=0}^{6} x^i e_i, \quad x^i \in R, \ i = 0,1,\ldots,6.$$

If x = 0, the element is called a purely imaginary Cayley number. All purely imaginary numbers form a 7-dimensional subspace $E^7 \subset C$. Let $X = \Sigma_{i=0}^{6} x^i e_i$, $Y = \Sigma_{i=0}^{6} y^i e_i \in E^7$. Then

$$X \cdot Y = -\langle X,Y \rangle I + X \times Y,$$

where

$$\langle X,Y \rangle = \sum_{i=0}^{6} x^i y^i$$

is the scalar product in E^7, and

$$X \times Y = \sum_{i \neq j} x^i y^j e_i \cdot e_j$$

is the vector product of X and Y.

The operation of vector product is bilinear, and $X \times Y$ is orthogonal to both X and Y, that is, $\langle X, X \times Y \rangle = \langle Y, X \times Y \rangle = 0$. We have moreover $X \times Y = -Y \times X$.

Consider the unit 6-dimensional sphere S^6 in E^7:

$$S^6 = \{X \in E^7 : \langle X,X \rangle = 1\}.$$

The scalar product in E^7 induces the natural tensor field g on S^6. The tangent space $T_X(S^6)$ at $X \in S^6$ can naturally be identified with the subspace of E^7 orthogonal to X. Define the endomorphism J_X on $T_X(S^6)$ by

$$J_X Y = X \times Y \quad \text{for } Y \in T_X(S^6).$$

Then

$$\begin{aligned} J_X^2 Y &= J_X(X \times Y) = X \times (X \times Y) \\ &= X \cdot (X \times Y) = X \cdot (X \cdot Y) + \langle X,Y \rangle X \\ &= (X \cdot X) \cdot Y = -\langle X,X \rangle Y = -Y, \end{aligned}$$

from which $J_X^2 = -I$. Thus the correspondence $X \longrightarrow J_X$ defines a tensor field J such that $J^2 = -I$. On the other hand, we have

$$g(J_X Y, J_X Z) = g(Y,Z), \quad Y,Z \in T_X(S^6).$$

Consequently, S^6 admits an almost Hermitian structure (J,g) (see §3).

3. HERMITIAN MANIFOLDS

Let M be an almost complex manifold with almost complex structure J. A *Hermitian metric* on M is a Riemannian metric g such that

$$g(JX,JY) = g(X,Y)$$

for any vector fields X and Y on M.

An almost complex manifold with a Hermitian metric is called an *almost Hermitian manifold* and a complex manifold with a Hermitian metric is called a *Hermitian manifold*.

PROPOSITION 3.1. Every almost complex manifold admits a Hermitian metric provided it is paracompact.

Proof. Since the manifold M is paracompact, we can take a Riemannian metric g. If we set

$$h(X,Y) = g(X,Y) + g(JX,JY)$$

for any vector fields X and Y on M, then h is a Hermitian matric on M. QED.

We assume that all manifolds M are always paracompact. We can easily see that a Hermitian metric g on an almost complex manifold M can be extended uniquely to a complex symmetric tensor field of covariand degree 2, also denoted by g, such that

(3.1) $g(\bar{Z},\bar{W}) = \overline{g(Z,W)}$ for any complex vector fields Z and W;

(3.2) $g(Z,\bar{Z}) > 0$ for any non-zero complex vector Z;

(3.3) $g(Z,\bar{W}) = 0$ for any vector field Z of type $(1,0)$ and any vector field W of type $(0,1)$.

Conversely, every complex symmetric tensor g satisfying (3.1), (3.2) and (3.3) is the natural extension of a Hermitian metric g on M.

Let M be an almost Hermitian manifold with almost complex structure J and Hermitian metric g. The *fundamental 2-form* Φ of M is defined by

$$\Phi(X,Y) = g(X,JY).$$

for all vector fields X and Y on M. Then we have

$$\Phi(JX,JY) = \Phi(X,Y).$$

Since g is positive-definite and J is non-singular at each point, it follows that

$$\Phi^p = \Phi \wedge \ldots\ldots \wedge \Phi \quad \text{(p times)}, \ 1 \le p \le n,\ 2n = \dim M,$$

is non-zero at each point. Thus we have

PROPOSITION 3.2. Let M be a real $2n$-dimensional almost complex manifold. Then M is orientable.

THEOREM 3.1. Let M be an almost complex manifold with almost complex structure J. Then M is a complex manifold if and only if M admits a linear connection ∇ such that $\nabla J = 0$ and $T = 0$, where T denote the torsion of ∇.

Proof. Suppose that there exists a linear connection ∇ on M satisfying $\nabla J = 0$ and $T = 0$. Then we have

$$[X,Y] = \nabla_X Y - \nabla_Y X, \qquad \nabla_X JY = J\nabla_X Y$$

for any vector fields X and Y on M. From these equations we obtain

$$\begin{aligned} N(X,Y) &= [JX,JY] - J[X,JY] - J[JX,Y] - [X,Y] \\ &= J\nabla_{JX}Y - J\nabla_{JY}X - J^2\nabla_X Y + J\nabla_{JY}X - J\nabla_{JX}Y + J^2\nabla_Y X - \nabla_X Y + \nabla_Y X \\ &= 0. \end{aligned}$$

Therefore, from Theorem 1.2, we see that M is a complex manifold.

Conversely, we suppose that M is a complex manifold, that is, $N = 0$. First of all, we can take a linear connection ∇ with vanishing torsion $T = 0$. We put

$$A(X,Y) = (\nabla_X J)Y - (\nabla_Y J)X, \quad S(X,Y) = (\nabla_X J)Y + (\nabla_Y J)X.$$

If we set

$$\nabla'_X Y = \nabla_X Y + \tfrac{1}{4}[A(X,JY) - JS(X,Y)],$$

then we can prove that ∇' is the desired connection. From the definition ∇' is obviously a linear connection on M. We shall prove that $\nabla' J = 0$ and $T' = 0$, where T' is the torsion of ∇'. For any vector fields X and Y on M we have

$$\begin{aligned} (\nabla'_X J)Y &= \nabla'_X JY - J\nabla'_X Y \\ &= \nabla_X JY - J\nabla_X Y - \tfrac{1}{4}[A(X,Y) + JS(X,JY) + JA(X,JY) + S(X,Y)] \\ &= (\nabla_X J)Y - \tfrac{1}{2}[(\nabla_X J)Y + J(\nabla_X J)JY] \\ &= \tfrac{1}{2}[(\nabla_X J)Y - J(\nabla_X J)JY]. \end{aligned}$$

On the other hand, we obtain

$$J(\nabla_X J)JY = J\nabla_X J^2 Y - J^2\nabla_X JY = -J\nabla_X Y + (\nabla_X J)Y + J\nabla_X Y = (\nabla_X J)Y.$$

Thus we have $\nabla'_X J = 0$.

We next prove that $T' = 0$. Since $T = 0$, we obtain

$$A(JX,Y) + A(X,JY) = -[X,Y] + [JX,JY] - J[JX,Y] - J[X,JY] = N(X,Y),$$

from which

$$\begin{aligned} T'(X,Y) &= \nabla'_X Y - \nabla'_Y X - [X,Y] \\ &= T(X,Y) + \tfrac{1}{4}[A(JX,Y) + A(X,JY)] \\ &= \tfrac{1}{4}N(X,Y). \end{aligned}$$

Consequently, if $N = 0$, then $T' = 0$. QED.

LEMMA 3.1. Let M be an almost Hermitian manifold with almost complex structure J and Hermitian metric g. Then the covariant derivative ∇ of the Riemannian connection defined by g, the fundamental 2-form Φ and the torsion N of J satisfy

$$2g((\nabla_X J)Y,Z) - g(JX,N(Y,Z)) = 3d\Phi(X,JY,JZ) - 3d\Phi(X,Y,Z)$$

for any vector fields X, Y and Z on M.

Proof. First of all, we have

$$(\nabla_X J)JY = -J(\nabla_X J)Y, \qquad (\nabla_X \Phi)(Y,Z) = g(Y,(\nabla_X J)Z),$$

$$N(Y,Z) = (\nabla_{JY} J)Z - (\nabla_{JZ} J)Y + J(\nabla_Z J)Y - J(\nabla_Y J)Z.$$

On the other hand, we obtain

$$\begin{aligned} 3d\Phi(X,Y,Z) &= X(\Phi(Y,Z)) + Y(\Phi(Z,X)) + Z(\Phi(X,Y)) \\ &\qquad - \Phi([X,Y],Z) - \Phi([Y,Z],X) - \Phi([Z,X],Y) \\ &= g(Y,(\nabla_X J)Z) + g(Z,(\nabla_Y J)X) + g(X,(\nabla_Z J)Y), \end{aligned}$$

$$3d\Phi(X,JY,JZ) = g(JY,(\nabla_X J)JZ)+g(JZ,(\nabla_{JY}J)X)+g(X,(\nabla_{JZ}J)JY).$$

From these equations we have

$$3d\Phi(X,Y,Z) - 3d\Phi(X,JY,JZ) = -2g((\nabla_X J)Y,Z) + g(Z,(\nabla_Y J)X)$$
$$+ g(X,(\nabla_Z J)Y) - g(JZ,(\nabla_{JY}J)X) - g(X,(\nabla_{JZ}J)JY)$$
$$= -2g((\nabla_X J)Y,Z) - g(J(\nabla_Y J)Z,JX) + g(JX,J(\nabla_Z J)Y)$$
$$+ g((\nabla_{JY}J)Z,JX) - g((\nabla_{JZ}J)Y,JX)$$
$$= -2g((\nabla_X J)Y,Z) + g(JX,N(Y,Z)).$$

Thus we have our equation. QED.

THEOREM 3.2. Let M be an almost complex manifold with almost complex structure J and Hermitian metric g. Let ∇ be the covariant differentiation of the Riemannian connection defined by g. Then the following conditions are equivalent:

(a) $\nabla J = 0$;

(b) $\nabla\Phi = 0$;

(c) The almost complex structure has no torsion and the fundamental 2-form Φ is closed, that is, $N = 0$ and $d\Phi = 0$.

Proof. We have seen that $(\nabla_X\Phi)(Y,Z) = g(Y,(\nabla_X J)Z)$ for any vector fields X, Y and Z on M. Thus $\nabla J = 0$ if and only if $\nabla\Phi = 0$. Hence (a) is equivalent to (b).

We suppose (b). Then $d\Phi = 0$ obviously. Moreover, by Lemma 3.1, we have $N = 0$.

Conversely, we suppose (c). Then Lemma 3.1 implies $\nabla J = 0$ and hence $\nabla\Phi = 0$. Hence (b) is equivalent to (c). QED.

A Hermitian metric g on an almost complex manifold M is called a *Kaehlerian metric* if the fundamental 2-form Φ is closed. An almost complex manifold M with a Kaehlerian metric is called an *almost Kaehlerian manifold*. A complex manifold M with a Kaehlerian metric is called a *Kaehlerian manifold*. In view of Theorem 3.2, a Hermitian manifold M is a Kaehlerian manifold if and only if $\nabla J = 0$.

An almost Hermitian manifold M with almost complex structure J is called a *nearly Kaehlerian manifold* (*almost Tachibana manifold*, K-*space*) if

$$(\nabla_X J)Y + (\nabla_Y J)X = 0$$

for any vector fields X and Y on M or equivalently

$$(\nabla_X J)X = 0$$

for any vector field X on M.

If we have

$$(\nabla_X J)Y + (\nabla_{JX} J)JY = 0$$

for any vector fields X and Y on M, then M is called a *quasi-Kaehlerian manifold*. Let $\{e_1,\dots,e_n,Je_1,\dots,Je_n\}$ be a basis of M. Then

$$\delta\Phi(X) = - \sum_{i=1}^{n} [(\nabla_{e_i}\Phi)(e_i,X) + (\nabla_{Je_i}\Phi)(Je_i,X)].$$

If $\delta\Phi = 0$, M is called an *almost semi-Kaehlerian manifold* and moreover if $N = 0$, M is called a *semi-Kaehlerian manifold*.

A 2n-dimensional manifold M with a 2-form (resp. a closed 2-form) Φ which is non-degenerate at each point of M is called an *almost symplectec* or *almost Hamiltonian manifold* (resp. a *symplectec* or *Hamiltonian manifold*).

4. KAEHLERIAN MANIFOLDS

Let M be a real 2n-dimensional Kaehlerian manifold with almost complex structure J and Kaehlerian metric g. We denote by R and S the Riemannian curvature tensor and the Ricci tensor of M, respectively.

PROPOSITION 4.1. For a Kaehlerian manifold M we have the following properties:

(a) $R(X,Y)J = JR(X,Y)$ and $R(JX,JY) = R(X,Y)$ for any vector fields X and Y on M;

(b) $S(JX,JY) = S(X,Y)$ and $S(X,Y) = \frac{1}{2}(\text{Trace of } JR(X,JY))$ for any vector fields X and Y on M.

Proof. (a) Since J is parallel, the first equation is clear. We prove the second equation. For any vector fields X, Y, Z and W on M we have

$$\begin{aligned} g(R(JX,JY)Z,W) &= g(R(W,Z)JY,JX) = g(JR(W,Z)Y,JX) \\ &= g(R(W,Z)Y,X) = g(R(X,Y)Z,W). \end{aligned}$$

Thus we have $R(JX,JY) = R(X,Y)$.

(b) Let $\{e_1,\dots,e_{2n}\}$ be an orthonormal basis of M. Then

$$\begin{aligned} S(JX,JY) &= \sum_i g(R(e_i,JX)JY,e_i) = \sum_i g(R(Je_i,JX)JY,Je_i) \\ &= \sum_i g(R(e_i,X)JY,Je_i) = \sum_i g(JR(e_i,X)Y,Je_i) \\ &= \sum_i g(R(e_i,X)Y,e_i) = S(X,Y). \end{aligned}$$

Thus we have the first formula of (b).

Using Bianchi's first identity, we obtain

$$\begin{aligned} S(X,Y) &= \sum_i g(R(e_i,X)Y,e_i) = -\sum_i g(JR(e_i,X)JY,e_i) \\ &= \sum_i [g(JR(X,JY)e_i,e_i) + g(JR(JY,e_i)X,e_i)] \\ &= \sum_i [g(JR(X,JY)e_i,e_i) + g(JR(JY,Je_i)X,Je_i)] \\ &= \sum_i [g(JR(X,JY)e_i,e_i) + g(R(Y,e_i)X,e_i)] \\ &= (\text{Trace of } JR(X,JY)) - S(X,Y). \end{aligned}$$

Hence we obtain the second formula of (b). QED.

PROPOSITION 4.2. The Ricci tensor S of a Kaehlerian manifold M satisfies

$$(\nabla_Z S)(X,Y) = (\nabla_X S)(Y,Z) + (\nabla_{JY} S)(JX,Z).$$

Proof. From the second equation of (b) in Proposition 4.1 and Bianchi's second identity we have

$$\begin{aligned}(\nabla_Z S)(X,Y) &= \tfrac{1}{2}\sum_i g(J(\nabla_Z R)(X,JY)e_i,e_i)\\ &= \tfrac{1}{2}\sum_i g(J(\nabla_X R)(Z,JY)e_i,e_i) + \tfrac{1}{2}\sum_i g(J(\nabla_{JY} R)(X,Z)e_i,e_i)\\ &= (\nabla_X S)(Y,Z) + (\nabla_{JY} S)(JX,Z).\end{aligned}$$

QED.

PROPOSITION 4.3. Let M be a real 2n-dimensional Kaehlerian manifold. If M is of constant curvature, then M is flat provided $n > 1$.

Proof. If M is of constant curvature c, then

$$R(X,Y)Z = c[g(Y,Z)X - g(X,Z)Y]$$

for any vector fields X, Y and Z on M. From the second formula of (a) in Proposition 4.1 we obtain

$$\begin{aligned}R(X,Y)Y &= c[g(Y,Y)X - g(X,Y)Y]\\ &= c[g(JY,Y)JX - g(JX,Y)JY] = R(JX,JY)Y,\end{aligned}$$

from which $(2n-1)cX = cX$. Hence we have $2(n-1)c = 0$. Since $n > 1$, we obtain $c = 0$. QED.

In view of Proposition 4.3, the notion of constant curvature for Kaehlerian manifolds is not essential. So we introduce the notion of constant holomorphic sectional curvature for Kaehlerian manifolds. To this purpose we prepare some algebraic results for quadrilinear mappings on a real vector space with almost complex structure.

Let V be a 2n-dimensional real vector space with a complex

structure J. We consider a quadrilinear mapping $B : V \times V \times V \times V \longrightarrow R$ with the following four conditions:

(a) $B(X,Y,Z,W) = -B(Y,X,Z,W) = -B(X,Y,W,Z)$,

(b) $B(X,Y,Z,W) = B(Z,W,X,Y)$,

(c) $B(X,Y,Z,W) + B(X,Z,W,Y) + B(X,W,Y,Z) = 0$,

(d) $B(JX,JY,Z,W) = B(X,Y,JZ,JW) = B(X,Y,Z,W)$.

We easily see that the Riemannian curvature tensor R of a Kaehlerian manifold satisfies the above conditions (a), (b), (c) and (d) (see §3 of Chapter I).

LEMMA 4.1. Let B and T be two quadrilinear mappings satisfying the conditions (a), (b), (c) and (d). If

$$B(X,JX,X,JX) = T(X,JX,X,JX)$$

for all $X \in V$, then $B = T$.

Proof. We may assume that $T = 0$; consider $B - T$ and 0 instead of B and T. We assume therefore that $B(X,JX,X,JX) = 0$ for all $X \in V$. Replacing X by $X + Y$, we obtain

$$(4.1) \qquad 2B(X,JY,X,JY) + B(X,JX,Y,JY) = 0.$$

On the other hand, (a), (c) and (d) imply

$$B(X,JX,Y,JY) - B(X,Y,X,Y) - B(X,JY,X,JY) = 0.$$

This, combined with (4.1), yields

$$(4.2) \qquad 3B(X,JY,X,JY) + B(X,Y,X,Y) = 0,$$

from which

$$(4.3)\qquad 3B(X,Y,X,Y) + B(X,JY,X,JY) = 0.$$

From (4.2) and (4.3) we obtain $B(X,Y,X,Y) = 0$. By Lemma 3.2 of Chapter I, we have $B = 0$. QED.

Let g be a Hermitian inner product on V. We set

$$B_0(X,Y,Z,W) = \tfrac{1}{4}[g(X,Z)g(Y,W) - g(X,W)g(Y,Z) + g(X,JZ)g(Y,JW) - g(X,JW)g(Y,JZ) + 2g(X,JY)g(Z,JW)].$$

Then B_0 satisfies (a), (b), (c) and (d). We also have

$$B_0(X,Y,X,Y) = \tfrac{1}{4}[g(X,X)g(Y,Y) - g(X,Y)^2 + 3g(X,JY)^2],$$

$$B_0(X,JX,X,JX) = g(X,X)^2.$$

Let p be a plane in V and X, Y be orthonormal basis for p. We set

$$K(p) = B(X,Y,X,Y).$$

Then K(p) depends only on p and is independent of the choice of an orthonormal basis for p. We assume that p is independent by J, that is, $Jp = p$. Then X,JX is an orthonormal basis for p for any unit vector X in p. From Lemma 4.1 we have

LEMMA 4.2. Let B be a quadrilinear mapping satisfying (a), (b), (c) and (d). If $K(p) = c$ for all J-invariant planes p, then $B = cB_0$.

We shall now apply these algebraic results to the Riemannian curvature tensor R of a Kaehlerian manifold M with almost complex structure J and metric g.

For each plane p in the tangent space $T_x(M)$, the sectional curvature K(p) is defined to be

$$K(p) = R(X,Y,X,Y) = g(R(X,Y)Y,X),$$

where $\{X,Y\}$ is an orthonormal basis for p. If p is invariant by J, then K(p) is called the *holomorphic sectional curvature* by p. The holomorphic sectional curvature K(p) is given by

$$K(p) = R(X,JX,X,JX) = g(R(X,JX)JX,X),$$

where X is a unit vector in p. From Lemma 4.1 we see that the holomorphic sectional curvature K(p) for all J-invariant planes p in $T_x(M)$ determine the Riemannian curvature tensor R at x of M. If K(p) is a constant for all J-invariant planes p in $T_x(M)$ and for all points $x \in M$, then M is called a *space of constant holomorphic sectional curvature* or a *complex space form*. Sometimes, a complex space form is defined to be a simply connected complete Kaehlerian manifold of constant holomorphic sectional curvature.

We prove the following theorem which is a Kaehlerian analogue of Schur's theorem.

THEOREM 4.1. Let M be a connected Kaehlerian manifold of complex dimension $n > 1$. If the holomorphic sectional curvature K(p), where p is a J-invariant plane in $T_x(M)$, depends only on x, then M is a complex space form.

Proof. We put

$$R_0(X,Y,Z,W) = \tfrac{1}{4}[g(X,Z)g(Y,W) - g(X,W)g(Y,Z) + g(X,JZ)g(Y,JW) - g(X,JW)g(Y,JZ) + 2g(X,JY)g(Z,JW)].$$

Then Lemma 4.2 implies that $R = cR_0$, where c is a function on M. Then the Ricci tensor S of M is given by

$$S = \tfrac{1}{2}(n+1)cg.$$

From Theorem 3.3 of Chapter I we see that c is constant provided $n > 1$.

QED.

In the course of the proof for Theorem 4.1 we have

THEOREM 4.2. A Kaehlerian manifold M is of constant holomorphic sectional curvature c if and only if

$$R(X,Y)Z = \tfrac{c}{4}[g(X,Z)Y-g(Y,Z)X+g(JX,Z)JY-g(JY,Z)JX+2g(JX,Y)JZ]$$

for any vector fields X, Y and Z on M.

PROPOSITION 4.4. If a Kaehlerian manifold M is of constant holomorphic sectional curvature, then M is an Einstein manifold.

Let M be a complex n-dimensional Kaehlerian manifold with complex structure J. We can choose a local field of orthonormal frames $e_1,\dots,e_n,e_{1*}=Je_1,\dots,e_{n*}=Je_n$ in M. With respect to the frame field of M chosen above, let $\omega^1,\dots,\omega^n,\omega^{1*},\dots,\omega^{n*}$ be the field of dual frames. Let $\omega = (\omega^i_j)$, $i,j = 1,\dots,2n$, be the connection form of M. Then we have

$$(4.4)\qquad \omega^a_b = \omega^{a*}_{b*},\quad \omega^a_{b*} = -\omega^{a*}_b,\quad \omega^a_b = -\omega^b_a,\quad \omega^a_{b*} = \omega^b_{a*},$$

where $a,b = 1,\dots,n$. We denote by $\Omega = (\Omega^i_j)$ the curvature form and write

$$\Omega^i_j = \tfrac{1}{2}\sum_{k,l} R^i_{jkl}\omega^k \wedge \omega^l.$$

We set

$$\xi_a = \tfrac{1}{2}(e_a - ie_{a*}),\qquad \xi_{\bar a} = \tfrac{1}{2}(e_a + ie_{a*}),$$

$$\theta^a = \omega^a + i\omega^{a*},\qquad \theta^{\bar a} = \omega^a - i\omega^{a*}.$$

Then $\{\xi_a\}$ form a complex basis of $T^{1,0}_x(M)$ and $\{\xi_{\bar a}\}$ form a complex basis of $T^{0,1}_x(M)$. The Kaehlerian metric g is given by

$$g = \sum_a \theta^a \otimes \theta^{\bar a}.$$

Moreover, we set

$$\theta^a_b = \omega^a_b + i\omega^{a*}_b, \qquad \theta^{\bar a}_{\bar b} = \omega^a_b - i\omega^{a*}_b,$$

$$\Psi^a_b = \Omega^a_b + i\Omega^{a*}_b, \qquad \Psi^{\bar a}_{\bar b} = \Omega^a_b - i\Omega^{a*}_b.$$

Then we obtain

$$d\theta^a = -\sum_b \theta^a_b \wedge \theta^b, \qquad \theta^a_b + \theta^{\bar b}_{\bar a} = 0,$$

$$d\theta^a_b = -\sum_c \theta^a_c \wedge \theta^c_b + \Psi^a_b, \qquad \Psi^a_b = \sum_{c,d} K^a_{bc\bar d}\theta^c \wedge \theta^{\bar d},$$

$$K^a_{bc\bar d} = \tfrac{1}{2}[R^a_{bcd} + R^a_{b*cd*} + i(R^a_{bcd*} - R^a_{b*cd})].$$

We see that M is of constant holomorphic sectional curvature c if and only if

$$K^a_{bc\bar d} = \tfrac{1}{4}c(\delta_{ac}\delta_{bd} + \delta_{ab}\delta_{cd}),$$

or

$$\Psi^a_b = \tfrac{1}{4}c(\theta^a \wedge \theta^{\bar b} + \delta_{ab}\sum_c \theta^c \wedge \theta^{\bar c}).$$

We now take a complex local coordinate system $\{z^1,\ldots,z^n\}$ in M. We set

$$Z_a = \partial/\partial z^a, \quad Z_{\bar a} = \bar Z_a = \partial/\partial \bar z^a, \quad a = 1,\ldots,n.$$

We extend a Hermitian metric g to a complex symmetric bilinear form in $T^c_x(M)$. We put

$$g_{AB} = g(Z_A, Z_B), \quad A,B = 1,\ldots,n,\bar 1,\ldots,\bar n.$$

Then we have $g_{ab} = g_{\bar a\bar b} = 0$ and $(g_{a\bar b})$ is an (n,n)-Hermitian matrix. We write

$$ds^2 = 2 \sum_{a,b} g_{a\bar{b}} dz^a d\bar{z}^b$$

for the metric g. Let Φ be the fundamental 2-form of M. For any complex vectors Z, W we may write

$$Z = \sum_a (dz^a(Z)Z_a + d\bar{z}^a(Z)\bar{Z}_a), \quad W = \sum_b (dz^b(W)Z_b + d\bar{z}^b(W)\bar{Z}_b).$$

Then

$$\Phi(Z,W) = -i \sum_{a,b} g_{a\bar{b}}(dz^a(Z)d\bar{z}^b(W) - dz^a(W)d\bar{z}^b(Z)),$$

from which

$$\Phi = -2i \sum_{a,b} g_{a\bar{b}} dz^a \wedge d\bar{z}^b.$$

Since the fundamental 2-form Φ is closed, we have

$$\partial g_{a\bar{b}}/\partial z^c = \partial g_{c\bar{b}}/\partial z^a, \quad \partial g_{a\bar{b}}/\partial \bar{z}^c = \partial g_{a\bar{c}}/\partial \bar{z}^b.$$

The covariant differentiation ∇, which is originally defined for real vector fields, is extended by complex linearity to the act on complex vector fields. Setting

$$\nabla_{Z_B} Z_C = \Gamma^A_{BC} Z_A,$$

we obtain

$$\bar{\Gamma}^A_{BC} = \Gamma^{\bar{A}}_{\bar{B}\bar{C}}$$

with the connection that $\bar{\bar{a}} = a$. From the fact that $JZ_a = iZ_a$, $JZ_{\bar{a}} = -iZ_{\bar{a}}$ and $\nabla J = 0$ we have

$$\Gamma^{a}_{B\bar{C}} = \Gamma^{\bar{a}}_{BC} = 0.$$

Since J has no torsion we find

$$\Gamma^{a}_{bc} = \Gamma^{a}_{cb}, \qquad \Gamma^{\bar{a}}_{\bar{b}\bar{c}} = \Gamma^{\bar{a}}_{\bar{c}\bar{b}}, \quad \text{and other } \Gamma^{A}_{BC} = 0.$$

Thus we see that the Γ^{A}_{BC}'s are determined by the metric as follows:

$$\sum_a g_{a\bar{d}}\Gamma^{a}_{bc} = \partial g_{\bar{d}b}/\partial z^c, \qquad \sum_a g_{\bar{a}d}\Gamma^{\bar{a}}_{\bar{b}\bar{c}} = \partial g_{d\bar{b}}/\partial \bar{z}^c.$$

We set

$$R(Z_C,Z_D)Z_B = \sum_A K^{A}_{BCD}Z_A, \quad K_{ABCD} = g(R(Z_C,Z_D)Z_B,Z_A),$$

$$K_{ABCD} = \sum_E g_{AE}K^{E}_{BCD}.$$

From Proposition 4.1 we see that $R(Z_C,Z_D)$ commutes with J and hence

$$K^{a}_{\bar{b}CD} = K^{\bar{a}}_{bCD} = 0, \qquad K_{abCD} = K_{\bar{a}\bar{b}CD} = 0,$$

from which

$$K^{A}_{Bcd} = K^{A}_{B\bar{c}\bar{d}} = 0, \qquad K_{ABcd} = K_{AB\bar{c}\bar{d}} = 0.$$

Therefore, only components of the following types can be different from zero:

$$K^{a}_{bc\bar{d}},\ K^{a}_{b\bar{c}d},\ K^{\bar{a}}_{\bar{b}c\bar{d}},\ K^{\bar{a}}_{\bar{b}\bar{c}d}, \qquad K_{a\bar{b}c\bar{d}},\ K_{a\bar{b}\bar{c}d},\ K_{\bar{a}bc\bar{d}},\ K_{\bar{a}b\bar{c}d}.$$

Putting $X = Z_c$, $Y = Z_{\bar{d}}$ and $Z = Z_b$ in $R(X,Y)Z = [\nabla_X,\nabla_Y]Z - \nabla_{[X,Y]}Z$, we obtain

$$K^a_{bc\bar{d}} = -\partial\Gamma^a_{bc}/\partial\bar{z}^d,$$

from which

$$K_{a\bar{b}c\bar{d}} = \frac{\partial^2 g_{a\bar{b}}}{\partial z^c \partial\bar{z}^d} - \sum_{e,f} g^{\bar{e}f} \frac{\partial g_{a\bar{f}}}{\partial z^c} \frac{\partial g_{\bar{b}e}}{\partial\bar{z}^d},$$

where $(g^{a\bar{b}})$ is the inverse matrix to $(g_{a\bar{b}})$ so that $\Sigma_b g^{a\bar{b}} g_{c\bar{b}} = \delta_{ac}$.

The components K_{AB} of the Ricci tensor of M are given by

$$K_{a\bar{b}} = -\sum_c \partial\Gamma^c_{ac}/\partial\bar{z}^b, \quad K_{\bar{a}b} = \bar{K}_{a\bar{b}}, \quad K_{ab} = K_{\bar{a}\bar{b}} = 0.$$

We denote by G the determinant of the matrix $(g_{a\bar{b}})$, that is, $G^2 = |g_{AB}| = |g_{a\bar{b}}|^2$. Then we have

$$\frac{\partial G}{\partial z^a} = G \sum_{b,c} g^{b\bar{c}} \frac{\partial g_{b\bar{c}}}{\partial z^a}.$$

Thus we obtain

$$\sum_c \Gamma^c_{ac} = \frac{\partial \log G}{\partial z^a},$$

from which

$$K_{a\bar{b}} = -\frac{\partial^2 \log G}{\partial z^a \partial\bar{z}^b}.$$

We now define the Ricci form associated to the Ricci tensor S of M by

$$\rho(X,Y) = S(X,JY)$$

for all vector fields X and Y on M. We then have the following expression

$$\rho = -2i \sum_{a,b} K_{a\bar{b}} dz^a \wedge d\bar{z}^b,$$

or

$$\rho = 2id'd''\log G.$$

We can see that every Kaehlerian metric can be locally written as $ds^2 = 2\sum_{a,b} g_{a\bar{b}}dz^a d\bar{z}^b$ with $g_{a\bar{b}} = (\partial^2 f/\partial z^a \partial \bar{z}^b)$ for some real function f (cf. Morrow-Kodaira [1; p.86]).

We shall now give some examples of Kaehlerian manifolds.

Example 4.1. Consider the complex n-space C^n with the metric

$$ds^2 = \sum_{j=1}^{n} dz^j d\bar{z}^j.$$

The fundamental 2-form Φ in this case is given by

$$\Phi = -i\sum_{j=1}^{n} dz^j \wedge d\bar{z}^j.$$

Clearly, Φ is closed, and so the metric defines a Kaehlerian structure on C^n. Thus C^n is a complete, flat Kaehlerian manifold.

Example 4.2. Let CP^n be the n-dimensional complex projective space with homogeneous coordinate system $\{z^0, z^1, \ldots, z^n\}$. For every index j, let U_j be the open subset of CP^n defined by $z^j \neq 0$. We set

$$t_j^k = z^k/z^j, \quad j,k = 0,1,\ldots,n.$$

On each U_j, we take $t_j^0, \ldots, \hat{t}_j^j, \ldots, t_j^n$ (where $\hat{t}_j^j$ indicates that t_j^j is deleted) as a local coordinate system and consider the function $f_j = \Sigma_{k=0}^n t_j^k \bar{t}_j^k$. Then

$$f_j = \sum_{i=0}^{n} t_j^i \bar{t}_j^i = \sum_{i=0}^{n} (t_k^i \bar{t}_k^i) t_j^k \bar{t}_j^k = f_k t_j^k \bar{t}_j^k$$

on $U_j \cap U_k$. Thus we have

$$\log f_j = \log f_k + \log t_j^k + \overline{\log t_j^k}.$$

Since t^k_j is holomorphic in $U_j \cap U_k$, we have $d''\log t^k_j = 0$, $d'\overline{\log t^k_j} = d''\overline{\log t^k_j} = 0$. From $d'd'' = -d''d'$ we obtain

$$d'd''\log f_j = d'd''\log f_k \quad \text{on } U_j \cap U_k.$$

By setting

$$\Phi = -4id'd''\log f_j \quad \text{on } U_j,$$

we obtain a globally defined closed (1,1)-form Φ on CP^n. On the other hand, we have

$$f_0 = \sum_{j=0}^{n} t^j_0\bar{t}^j_0 = 1 + \sum_{\alpha=1}^{n} t^\alpha\bar{t}^\alpha,$$

$$\Phi = -4i \sum_{\alpha,\beta=1}^{n} \frac{\partial^2 \log f_0}{\partial t^\alpha \partial \bar{t}^\beta} dt^\alpha \wedge d\bar{t}^\beta$$

where we have put $t^\alpha = t^\alpha_0$, $\alpha = 1,\dots,n$. Thus we have

$$\Phi = -4i\frac{\Sigma dt^\alpha \wedge d\bar{t}^\alpha + \Sigma t^\alpha\bar{t}^\alpha \Sigma dt^\alpha \wedge d\bar{t}^\alpha - \Sigma\bar{t}^\alpha dt^\alpha \wedge \Sigma t^\alpha d\bar{t}^\alpha}{(1 + \Sigma t^\alpha\bar{t}^\alpha)^2}.$$

The associated metric tensor g is given by

$$ds^2 = 4\frac{(1 + \Sigma t^\alpha\bar{t}^\alpha)(\Sigma dt^\alpha d\bar{t}^\alpha) - (\Sigma\bar{t}^\alpha dt^\alpha)(\Sigma t^\alpha d\bar{t}^\alpha)}{(1 + \Sigma t^\alpha\bar{t}^\alpha)^2}.$$

This metric is sometimes called the *Fubini-Study metric*.

Example 4.3. Let D^n be the open unit ball in C^n defined by

$$D^n = \{(z^1,\dots,z^n) : \sum z^\alpha\bar{z}^\alpha < 1\}.$$

We set

$$\Phi = 4id'd''(1 - \sum z^\alpha \bar{z}^\alpha).$$

The associated metric g is given by

$$ds^2 = 4\frac{(1 - \Sigma z^\alpha \bar{z}^\alpha)(\Sigma dz^\alpha d\bar{z}^\alpha) + (\Sigma \bar{z}^\alpha dz^\alpha)(\Sigma z^\alpha d\bar{z}^\alpha)}{(1 - \Sigma z^\alpha \bar{z}^\alpha)^2}.$$

D^n is obviously a Kaehlerian manifold.

Example 4.4. Let CP^n be the complex projective space with homogeneous coordinates $z^0, z^1, \ldots, z^n$. The *complex quadric* Q^{n-1} is a complex hypersurface of CP^n defined by the equation

$$(z^0)^2 + (z^1)^2 + \ldots + (z^n)^2 = 0.$$

Then Q^{n-1} is a Kaehlerian manifold. Moreover, Q^{n-1} is an Einstein manifold. (For complex submanifolds of Kaehlerian manifolds see §1 of Chapter IV.)

In the next place, we give examples of complex manifolds which do not admit any Kaehlerian metric.

Example 4.5. We notice that the even-dimensional Betti numbers of a compact Kaehlerian manifold M are all positive (cf. Hodge [1]). For the complex manifold $S^{2p+1} \times S^{2q+1}$ in Example 2.7 we see that the Betti number B_{2k} $(1 \leq k \leq p+q)$ vanish. Therefore, $S^{2p+1} \times S^{2q+1}$ cannot carry any Kaehlerian metric except for $p = q = 0$ (Calabi-Eckmann [1]).

THEOREM 4.2. (1) For any positive number c, the complex projective space CP^n carries a Kaehlerian metric of constant holomorphic sectional curvature c. With respect to an inhomogeneous coordinate system $z^1, \ldots, z^n$ the metric is given by

$$ds^2 = \frac{4}{c}\frac{(1 + \Sigma z^\alpha \bar{z}^\alpha)(\Sigma dz^\alpha d\bar{z}^\alpha) - (\Sigma \bar{z}^\alpha dz^\alpha)(\Sigma z^\alpha d\bar{z}^\alpha)}{(1 + \Sigma z^\alpha \bar{z}^\alpha)^2}.$$

(2) For any negative number c, the open unit ball D^n in C^n carries a complete Kaehlerian metric of constant holomorphic sectional curvature c. With respect to the coordinate system $z^1,\ldots,z^n$ of C^n the metric is given by

$$ds^2 = -\frac{4}{c}\frac{(1 - \Sigma z^\alpha \bar{z}^\alpha)(\Sigma dz^\alpha d\bar{z}^\alpha) + (\Sigma \bar{z}^\alpha dz^\alpha)(\Sigma z^\alpha d\bar{z}^\alpha)}{(1 - \Sigma z^\alpha \bar{z}^\alpha)^2} .$$

Proof. First of all, we see that $(g_{\alpha\bar{\beta}})$ of (1) and (2) are given by

$$(\pm\tfrac{1}{2}c)(1 \pm \Sigma z^\gamma \bar{z}^\gamma)^2 g_{\alpha\bar{\beta}} = (1 \pm \Sigma z^\gamma \bar{z}^\gamma)\delta_{\alpha\beta} \mp \bar{z}^\alpha z^\beta,$$

respectively. We differentiate this identity with respect to $\partial/\partial z^\gamma$ and $\partial^2/\partial z^\gamma \bar{z}^\delta$ and set $z^1 = \ldots = z^n = 0$. We then have

$$g_{\alpha\bar{\beta}} = (\pm\frac{2}{c})\delta_{\alpha\beta}, \qquad \partial g_{\alpha\bar{\beta}}/\partial z^\gamma = 0,$$

$$\frac{\partial^2 g_{\alpha\bar{\beta}}}{\partial z^\gamma \partial \bar{z}^\delta} = -\frac{2}{c}(\delta_{\alpha\beta}\delta_{\gamma\delta} + \delta_{\alpha\delta}\delta_{\beta\gamma}).$$

Thus we have

$$K_{\alpha\bar{\beta}\gamma\bar{\delta}} = \frac{\partial^2 g_{\alpha\bar{\beta}}}{\partial z^\gamma \partial \bar{z}^\delta} - \sum_{\tau,\varepsilon} g^{\bar{\varepsilon}\tau}\frac{\partial g_{\alpha\bar{\varepsilon}}}{\partial z^\gamma}\frac{\partial g_{\bar{\beta}\tau}}{\partial \bar{z}^\delta}$$

$$= -\tfrac{1}{2}c(g_{\alpha\bar{\beta}}g_{\gamma\bar{\delta}} + g_{\alpha\bar{\delta}}g_{\bar{\beta}\gamma}),$$

from which we see that the metric ds^2 is of constant holomorphic sectional curvature c at the origin $z^1 = \ldots = z^n = 0$. Since both CP^n and D^n admit a transitive group of holomorphic isometric transformations, ds^2 is of constant holomorphic sectional curvature c and is complete. QED.

Now we can prove the following theorem (Hawley [1], Igusa [1]).

THEOREM 4.3. A simply connected complete Kaehlerian manifold M of constant holomorphic sectional curvature c can be identified with the complex projective space CP^n, the open unit ball D^n in C^n or C^n according as $c > 0$, $c < 0$ or $c = 0$.

Proof. From Theorem 4.2 it sufficies to show that any two simply connected complete Kaehlerian manifolds of constant holomorphic sectional curvature c are holomorphically isometric to each other. Let M and M' be two such manifolds. We choose a point o in M and a point o' in M'. Then any linear isomorphism $F : T_o(M) \longrightarrow T_{o'}(M')$ preserving both the metric and the almost complex structure maps the curvature tensor of M at o into the curvature tensor of M' at o'. Then we can see that there exists a unique affine isomorphism f of M such that $f(o) = o'$ and the differential of f at o is F. For any point x of M we take a curvae τ from o joining to x. We put $x' = f(x)$ and $\tau' = f(\tau)$. Since the parallel displacement along τ' corresponds to that along τ under f and since the metric tensors and almost complex structures of M and M' are all parallel, the affine isomorphism f maps the metric tensor and the almost complex structure of M into those of M'. Moreover, Proposition 1.1 shows that f is holomorphic.

5. NEARLY KAEHLERIAN MANIFOLDS

Let M be an almost Hermitian manifold with almost complex structure J and Riemannian metric tensor field g. Then

$$J^2 = -I, \qquad g(JX,JY) = g(X,Y)$$

for any vector fields X and Y on M. We denote by ∇ the operator of covariant differentiation with respect to g in M. If the almost complex structure J of M satisfies

$$(\nabla_X J)Y + (\nabla_Y J)X = 0$$

for any vector fields X and Y on M, then the manifold M is called a nearly Kaehlerian manifold (Tachibana space, K-space). The condition above is equivalent to

$$(\nabla_X J)X = 0.$$

We assume that M is a nearly Kaehlerian manifold. Then

$$(\nabla_X J)Y + (\nabla_{JX} J)JY = 0, \qquad (\nabla_X J)JY = -J(\nabla_X J)Y.$$

Let N be the torsion tensor field of J defined by (1.5). By a simple computation we have

$$N(X,Y) = -4J(\nabla_X J)Y,$$

and

$$N(X,Y) + N(Y,X) = 0, \quad g(N(X,Y),Z) + g(N(X,Z),Y) = 0,$$

$$N(JX,JY) + N(X,Y) = 0, \quad g(N(X,JY),JZ) + g(N(X,Y),Z) = 0.$$

Let T be a tensor field of type (0,2). If T satisfies T(JX,JY) = T(X,Y) for any vector fields X and Y ($T_{sr}J_j^s J_i^r = T_{ji}$), then T is said to be *hybrid* (with respect to j and i), and if T(JX,JY) = -T(X,Y) ($T_{sr}J_j^s J_i^r = -T_{ji}$), then T is said to be *pure* (with respect to j and i).

We see that $N_{jih} = g(N(e_j,e_i),e_h)$ is skew-symmetric and pure with respect to j, i and i, h. It is well known that if a tensor field T = (T_{ji}) is pure and a tensor field S = (S_{ji}) is hybrid with respect to j and i, then $T_{ji}S^{ji} = 0$.

First of all, we have

PROPOSITION 5.1. If the Nijenhuis torsion N of a nearly Kaehlerian manifold M vanishes, then M is a Kaehlerian manifold.

Let R be the Riemannian curvature tensor field of M. Then we have

$$g(R(X,Y)Z,W) = g(R(X,Y)JZ,JW) - g((\nabla_X J)Y,(\nabla_Z J)W).$$

Thus we also have

$$g(R(X,Y)Y,X) = g(R(X,Y)JY,JX) + g((\nabla_X J)Y,(\nabla_X J)Y),$$

$$g(R(X,Y)Z,W) = g(R(JX,JY)JZ,JW).$$

We denote by Q the Ricci operator of M, that is, $QX = \Sigma_i R(X,e_i)e_i$, $\{e_i\}$ being an orthonormal basis of M. We put

$$g(Q^*X,Y) = -\tfrac{1}{2}\sum_i g(R(X,JY)e_i,Je_i).$$

We call Q^* the Ricci * operator of M. We now define the tensor field S by

$$S = Q - Q^*.$$

The we have

$$g(SX,Y) = \sum_i g((\nabla_X J)e_i,(\nabla_Y J)e_i), \qquad \mathrm{Tr}S = \text{constant} > 0.$$

In the following we put $R_{ijkl} = g(R(e_i,e_j)e_k,e_l) = g(R(e_k,e_l)e_i,e_j)$, that is, R_{ijkl} are the components of R. But R_{ijkl} differ from the components of the curvature tensor of Chapter I by sign. We also put $R_{ij} = g(Qe_i,e_j)$, $R^*_{ij} = g(Q^*e_i,e_j)$ and $S_{ij} = g(Se_i,e_j)$. Then R_{ij}, R^*_{ij} and S_{ij} are symmetric and hybrid with respect to i and j. We now give the equations above by using the components.

We have

$$J_j^i J_i^h = -\delta_j^h, \quad g_{sr}J_j^s J_i^r = g_{ji}, \quad \nabla_j J_i^h + \nabla_i J_j^h = 0, \quad \nabla_j J_i^j = 0,$$

$$\nabla_j J_i^h + J_j^b J_i^a \nabla_b J_a^h = 0, \quad J_i^a \nabla_j J_a^h = -J_a^h \nabla_j J_i^a,$$

where ∇_j denotes the operator of covariant differentiation. The Nijenhuis torsion $N_{ji}{}^h = J_j^t(\partial_t J_i^h - \partial_i J_t^h) - J_i^t(\partial_t J_j^h - \partial_j J_t^h)$ is given by

$$N_{ji}{}^h = -4(\nabla_j J_i^r)J_r^h.$$

Moreover, we obtain

(5.1) $$R_{ijkl} = R_{ijab}J_k^a J_l^b - (\nabla_i J_j^a)\nabla_k J_{la},$$

(5.2) $$R_{ijkl} = R_{abcd}J_i^a J_j^b J_k^c J_l^d,$$

(5.3) $$S_{ij} = R_{ij} - R^*_{ij} = \nabla_i J_{ab}(\nabla_j J^{ab}), \quad R^*_{ij} = -\tfrac{1}{2}R_{iabc}J_j^a J^{bc},$$

(5.4) $$\mathrm{Tr}S = \nabla_i J_{jh}(\nabla^i J^{jh}) = r - r^* = \text{constant} > 0,$$

where $r = \mathrm{Tr}Q = g^{ij}R_{ij}$, that is, r is the scalar curvature of M and $r^* = \mathrm{Tr}Q^* = g^{ij}R^*_{ij}$. Since we have $T^{ijk}\nabla_i R_{iklh} = 0$ for any skew-symmetric tensor T^{ijk}, we obtain

(5.5) $$(\nabla^i J^{jk})\nabla_i R_{jklh} = 0.$$

From $\nabla_i r^* = 2\nabla^j R^*_{ji}$ we have

(5.6) $$\nabla^j(R_{ij} - R^*_{ij}) = \tfrac{1}{2}\nabla_i(r - r^*) = 0.$$

LEMMA 5.1. In a nearly Kaehlerian manifold M, we have

$$S^{ji}(R_{kjih} - 5R_{kjba}J_i^b J_h^a) = 0.$$

Proof. From (5.1) and (5.2) we have

$$2(\nabla_h J^{ji})R_{jist} = -S_h^r \nabla_s J_{tr}.$$

Applying ∇^h to this equation, we obtain, by (5.5) and (5.6),

$$2(\nabla^h\nabla_h J^{ji})R_{jist} = -S_h^r\nabla^h\nabla_s J_{tr},$$

or by Ricci s identity and $S^{hr}\nabla_s\nabla_h J_{tr} = 0$,

$$2(\nabla^h\nabla_h J^{ji})R_{jist} = S_h^r(R^h{}_{sat}J_r^a - R^h{}_{sar}J_t^a).$$

Transvecting the equation above with J_k^t, we have

$$2J_k^t(\nabla^h\nabla_h J^{ji})R_{jist} = S^{hr}(R_{hskr} - R_{hsta}J_k^tJ_r^a).$$

Since we have

$$\nabla^t\nabla_t J_j^h = J^{hl}(R_{jl} - R^*{}_{lj}), \tag{5.7}$$

the above equation reduces to

$$2J_k^tJ_a^jS^{ai}R_{jist} = S^{hr}(R_{hskr} - R_{hsta}J_k^tJ_r^a),$$

or by Bianchi's identity,

$$2J_k^tJ_a^jS^{ai}(R_{itjs} + R_{tjis}) = S^{hr}(R_{hskr} - R_{hsta}J_k^tJ_r^a).$$

Using $J_a^jS^{ai} = -J_a^iS^{aj}$, we have

$$4J_k^tJ_a^jR_{tjis}S^{ai} = S^{hr}(R_{hskr} - R_{hsta}J_k^tJ_r^a).$$

Therefore, we have our equation. QED.

We prove the following theorem (Watanabe-Takamatsu [1]).

THEOREM 5.1. Let M be a nearly Kaehlerian manifold. If M is non-Kaehlerian and if

$$S_{ji} = R_{ji} - R^*{}_{ji} = ag_{ji},$$

then M is an Einstein manifold.

Proof. From Lemma 5.1 we have $R_{ji} = 5R^*_{ji}$. Thus we have $r = 5r^*$. Since $r - r^* = na$, we see that $R_{ji} = (r/n)g_{ji}$. Consequently, M is an Einstein manifold. QED.

In the following we put

$$O^{ml}_{ij} = \tfrac{1}{2}(\delta^m_i\delta^l_j - J^m_iJ^l_j), \qquad {}^*O^{ml}_{ij} = \tfrac{1}{2}(\delta^m_i\delta^l_j + J^m_iJ^l_j).$$

Then we obtain

$$O^{ab}_{ij}R_{ab} = 0, \qquad O^{ab}_{ij}R^*_{ab} = 0, \qquad {}^*O^{ab}_{ij}\nabla_aJ_b = 0.$$

From these equations we obtain

$$J^r_j(5R^*_{ri} - R_{ri})\nabla_hJ^{ji} = 0.$$

Applying ∇^h, we find

$$(\nabla^h\nabla_hJ^{ji})J^r_j(5R^*_{ri} - R_{ri}) = 0.$$

Thus (5.7) implies the following

LEMMA 5.2. In a nearly Kaehlerian manifold M, we have

$$(R_{ji} - R^*_{ji})(5R^{*ji} - R^{ji}) = 0.$$

We also have (see Takamatsu [1])

$$R^*_{ji}(R^{ji} - R^{*ji}) = R_{kjih}O^{ih}_{ts}R^{kjts}. \tag{5.8}$$

From Lemma 5.2 and (5.8) we get

$$(R_{ji} - R^*_{ji})(R^{ji} - R^{*ji}) = 4R_{kjih}O^{ih}_{ts}R^{kjts}. \tag{5.9}$$

We now put, for arbitrary given constants a and b,

$$(5.10)\quad T_{kjih} = R_{kjih} - a(g_{kh}S_{ji} - g_{jh}S_{ki} + g_{ji}S_{kh} - g_{ki}S_{jh}) + b(r - r^*)(g_{kh}g_{ji} - g_{jh}g_{ki})$$

and

$$(5.11)\quad U_{kjih} = O_{ih}^{ts}T_{kjts}.$$

LEMMA 5.3. In an n-dimensional nearly Kaehlerian manifold M, we have

$$U_{kjih}U^{kjih} = R^{kjih}O_{ih}^{ts}R_{kjts} + 2[(n-4)a^2 - 2a]S_{ji}S^{ji} + [2a^2 + 2b - 4(n-2)ab + n(n-2)b^2](r - r^*)^2$$

for arbitrary constants a and b.

Proof. First of all we have

$$\begin{aligned} U_{kjih}U^{kjih} &= T^{kjih}O_{ih}^{ts}T_{kjts} \\ &= R^{kjih}O_{ih}^{ts}T_{kjts} - ag^{kh}S^{ji}O_{ih}^{ts}T_{kjts} + ag^{jh}S^{ki}O_{ih}^{ts}T_{kjts} \\ &\quad - ag^{ji}S^{kh}O_{ih}^{ts}T_{kjts} + ag^{ki}S^{jh}O_{ih}^{ts}T_{kjts} \\ &\quad + b(r-r^*)g^{kh}g^{ji}O_{ih}^{ts}T_{kjts} - b(r-r^*)g^{jh}g^{ki}O_{ih}^{ts}T_{kjts} \\ &= R^{kjih}O_{ih}^{ts}T_{kjts} - 4ag^{kh}S^{ji}O_{ih}^{ts}T_{kjts} \\ &\quad + 2b(r-r^*)g^{kh}g^{ji}O_{ih}^{ts}T_{kjts}. \end{aligned}$$

On the other hand, we have

$$R^{kjih}O_{ih}^{ts}T_{kjts} = R^{kjih}O_{ih}^{ts}R_{kjts} - 2aS_{ji}S^{ji} + b(r - r^*)^2,$$

$$-ag^{kh}S^{ji}O_{ih}^{ts}T_{kjts} = \tfrac{1}{2}[(n-4)a^2-a]S_{ji}S^{ji} + \tfrac{1}{2}[a^2-(n-2)ab](r - r^*)^2.$$

From these equations we obtain our equation. QED.

From (5.9) and Lemma 5.3 we find

$$U_{kjih}U^{kjih} = AS_{ji}S^{ji} + B(r - r^*)^2, \tag{5.12}$$

where

$$A = 2(n-4)a^2 - 4a + \tfrac{1}{4}, \quad B = 2a^2 + 2b - 4(n-2)ab + n(n-2)b^2.$$

LEMMA 5.4. In an n-dimensional nearly Kaehlerian manifold M, if $A + nB = 0$, then

$$U_{kjih}U^{kjih} = A[S_{ji} - \frac{(r-r^*)}{n} g_{ji}][S^{ji} - \frac{(r-r^*)}{n} g^{ji}].$$

We notice that there exist real numbers a and b satisfying $A + nB = 0$ if and only if $n \geq 6$.

When $n = 6$, there exist constant a and b, for example $a = 1/2$, $b = 1/8$, satisfying $A + nB = 0$. If we put $n = 6$, $a = 1/2$ and $b = 1/8$, then

$$U_{kjih}U^{kjih} = -\frac{3}{4}[S_{ji} - \frac{(r-r^*)}{n} g_{ji}][S^{ji} - \frac{(r-r^*)}{n} g^{ji}].$$

Consequently, we obtain (Takamatsu [1])

THEOREM 5.2. In a 6-dimensional nearly Kaehlerian manifold M, we have

$$S_{ji} = R_{ji} - R^*_{ji} = \frac{1}{6}(r - r^*)g_{ji}.$$

From Theorems 5.2 and 5.2 we have (Matsumoto [1])

THEOREM 5.3. A 6-dimensional nearly Kaehlerian manifold M is an Einstein manifold.

PROPOSITION 5.2. Let M be an Einstein nearly Kaehlerian manifold. Then the scalar curvature r of M is a positive constant.

Proof. From Lemma 5.2 we have

$$5S_{ji}S^{ji} = 4R_{ji}S^{ji},$$

from which

$$S_{ji}S^{ji} = \frac{4r}{5n}(r - r^*).$$

Using (5.4), we see that $r > 0$. QED.

In the sequel, we shall study a nearly Kaehlerian manifold M of constant holomorphic sectional curvature. The holomorphic sectional curvature H(X) with respect to a unit vector X is given by

$$H(X) = g(R(X,JX)JX,X).$$

We put

$$B(X,Y,Z,W) = g(R(X,Y)Z,W) - \tfrac{1}{4}[g((\nabla_X J)W,(\nabla_Y J)Z) - g((\nabla_X J)Z,(\nabla_Y J)W) - 2g((\nabla_X J)Y,(\nabla_Z J)W)].$$

Then B is a quadrilinear mapping satisfying the four conditions (a), (b), (c) and (d) in §4. We set

$$B_0(X,Y,Z,W) = \tfrac{1}{4}[g(Y,Z)g(X,W) - g(X,Z)g(Y,W) + g(Y,JZ)g(X,JW) - g(X,JZ)g(Y,JW) - 2g(X,JY)g(Z,JW)].$$

Then

$$B_0(X,JX,X,JX) = g(X,X)^2.$$

Suppose that M is of constant holomorphic sectional curvature c. Then we have

$$B(X,JX,X,JX) = g(R(X,JX)JX,X) = c$$

for any unit vector X. Thus Lemma 4.2 implies that $B = cB_0$. Hence we have (Sawaki-Watanabe-Sato [1])

THEOREM 5.3. Let M be a nearly Kaehlerian manifold. If M is of constant holomorphic sectional curvature c, then

$$\begin{aligned} g(R(X,Y)Z,W) = {} & \tfrac{1}{4}c[g(Y,Z)g(X,W) - g(X,Z)g(Y,W) + g(Y,JZ)g(X,JW) \\ & - g(X,JZ)g(Y,JW) - 2g(X,JY)g(Z,JW)] \\ & + \tfrac{1}{4}[g((\nabla_X J)W,(\nabla_Y J)Z) - g((\nabla_X J)Z,(\nabla_Y J)W) \\ & - 2g((\nabla_X J)Y,(\nabla_Z J)W)]. \end{aligned}$$

We now prove the following theorem (Takamatsu-Sato [1]).

THEOREM 5.4. There does not exist any dimensional, except 6-dimensional, non-Kaehlerian nearly Kaehlerian manifold of constant holomorphic sectional curvature.

To prove Theorem 5.4, we prepare some lemmas. Let M be an n-dimensional nearly Kaehlerian manifold of constant holomorphic sectional curvature c. We suppose that M is non-Kaehlerian.

From the assumption we have

$$\begin{aligned} R_{kjih} = {} & \tfrac{1}{4}c(g_{kh}g_{ji} - g_{ki}g_{jh} + J_{kh}J_{ji} - J_{ki}J_{jh} - 2J_{kj}J_{ih}) \\ & + \tfrac{1}{4}[\nabla_k J_h^r(\nabla_j J_{ir}) - \nabla_k J_i^r(\nabla_j J_{hr}) - 2(\nabla_k J_j^r)\nabla_i J_{hr}]. \end{aligned}$$

Thus we have

$$R_{ji} + 3R^*_{ji} = (n+2)cg_{ji}.$$

Since $R_{ji} = 5R^*_{ji}$, we have $R_{ji} = (r/n)g_{ji}$. Thus M is Einstein. We also have $c = (8r)/(5n^2+10n)$ and hence $c > 0$.

By a straightforward computation we have

LEMMA 5.5. Let M be an n-dimensional non-Kaehlerian nearly Kaehlerian manifold of constant holomorphic sectional curvature. Then

$$R_{kjih}R^{kjih} = \frac{(6n+44)}{25n(n+2)} r^2 = \text{constant}.$$

LEMMA 5.6. We have the following equations:

$$J_i^b J_h^a R^{kjih} R_{kjba} = \frac{(-2n+28)}{25n(n+2)} r^2, \tag{5.13}$$

$$J_j^b J_h^a R^{kjih} R_{kbia} = \frac{4}{25n} r^2, \tag{5.14}$$

$$J_h^b J_j^a R^{kjih} R_{kbia} = \frac{(5n-6)}{25n(n+2)} r^2. \tag{5.15}$$

Proof. We prove (5.13) only. From (5.1) and (5.3) we have

$$(R_{kjih} - J_i^b J_h^a R_{kjba})(R^{kjih} - J_d^i J_c^h R^{kjdc}) = S_{rs}S^{rs},$$

from which

$$R_{kjih}R^{kjih} - J_i^b J_h^a R_{kjba} R^{kjih} = \frac{8}{25n} r^2.$$

From this and Lemma 5.5 we have (5.13). QED.

LEMMA 5.7. We have the following equations:

$$\nabla^s J_{ka}(\nabla_s J_{jb})\nabla^t J^{bh}(\nabla_t J^{ai})R^{kj}{}_{ih} = -\frac{128}{125n^2(n+2)} r^3, \tag{5.16}$$

(5.17) $$\nabla^t J^{bh}(\nabla_t J^{ai}) R^{kj}{}_{ih} R_{kjba} = -\frac{64}{125n^2(n+2)} r^3,$$

(5.18) $$\nabla^t J^b_h(\nabla_t J_{ja}) R^{kjih} R_{kbi}{}^a = -\frac{96}{125n^2(n+2)} r^3,$$

(5.19) $$\nabla^t J^b_j(\nabla_t J_{ah}) R^{kjih} R_{kbi}{}^a = -\frac{128}{125n^2(n+2)} r^3.$$

Proof. We prove (5.16) only. First of all we have

$$\begin{aligned}
&\nabla^s J_{ka}(\nabla_s J_{jb}) \nabla^t J^{bh}(\nabla_t J^{ai}) R^{kj}{}_{ih} \\
&= \tfrac{1}{4}c(\delta^k_h \delta^j_i - \delta^k_i \delta^j_h + J^k_h J^j_i - J^k_i J^j_h - 2J^{kj} J_{ih}) \\
&\qquad \times \nabla^s J_{ka}(\nabla_s J_{jb}) \nabla^t J^{bh}(\nabla_t J^{ai}) \\
&+ \tfrac{1}{4}[\nabla^r J^k_h(\nabla_r J^j_i) - \nabla^r J^k_i(\nabla_r J^j_h) - 2\nabla^r J^{kj}(\nabla_r J_{ih})] \\
&\qquad \times \nabla^s J_{ka}(\nabla_s J_{jb}) \nabla^t J^{bh}(\nabla_t J^{ai}).
\end{aligned}$$

In the equation above, calculating term by term, we get the following identities:

$$\delta^k_h \delta^j_i \nabla^s J_{ka}(\nabla_s J_{jb}) \nabla^t J^{bh}(\nabla_t J^{ai}) = \nabla^s J_{ka}(\nabla_s J_{jb}) \nabla^t J^{bk}(\nabla_t J^{aj}) = 0,$$

because, $\nabla^t J^{bk}(\nabla_t J^{aj})$ is hybrid with respect to k, a, but $\nabla^s J_{ka}$ is pure with respect to k, a,

$$\begin{aligned}
-\delta^k_i \delta^j_h \nabla^s J_{ka}(\nabla_s J_{jb}) \nabla^t J^{bh}(\nabla_t J^{ai}) &= -\nabla^s J_{ka}(\nabla_s J_{jb}) \nabla^t J^{bj}(\nabla_t J^{ak}) \\
&= -S_{st} S^{st} = -\frac{16}{25n} r^2,
\end{aligned}$$

$$J^k_h J^j_i \nabla^s J_{ka}(\nabla_s J_{jb}) \nabla^t J^{bh}(\nabla_t J^{ai}) = \nabla^s J_{ah}(\nabla_s J_{bi}) \nabla^t J^{bh}(\nabla_t J^{ai}) = 0,$$

$$-J^k_i J^j_h \nabla^s J_{ka}(\nabla_s J_{jb}) \nabla^t J^{bh}(\nabla_t J^{ai}) = -S_{st} S^{st} = -\frac{16}{25n} r^2,$$

$$-2J^{kj}J_{ih}\nabla^s J_{ka}(\nabla_s J_{jb})\nabla^t J^{bh}(\nabla_t J^{ai}) = -2S_{aj}S^{aj} = -\frac{32}{25n}r^2,$$

$$-\nabla^r J_i^k(\nabla_r J_h^j)\nabla^s J_{ka}(\nabla_s J_{jb})\nabla^t J^{bh}(\nabla_t J^{ai}) = 0,$$

$$-2\nabla^r J^{kj}(\nabla_r J_{ih})\nabla^s J_{ka}(\nabla_s J_{jb})\nabla^t J^{bh}(\nabla_t J^{ai}) = 0$$

and

$$\nabla^r J_h^k(\nabla_r J_i^j)\nabla^s J_{ka}(\nabla_s J_{jb})\nabla^t J^{bh}(\nabla_t J^{ai}) = 0,$$

because, $\nabla^r J_h^k(\nabla_r J_i^j)\nabla^s J_{ka}(\nabla_s J_{jb})\nabla^t J^{bh}$ is symmetric with respect to i, a, but $\nabla_t J^{ai}$ is skew-symmetric with respect to i, a. From these equations we obtain (5.16). QED.

Proof of Theorem 5.4. From Lemma 5.5 we obatin

$$\nabla^b R^{kjih}(\nabla_b R_{kjih}) + R^{kjih}\nabla^b\nabla_b R_{kjih} = 0. \tag{5.20}$$

Using Bianchi's and Ricci's identities, we have

$$\begin{aligned} R^{kjih}\nabla^b\nabla_b R_{kjih} &= -R^{kjih}\nabla^b(\nabla_k R_{jbih} + \nabla_j R_{bkih}) \\ &= 2R^{kjih}\nabla^b\nabla_k R_{bjih} \\ &= 2R^{kjih}(\nabla_k\nabla^b R_{bjih} - R^b{}_{kb}{}^a R_{ajih} - R^b{}_{kj}{}^a R_{baih} \\ &\qquad - R^b{}_{ki}{}^a R_{bjah} - R^b{}_{kh}{}^a R_{bjia}). \end{aligned}$$

Since we have $\nabla^b R_{bjih} = 0$, we obtain, by Bianchi's identity,

$$\begin{aligned} R^{kjih}\nabla^b\nabla_b R_{kjih} = \frac{2r}{n}R^{kjih}R_{kjih} + R^{kjih}R_{kjba}R^{ba}{}_{ih} \\ + 4R^{kjih}R_{kbi}{}^a R^b{}_{jah}. \end{aligned} \tag{5.21}$$

On the other hand, we have

$$(5.22) \quad R^{kjih}R_{kjba}R^{ba}{}_{ih} = \tfrac{1}{4}c(\delta_h^b\delta_i^a - \delta_i^b\delta_h^a + J_h^bJ_i^a - J_i^bJ_h^a$$
$$- 2J^{ba}J_{ih})R^{kjih}R_{kjba} + \tfrac{1}{4}[\nabla^t J_h^b(\nabla_t J_i^a) - \nabla^t J_i^b(\nabla_t J_h^a)$$
$$- 2\nabla^t J^{ba}(\nabla_t J_{ih})]R^{kjih}R_{kjba}.$$

We calculate each term of the equation above. By Lemma 5.5 we have

$$(\delta_h^b\delta_i^a - \delta_i^b\delta_h^a)R^{kjih}R_{kjba} = -2R^{kjih}R_{kjih} = -\frac{2(6n+44)}{25n(n+2)}r^2.$$

From Lemma 5.6 we obtain

$$(J_h^bJ_i^a - J_i^bJ_h^a - 2J^{ba}J_{ih})R^{kjih}R_{kjba} = \frac{2(2n-28)}{25n(n+2)}r^2 - \frac{8}{25n}r^2.$$

We also have, by Lemma 5.7,

$$\nabla^t J_h^b(\nabla_t J_i^a)R^{kjih}R_{kjba} = -\frac{64}{125n^2(n+2)}r^3,$$

$$-\nabla^t J_i^b(\nabla_t J_h^a)R^{kjih}R_{kjba} = -\frac{64}{125n^2(n+2)}r^3,$$

$$-2\nabla^t J^{ba}(\nabla_t J_{ih})R^{kjih}R_{kjba} = -\tfrac{1}{2}S^{tr}(\nabla_r J_{kj})S_{ts}\nabla^s J^{kj}$$
$$= -\frac{32}{125n^2}r^3.$$

Consequently, we obtain

$$(5.23) \qquad R^{kjih}R_{kjba}R^{ba}{}_{ih} = -\frac{(8n^2+96n+416)}{125n^2(n+2)^2}r^3.$$

Similarly, we have

$$(5.24) \qquad R^{kjih}R_{kbi}{}^{a}R^{b}{}_{jah} = -\frac{(n^2+164n+260)}{125n^2(n+2)^2}r^3.$$

Substituting (5.23) and (5.24) into (5.21), and using Lemma 5.5, we obtain

$$R^{kjih}\nabla^b\nabla_b R_{kjih} = \frac{48(n-6)}{125n^2(n+2)} r^3. \tag{5.25}$$

From (5.20) and (5.25) we get

$$\nabla^a R^{kjih}(\nabla_a R_{kjih}) + \frac{48(n-6)}{125n^2(n+2)} r^3 = 0.$$

Thus, if $n \geq 6$, then $\nabla_a R_{kjih} = 0$ and $n = 6$ because of $r > 0$. When $n = 4$, we can prove the following (A. Gray [1])

THEOREM 5.5. Any 4-dimensional nearly Kaehlerian manifold is a Kaehlerian manifold.

Proof. In (5.12), we put $n = 4$, $a = 1/8$ and $b = -1/16$. Then we have $S_{ji} = R_{ji} - R^*_{ji} = 0$, and hence $r = r^*$. Thus (5.4) implies that $\nabla_j J_{ih} = 0$. This means that the manifold is Kaehlerian. QED.

If $n = 2$, we easily see that the torsion N of J vanishes. From these considerations we have Theorem 5.4. QED.

Example 5.1. Let S^6 be the 6-dimensional unit sphere defined in Example 2.8. Then S^6 admits a nearly Kaehlerian structure (Fukami-Ishihara [1]). Any 6-dimensional non-Kaehlerian nearly Kaehlerian manifold of constant holomorphic sectional curvature is of constant curvature (Tanno [7]).

6. QUATERNION KAEHLERIAN MANIFOLDS

In this section we shall study quaternion manifolds by using tensor calculas. The results in this section have been proved by S. Ishihara [3].

Let M be an n-dimensional manifold with a 3-dimensional vector bundle V consisting of tensors of type (1,1) over M satisfying the following condition:

(a) In any coordinate neighborhood U of M, there exists a local basis {F,G,H} of V such that

$$F^2 = -I, \quad G^2 = -I, \quad H^2 = -I,$$
$$(6.1)$$
$$GH = -HG = F, \quad HF = -FH = G, \quad FG = -GF = H,$$

I denoting the identity tensor of type (1,1) in M.

Such a local basis {F,G,H} is called a *canonical local basis* of the bundle V in U. Then the bundle V is called an *almost quaternion structure* in M, and M with V an *almost quaternion manifold*, which will be denoted by (M,V). An almost quaternion manifold M is of dimension $n = 4m$ ($m \geq 1$). In an almost quaternion manifold M, we take coordinate neighborhoods U and U' such that $U \cap U' \neq \emptyset$. Let {F,G,H} and {F',G',H'} be canonical basis of V in U and U' respectively. Then

$$F' = s_{11}F + s_{12}G + s_{13}H,$$

$$G' = s_{21}F + s_{22}G + s_{23}H,$$

$$H' = s_{31}F + s_{32}G + s_{33}H,$$

where s_{ab} (a,b = 1,2,3) are functions in $U \cap U'$. From (6.1) we see that (s_{ab}) is an element of the proper orthogonal group SO(3) of dimension 3. Thus every almost quaternion manifold M is orientable.

If, in an almost quaternion manifold (M,V), there is a global basis {F,G,H} of V which satisfies (6.1), then (M,V) is traditionally called an almost quaternion manifold. Such a global basis {F,G,H} of V is called a *canonical global basis* of V.

Let (M,V) be an almost quaternion manifold with a canonical local basis of V in a coordinate neighborhood U. We now assume that there exists a system of coordinates (x^h) in each U with respect to which F, G and H have components of the form

$$(6.2) \qquad F = \begin{pmatrix} 0 & -E & 0 & 0 \\ E & 0 & 0 & 0 \\ 0 & 0 & 0 & -E \\ 0 & 0 & E & 0 \end{pmatrix}, \quad G = \begin{pmatrix} 0 & 0 & -E & 0 \\ 0 & 0 & 0 & E \\ E & 0 & 0 & 0 \\ 0 & -E & 0 & 0 \end{pmatrix}, \quad H = \begin{pmatrix} 0 & 0 & 0 & -E \\ 0 & 0 & -E & 0 \\ 0 & E & 0 & 0 \\ E & 0 & 0 & 0 \end{pmatrix}$$

where E denotes the identity (m,m)-matrix. In such a case, the given almost quaternion structure V is said to be *integrable*.

In any almost quaternion manifold (M,V), there is a Riemannian metric tensor field g such that

$$g(\phi X,Y) + g(X,\phi Y) = 0$$

for any cross-section ϕ and any vector fields X, Y of M. An almost quaternion structure V with such a Riemannian metric g is called an *almost quaternion metric structure*. A manifold M with an almost quaternion metric structure $\{g,V\}$ is called an *almost quaternion metric manifold*, which will be denoted by (M,g,V).

Let $\{F,G,H\}$ be a canonical local basis of V of an almost quaternion manifold (M,g,V). Since each of F, G and H is almost Hermitian with respect to g, putting

$$\Phi(X,Y) = g(FX,Y), \quad \Psi(X,Y) = g(GX,Y), \quad \Theta(X,Y) = g(HX,Y)$$

for any vector fields X and Y, we see that Φ, Ψ and Θ are local 2-forms. Then

$$\Omega = \Phi \wedge \Phi + \Psi \wedge \Psi + \Theta \wedge \Theta$$

is a 4-form defined globally in M. Moreover, we see that

$$\Lambda = F \otimes F + G \otimes G + H \otimes H$$

is a global tensor field of type (2,2) in M.

We now assume that the Riemannian connection ∇ of (M,g,V) satisfies the following conditions:

(b) If ϕ is a cross-section (local or global) of the bundle V, then $\nabla_X\phi$ is also a cross-section of V, X being an arbitrary vector field in M.

From (6.1) we see that the condition (b) is equivalent to the following condition:

(b') If F,G,H is a canonical local basis of V, then

$$(6.3)\qquad \begin{aligned} \nabla_X F &= \quad r(X)G - q(X)H, \\ \nabla_X G &= -r(X)F + p(X)H, \\ \nabla_X H &= q(X)F - p(X)G \end{aligned}$$

for any vector field X, where p, q and r are certain local 1-forms.

If an almost quaternion metric manifold M satisfies the condition (b) or (b'), then M is called a *quaternion Kaehlerian manifold* and an almost quaternion structure of M is called a *quaternion Kaehlerian structure*.

THEOREM 6.1. An almost quaternion metric manifold M is a quaternion Kaehlerian manifold if and only if $\nabla\Omega = 0$ or $\nabla\Lambda = 0$.

Proof. If M is a quaternion Kaehlerian manifold, then (6.3) implies $\nabla\Omega = 0$ and $\nabla\Lambda = 0$. Conversely, if $\nabla\Omega = 0$ or $\nabla\Lambda = 0$ we see that F, G and H satisfy (6.3). QED.

Let M = (M,g,V) be a quaternion Kaehlerian manifold of dimension n = 4m with canonical local basis {F,G,H} of V. Let R be the Riemannian curvature tensor of M. From (6.3) we have

$$\begin{aligned} (R(X,Y)F)Z &= ((\nabla_X\nabla_Y - \nabla_Y\nabla_X - \nabla_{[X,Y]})F)Z \\ &= dr(X,Y)GZ + (p \wedge r)(X,Y)HZ - dq(X,Y)HZ \\ &\qquad + (p \wedge r)(X,Y)GZ \\ &= (dr(X,Y) + (p \wedge q)(X,Y))GZ \\ &\qquad - (dq(X,Y) + (r \wedge p)(X,Y))HZ, \end{aligned}$$

from which

$$(6.4)\qquad (R(X,Y)F) = R(X,Y)F - FR(X,Y) = [R(X,Y),F]$$
$$= C(X,Y)G - B(X,Y)H,$$

where we have put

$$B = dq + r \wedge p, \quad C = dr + p \wedge q.$$

Similarly, we obtain

$$(6.5)\qquad (R(X,Y)G) = [R(X,Y),G] = -C(X,Y)F + A(X,Y)H,$$

$$(6.6)\qquad (R(X,Y)H) = [R(X,Y),H] = B(X,Y)F - A(X,Y)G,$$

where we have put

$$A = dp + q \wedge r.$$

From (6.4), (6.5) and (6.6) we have

$$g(R(X,Y)FZ,FW)-g(R(X,Y)Z,W) = c(X,Y)g(Z,HW)+B(X,Y)g(Z,GW),$$

$$(6.7)\qquad g(R(X,Y)GZ,GW)-g(R(X,Y)Z,W) = A(X,Y)g(Z,FW)+C(X,Y)g(Z,HW),$$

$$g(R(X,Y)HZ,HW)-g(R(X,Y)Z,W) = B(X,Y)g(Z,GW)+A(X,Y)g(Z,FW).$$

From the second equation of (6.7) we find

$$2\sum_{i=1}^{4m} g(R(X,Y)e_i,Fe_i) = 4mA(X,Y),$$

where $\{e_i\}$ is an orthonormal basis of M. Thus we have

$$(6.8)\qquad A(X,Y)=\frac{-1}{2m}\mathrm{Tr}FR(X,Y),\quad B(X,Y)=\frac{-1}{2m}\mathrm{Tr}GR(X,Y),\quad C(X,Y)=\frac{-1}{2m}\mathrm{Tr}HR(X,Y),$$

where Tr is the trace. From (6.8) and the Bianchi's 1'st identity

$$\begin{aligned}\sum g(R(X,e_i)Fe_i,Y) &= \tfrac{1}{2}\sum[g(R(X,e_i)Fe_i,Y)-g(R(X,Fe_i)e_i,Y)] \\ &= \tfrac{1}{2}\sum[g(R(X,e_i)Fe_i,Y)+g(R(Fe_i,X)e_i,Y)] \\ &= \tfrac{1}{2}\sum g(R(Fe_i,e_i)X,Y) = -\tfrac{1}{2}\sum g(R(X,Y)e_i,Fe_i) \\ &= -mA(X,Y).\end{aligned}$$

Similarly, we have

$$(6.9)\qquad \sum g(R(X,e_i)Fe_i,Y) = -mA(X,Y), \quad \sum g(R(X,e_i)Ge_i,Y) = -mB(X,Y),$$

$$\sum g(R(X,e_i)He_i,Y) = -mC(X,Y).$$

We denote by S the Ricci tensor of M. Then (6.7) and (6.9) imply

$$(6.10)\qquad \begin{aligned} S(X,Y) &= -mA(X,FY) - B(X,GY) - C(X,HY), \\ S(X,Y) &= -A(X,FY) - mB(X,GY) - C(X,HY), \\ S(X,Y) &= -A(X,FY) - B(X,GY) - mC(X,HY).\end{aligned}$$

Thus, if $m = 1$, then

$$(6.11)\qquad S(X,Y) = -A(X,FY) - B(X,GY) - C(X,HY),$$

and if $m > 1$, then

$$(6.12)\qquad S(X,Y) = -(m+2)A(X,FY) = -(m+2)B(X,GY) = -(m+2)C(X,HY).$$

From (6.7) and (6.12) we find, for $m > 1$,

$$g(R(X,Y)FZ,FW)-g(R(X,Y)Z,W)=\frac{1}{m+2}[g(Z,HW)S(X,HY) + g(Z,GW)S(X,GY)],$$

$$(6.13) \qquad g(R(X,Y)GZ,GW)-g(R(X,Y)Z,W)=\frac{1}{m+2}[g(Z,FW)S(X,FY)+g(Z,HW)S(X,HY)],$$

$$g(R(X,Y)HZ,HW)-g(R(X,Y)Z,W)=\frac{1}{m+2}[g(Z,GW)S(X,GY)+g(Z,FW)S(X,FY)].$$

Since A, B and C are all skew-symmetric, using (6.12) we find, for $m > 1$,

$$(6.14) \qquad S(FX,FY) = S(X,Y), \quad S(GX,GY) = S(X,Y), \quad S(HX,HY) = S(X,Y),$$

from which

$$(6.15) \qquad S(FX,Y) = -S(X,FY), \quad S(GX,Y) = -S(X,GY), \quad S(HX,Y) = -S(X,HY).$$

On the other hand, using (6.3), we find

$$(6.16) \qquad (\nabla_X S)(Y,FZ) + (\nabla_X S)(Z,FY) = 0.$$

Thus, if $m > 1$, then

$$(6.17) \qquad (\nabla_X S)(Y,Z) = (\nabla_X S)(FZ,FY).$$

LEMMA 6.1. For any quaternion Kaehlerian manifold M of dimension $n = 4m$ $(m > 1)$, the Ricci tensor S of M is parallel.

Proof. From (6.3) and (6.12) we have

$$(\nabla_X S)(Y,FZ) = (m+2)[(\nabla_X A)(Y,Z)+q(X)C(Y,Z)-r(X)B(Y,Z)],$$

from which

$$(6.18) \qquad (\nabla_X S)(Y,FZ) + (\nabla_Y S)(Z,FX) + (\nabla_Z S)(X,FY) = (m+2)(dA + q \wedge C - r \wedge B)(X,Y,Z) = 0.$$

Thus we have

$$-(\nabla_X S)(Y,Z) + (\nabla_Y S)(FZ,FX) + (\nabla_{FZ} S)(X,FY) = 0.$$

Substituting (6.17) into the equation above, we find

$$(\nabla_Y S)(Z,X) - (\nabla_X S)(Y,Z) = -(\nabla_{FZ} S)(X,FY).$$

Since we have $(\nabla_X S)(Y,Z) = (\nabla_X S)(GY,GZ)$, we obtain

$$(\nabla_Y S)(Z,X) - (\nabla_X S)(Y,Z) = (\nabla_{FZ} S)(GX,HY).$$

Similarly we obtain

$$(\nabla_Y S)(Z,X) - (\nabla_X S)(Y,Z) = (\nabla_{GZ} S)(HX,FY) = (\nabla_{HZ} S)(FX,GY).$$

Combining these equations we have

$$(6.18) \qquad (\nabla_{FZ} S)(GX,HY) = (\nabla_{GZ} S)(HX,FY) = (\nabla_{HZ} S)(FX,GY).$$

In particular, we have $(\nabla_{FZ} S)(GX,HY) = (\nabla_{HZ} S)(FX,GY)$. Changing Z, X, Y by GZ, HX, FY respectively, we obtain

$$(6.19) \qquad (\nabla_{HZ} S)(FX,GY) = -(\nabla_{FZ} S)(GX,HY).$$

From (6.18) and (6.19) we see that $(\nabla_{FZ} S)(GX,HY) = 0$ and hence $(\nabla_Z S)(X,Y) = 0$. Consequently, the Ricci tensor S of M is parallel. QED.

LEMMA 6.2. Let (M,g,V) be a quaternion Kaehlerian manifold of dimension ≥ 8. Then the Riemannian manifold (M,g) is irreducible if (M,g) has non-vanishing Ricci tensor.

Proof. Suppose that M is reducible and has non-vanishing Ricci tensor. Since M is not flat and the Ricci tensor is parallel, we can take a coordinate neighborhood U of M such that the Riemannian manifold (U,g) is decomposed into a Riemannian product of certain number of Riemannian manifolds $(W_1,g_1),\ldots,(W_p,g_p)$, $p \geq 2$, in such a way that

$$g(X,Y) = \sum_{A=1}^{p} g_A(\pi_A X,\pi_A Y), \qquad S(X,Y) = \sum_{A=1}^{p} c_A g_A(\pi_A X,\pi_A Y),$$

where c_A are constants such that $c_1 < \ldots < c_p$ and $\pi_A\colon U \longrightarrow W_A$ are the natural projections which denote at the same time their differential mappings. Since $g(FX,FY) = g(X,Y)$ and $S(FX,FY) = S(X,Y)$, we have

$$\sum_{A=1}^{p} g_A(\pi_A FX,\pi_A FY) = \sum_{A=1}^{p} g_A(\pi_A X,\pi_A Y),$$
$$\sum_{A=1}^{p} c_A g_A(\pi_A FX,\pi_A FY) = \sum_{A=1}^{p} c_A g_A(\pi_A X,\pi_A Y).$$

From these equations we obtain

$$g_A(\pi_A FX,\pi_A FY) = g_A(\pi_A X,\pi_A Y), \quad A = 1,\ldots,p.$$

From now on for simplicity we assume that $p = 2$, i.e., that $(U,g) = (W_1,g_1) \times (W_2,g_2)$. Let $(y^\alpha) = (y^1,\ldots,y^r)$ and $(z^\lambda) = (z^{r+1},\ldots,z^n)$ be coordinate systems in W_1 and W_2 respectively. Then $(x^h) = (y^\alpha,z^\lambda)$ are naturally coordinates in U. With respect to $(x^h) = (y^\alpha,z^\lambda)$, g has components of the form

$$(g_{ji}) = \begin{pmatrix} g_{\gamma\beta} & 0 \\ 0 & g_{\mu\nu} \end{pmatrix},$$

where $(g_{\gamma\beta})$ and $(g_{\mu\nu})$ are respectively the components of g_1 and g_2, $(g_{\gamma\beta})$ and $(g_{\mu\nu})$ being independent of the variables z^λ and y^α respectively (see Chapter VIII). We then have $F_\mu^\alpha = 0$ and hence

$A_{\mu\lambda} = B_{\mu\lambda} = C_{\mu\lambda} = 0$. Similarly, we find $A_{\beta\alpha} = B_{\beta\alpha} = C_{\beta\alpha} = 0$ and $A_{\lambda\alpha} = B_{\lambda\alpha} = C_{\lambda\alpha} = 0$. Consequently, we obtain A = B = C, which implies that S = 0 by (6.12). This is a contradiction. Therefore, M is irreducible. QED.

From Lemmas 6.1 and 6.2 we have the following theorems.

THEOREM 6.2. Any quaternion Kaehlerian manifold of dimension ≥ 8 is an Einstein manifold.

THEOREM 6.3. If a quaternion Kaehlerian manifold M of dimension ≥ 8 has non-vanishing scalar curvature, then M is irreducible.

THEOREM 6.4. If a quaternion Kaehlerian manifold M of dimension ≥ 8 has zero scalar curvature, then M is locally a Riemannian product of a flat quaternion Kaehlerian manifold and irreducible quaternion Kaehlerian manifolds with vanishing Ricci tensor.

In the next place, we assume that a quaternion Kaehlerian manifold (M,g,V) is of constant curvature c. Then we have

$$g(R(X,Y)Z,W) = c[g(Y,Z)g(X,W) - g(X,Z)g(Y,W)].$$

From this we have

$$g(R(X,Y)FZ,FW) = c[g(Y,FZ)g(X,FW) - g(X,FZ)g(Y,FW)].$$

Using (6.8), we obtain

$$A(X,Y) = -\frac{c}{m}g(FX,Y), \quad B(X,Y) = -\frac{c}{m}g(GX,Y), \quad C(X,Y) = -\frac{c}{m}g(HX,Y).$$

Substituting these equations into the first equation of (6.7), we have

$$c[g(Y,FZ)g(X,FW)-g(X,FZ)g(Y,FW)]-c[g(Y,Z)g(X,W)-g(X,Z)g(Y,W)]$$
$$= -\frac{c}{m}[g(HX,Y)g(Z,HW)+g(GX,Y)g(Z,GW)],$$

from which

$$c(2m+1)(m-1)g(Y,Z) = 0.$$

Thus, if $c \neq 0$, we get $m = 1$. Consequently, we have

THEOREM 6.5. If a quaternion Kaehlerian manifold M is of non-zero constant curvature, then M is of dimension 4.

Since any quaternion Kaehlerian manifold M is Einstein, if M is conformally flat, then it is of constant curvature. Thus, from Theorem 6.5 we have

THEOREM 6.6. If a quaternion Kaehlerian manifold of dimension ≥ 8 is conformally flat, then it is of zero curvature.

In the following we shall define Q-sectional curvatures in a quaternion Kaehlerian manifold.

We define the 4-dimensional subspace $Q(X)$ of the tangent space $T_x(M)$ of M at x by

$$Q(X) = \{Y : Y = aX+bFX+cGX+dHX\}$$

for a vector X of M, a, b, c and d being arbitrary real numbers. $Q(X)$ is called the *Q-section* determined by X. We denote by $K(X,Y)$ the sectional curvature of M with respect to the section spanned by X and Y. Since M is Einstein, using (6.13), we obtain, for a unit vector X,

$$\begin{aligned} &K(X,FX)=g(R(X,FX)FX,X) = \frac{k}{4m(m+2)} - g(R(X,FX)GX,HX), \\ (6.20)\quad &K(X,GX)=g(R(X,GX)GX,X) = \frac{k}{4m(m+2)} - g(R(X,GX)HX,FX), \\ &K(X,HX)=g(R(X,HX)HX,X) = \frac{k}{4m(m+2)} - g(R(X,HX)FX,GX), \end{aligned}$$

k being the scalar curvature of M.

Now we suppose that for any $Y,Z \in Q(X)$ the sectional curvature $K(Y,Z)$ is a constant $\rho(X)$, which will be called the *Q-sectional curvature* of M with respect to X at $x \in M$. Then, putting $Y = X$, $Z = FX$,

$Y = X$, $Z = GX$ and $Y = X$, $Z = HX$, we have respectively

$$K(X,FX) = K(X,GX) = K(X,HX) = \rho(X). \tag{6.21}$$

Since $K(Y,Z)$ is constant for any $Y,Z \in Q(X)$, (6.20) and (6.21) imply

$$\rho(X) = \frac{k}{4m(m+2)}, \tag{6.22}$$

$$g(R(X,FX)GX,HX) = g(R(X,GX)HX,FX) = g(R(X,HX)FX,GX) = 0, \tag{6.23}$$

where we have used the Bianchi's 1'st identity.

Next, from the assumption we find $K(X,aFX+bGX) = \rho(X)$, $a^2+b^2 \neq 0$, which together with (6.21) implies $g(R(X,FX)X,GX) = 0$. Similarly,

$$g(R(X,FX)X,GX) = g(R(X,GX)X,HX) = g(R(X,HX)X,FX) = 0. \tag{6.24}$$

Thus, using (6.23) and (6.24), we find

$$R(Y,Z)Y - \rho(X)Z \in Q^{\perp}(X) \tag{6.25}$$

for any $Y,Z \in Q(X)$, where $Q^{\perp}(X)$ denotes the orthogonal complement of $Q(X)$ in $T_x(M)$.

Conversely, if we assume that (6.25) holds for any $Y,Z \in Q(X)$ at $x \in M$, then the sectional curvature $K(Y,Z)$ is constant for any $Y,Z \in Q(X)$. In the case above, $\rho(X)$ is called the *Q-sectional curvature* of M with respect to X at x, and the Q-section is said to have Q-sectional curvature $\rho(X)$.

Let (M,g,V) be a quaternion Kaehlerian manifold and assume that each Q-section $Q(X)$ at each point x of M, X being an arbitrary vector of M at x, has a Q-sectional curvature $\rho(X)$. Moreover, if we suppose that the Q-sectional curvature $\rho(X)$ is a constant $c = c(x)$ independent of X at each point x, then we asy that M is of *constant* Q-*sectional curvature* $c(x)$. Since M is Einstein, from (6.22), we see that $c(x)$ is constant on M.

In the sequel we shall determine the form of the curvature tensor

R of a quaternion Kaehlerian manifold of constant Q-sectional curvature c. Let (M,g,V) be a quaternion Kaehlerian manifold of constant Q-sectional curvature c of dimension ≥ 8. In the following, we shall calculate equations by using components of tensors. But, we notice that the components $R_{ijkl} = g(R(e_i,e_j)e_k,e_l)$ of the curvature tensor R differ from that of Chapter I by sign.

Since M is Einstein, we may put

$$A_{kj} = -4aF_{kj}, \qquad B_{kj} = -4aG_{kj}, \qquad C_{kj} = -4aH_{kj},$$

a being a certain constant. Substituting these equations into (6.7), we get

$$-R_{kjts}F_i^tF_h^s + R_{kjih} = -4a(G_{kj}G_{ih} + H_{kj}H_{ih}), \tag{6.26}$$

from which

$$-R_{vjsh}F_k^vF_i^s - R_{vjsi}F_k^vF_h^s = 4a(G_{kj}G_{ih} + H_{kj}H_{ih}), \tag{6.27}$$

$$-R_{ktsh}F_j^tF_i^s - R_{ktsi}F_j^tF_h^s = -4a(G_{kj}G_{ih} + H_{kj}H_{ih}). \tag{6.28}$$

From the assumption we have K(X,FX) = c for any vector X, taking account of the definition of sectional curvatures, we get

$$R_{vjsh}F_k^vF_i^sX^kX^jX^iX^h = -cg_{jh}g_{ki}X^kX^jX^iX^h,$$

X^h being the components of X. Since X is arbitrary taken, from the equation above we obtain

$$\begin{aligned} R_{vjsh}F_k^vF_i^s &+ R_{vish}F_j^vF_k^s + R_{vksh}F_i^vF_j^s + R_{vjsi}F_k^vF_h^s \\ &+ R_{vhsi}F_j^vF_k^s + R_{vksi}F_h^vF_j^s + R_{vksh}F_j^vF_i^s + R_{vish}F_k^vF_j^s \\ &+ R_{vjsh}F_i^vF_k^s + R_{vksi}F_j^vF_h^s + R_{vhsi}F_k^vF_j^s + R_{vjsi}F_h^vF_k^s \\ &= -4c(g_{kj}g_{ih} + g_{ji}g_{kh} + g_{ik}g_{jh}), \end{aligned}$$

which, together with (6.27), implies

$$R_{vjsh}F^v_kF^s_i + R_{vish}F^v_jF^s_k + R_{vksh}F^v_iF^s_j$$
$$= -c(g_{kj}g_{ih} + g_{ji}g_{kh} + g_{ik}g_{jh}) - 4a(G_{kj}G_{ih} + H_{kj}H_{ih})$$
$$- 2a(G_{kj}G_{ih} + G_{ji}G_{kh} + G_{ik}G_{jh} + H_{kj}H_{ih} + H_{ji}H_{kh} + H_{ik}H_{jh}).$$

If we transvect the above equation with $F^k_jF^i_p$, then we obtain

$$R_{qjph} - R_{vuqh}F^v_pF^u_j - R_{pvuh}F^v_qF^u_j$$
$$= -c(F_{qj}F_{ph} - F_{qh}F_{jp} + g_{qp}g_{jh})$$
$$- 2a(G_{qj}G_{ph} + G_{qh}G_{jp} + G_{qp}G_{jh} + H_{qj}H_{ph} + H_{qh}H_{jp} + H_{qp}H_{jh}).$$

Substituting

$$R_{vuqh}F^v_qF^u_j = R_{jpqh} + 4a(G_{jp}G_{qh} + H_{jp}H_{qh}),$$

which is a consequence of (6.26), in the above equation we get

$$\text{(6.29)} \quad R_{qiph} - R_{jqph} - R_{pvuh}F^v_qF^u_j - 4a(G_{jp}G_{qh} + H_{jp}H_{qh})$$
$$= -c(F_{qj}F_{ph} - F_{qh}F_{jp} + g_{qp}g_{jh})$$
$$- 2a(G_{qj}G_{ph} + G_{qh}G_{jp} + G_{qp}G_{jh}$$
$$+ H_{qj}H_{ph} + H_{qh}H_{jp} + H_{qp}H_{jh}).$$

On the other hand, (6.26) implies

$$\text{(6.30)} \quad R_{pvuh}F^v_qF^u_j - R_{puvh}F^v_qF^u_j = (R_{pvuh} - R_{puvh})F^v_qF^u_j$$
$$= -R_{vuph}F^v_qF^u_j = -R_{qjph} - 4a(G_{qj}G_{ph} + H_{qj}H_{ph}).$$

Taking the skew-symmetric parts of the both sides of (6.29) with respect to q and j, and substituting (6.30) in the resulting equation we obtain

$$2R_{qjph} - R_{jpqh} + R_{qpjh} + R_{qjph} + 4a(G_{qj}G_{ph} + H_{qj}H_{ph})$$
$$- 4a(G_{jp}G_{qh} - G_{qp}G_{jh} + H_{jp}H_{qh} - H_{qp}H_{jh})$$
$$= -c(2F_{qj}F_{ph} - F_{qh}F_{jp} + F_{jh}F_{qp} + g_{qp}g_{jh} - g_{jp}g_{qh})$$
$$- 4a(G_{qj}G_{ph} + H_{qj}H_{ph}),$$

from which

$$R_{qjph} = \tfrac{1}{4}c(g_{qh}g_{jp} - g_{jh}g_{qp} + F_{qh}F_{jp} - F_{jh}F_{qp} - 2F_{qj}F_{ph})$$
$$+ a(G_{qh}G_{jp} - G_{jh}G_{qp} - 2G_{qj}G_{ih} + H_{qh}H_{jp} - H_{jh}H_{qj} - 2H_{qj}H_{ih}).$$

If we now take account of $K(X,FX) = c$, then we have $a = \frac{1}{4}c$ and hence

$$R_{kjih} = \tfrac{1}{4}c(g_{kh}g_{ji} - g_{jh}g_{ki} + F_{kh}F_{ji} - F_{jh}F_{ki} - 2F_{kj}F_{ih}$$
$$+ G_{kh}G_{ji} - G_{jh}G_{ki} - 2G_{kj}G_{ih} + H_{kh}H_{ji} - H_{jh}H_{ki} - 2H_{kj}H_{ih}). \tag{6.31}$$

Summing up, we have

THEOREM 6.7. A quaternion Kaehlerian manifold M of dimension $n \geq 8$ is of constant Q-sectional curvature c if and only if

$$R(X,Y)Z = \tfrac{1}{4}c[g(Y,Z)X-g(X,Z)Y+g(FY,Z)FX-g(FX,Z)FY-2g(FX,Y)FZ$$
$$+g(GY,Z)GX-g(GX,Z)GY-2g(GX,Y)GZ+g(HY,Z)HX-g(HX,Z)HY$$
$$-2g(HX,Y)HZ]$$

for any vector fields X, Y and Z on M.

Example 6.1. Let HP^m be a quaternion projective space of dimension 4m. HP^m is of constant Q-sectional curvature 4. Let S^{4m+3} be a unit sphere of dimension 4m+3. Then we can consider the Hopf fibering $S^{4m+3} \longrightarrow HP^m$. S^{4m+3} admits a Sasakian 3-structure.

EXERCISES

A. CONSTANCY OF THE HOLOMORPHIC SECTIONAL CURVATURE: Let M be a Kaehlerian manifold with almost complex structure J. Nomizu [5] gave the condition for constancy of the holomorphic sectional curvature of the curvature tensor of a Kaehlerian manifold M.

THEOREM 1. The curvature tensor R at a point of a Kaehlerian manifold M has constant holomorphic sectional curvature if and only if it has the following property:

(A) If $g(X,Y) = g(JX,Y) = 0$, then $g(R(X,JX)JX,Y) = 0$.

A subspace S of the tangent space of M is holomorphic if $JS = S$. S is said to be totally real if it satisfies the condition: $g(JX,Y) = 0$ for all $X,Y \in S$.

(B) Axiom of holomorphic 2k-planes. For any 2k-dimensional holomorphic subspace S of $T_x(M)$, there exists a 2k-dimensional totally geodesic submanifold V of M containing x such that $T_x(V) = S$.

(C) Axiom of totally real k-planes. For any k-dimensional totally real subspace S of $T_x(M)$, there exists a k-dimensional totally geodesic submanifold V of M containing x such that $T_x(V) = S$.

As applications of Theorem 1, Nomizu [5] proved the following theorems.

THEOREM 2. If a Kaehlerian manifold M of dimension 2n satisfies the axiom of holomorphic 2k-planes for some k, $1 \le k \le n-1$, then M has constant holomorphic sectional curvature.

THEOREM 3. If a Kaehlerian manifold M of dimension 2n satisfies the axiom of totally real k-planes for some k, $2 \le k \le n$, then M has constant holomorphic sectional curvature.

For constancy of the holomorphic sectional curvature, see also Chen-Ogiue [1], Harada [2] and Yamaguchi-Kon [1].

B. COMPACT KAEHLERIAN MANIFOLD WITH $R(X,Y)R = 0$: On a Kaehlerian manifold M, Ogawa [1] studied the condition $R(X,Y)R = 0$ and prove the following

THEOREM Let M be a compact Kaehlerian manifold with constant scalar curvature. If M satisfies the condition $R(X,Y)R = 0$, then M is locally symmetric.

To prove the theorem above, we used the integral formula of Lichnerowicz [4].

C. BOCHNER CURVATURE TENSOR: Let M be a complex n-dimensional Kaehlerian manifold. We denote by R, Q and r the Riemannian curvature tensor, the Ricci operator and the scalar curvature of M respectively. The Bochner curvature tensor B of M is defined by

$$\begin{aligned}B(X,Y)Z = R(X,Y)Z - \frac{1}{2n+4}[&g(Y,Z)QX-g(QX,Z)Y+g(JY,Z)QJX-g(QJX,Z)JY\\&+g(QY,Z)X-g(X,Z)QY+g(QJY,Z)JX-g(JX,Z)QJY-2g(JX,QY)JZ\\&-2g(JX,Y)QJZ] + \frac{r}{(2n+2)(2n+4)}[g(Y,Z)X-g(X,Z)Y\\&+g(JY,Z)JX-g(JX,Z)JY-2g(JX,Y)JZ].\end{aligned}$$

Matsumoto-Tanno [1] proved the following theorems.

THEOREM 1. Let M be a Kaehlerian space with parallel Bochner curvature tensor. Then M is a locally symmetric space or a space with vanishing Bochner curvature tensor.

THEOREM 2. If a Kaehlerian space with vanishing Bochner curvature tensor has constant scalar curvature, then M is one of the following spaces:

(1) M is a space of constant holomorphic sectional curvature;

(2) M is a locally product space of two spaces of constant holomorphic sectional curvatures c and -c.

D. NONNEGATIVE SECTIONAL CURVATURE: A. Gray [4] proved the following

THEOREM 1. Let M be a compact Kaehlerian manifold with nonnegative sectional curvature. If M has constant scalar curvature, then M is locally symmetric.

It is impossible to replace "locally symmetric" by "symmetric" in the conclusion to Theorem 1. Indeed, Auslander [1] has constructed compact flat Kaehlerian manifold which are not homogeneous.

Furthermore we have by Theorem 1 the following

THEOREM 2. If in addition to the hypotheses of Theorem 1, M has positive Ricci curvature, then M is globally symmetric. More generally, if M satisfies the hypotheses of Theorem 1 and has positive first Chern class, then M is globally symmetric.

We also have the following (Berger [1], Bishop-Goldberg [1])

THEOREM 3. Let M be a compact Kaehlerian manifold with positive sectional curvature and constant scalar curvature. Then M is isometric to a complex projective space.

E. COMPLEX SPACE FORMS AND CHERN CLASSES: Let M be a Kaehlerian manifold of complex dimension n. We denote by c_k the k-th Chern class of M. Let $\omega^1,\ldots,\omega^n$ be a local field of unitary coframes. Then the Kaehlerian metric of M is written as $g = \Sigma(\omega^\alpha \otimes \bar{\omega}^\alpha + \bar{\omega}^\alpha \otimes \omega^\alpha)$ and the fundamental 2-form is given by $\Phi = \frac{1}{2}\sqrt{-1}\sum\omega^\alpha \wedge \bar{\omega}^\alpha$. Let $\Omega^\alpha_\beta = \Sigma R^\alpha_{\beta\gamma\bar{\delta}}\omega^\gamma \wedge \bar{\omega}^\delta$ be the curvature form of M.

If we define a closed 2k-form r_k by

$$r_k = \frac{(-1)^k}{(2\pi\sqrt{-1})^k k!}\sum\delta^{\alpha_1\ldots\alpha_k}_{\beta_1\ldots\beta_k}\Omega^{\beta_1}_{\alpha_1} \wedge \cdots \wedge \Omega^{\beta_k}_{\alpha_k},$$

then the k-th Chern class c_k of M is represented by r_k. In particular, c_1 and c_2 are represented by

$$r_1 = \frac{\sqrt{-1}}{2\pi}\sum\Omega^\alpha_\alpha, \qquad r_2 = -\frac{1}{8\pi^2}\sum(\Omega^\alpha_\alpha \wedge \Omega^\beta_\beta - \Omega^\alpha_\beta \wedge \Omega^\beta_\alpha),$$

respectively. Chen-Ogiue [3] proved the following theorems.

THEOREM 1. If M is an n-dimensional complex space form ($n \geq 2$), then $c_2 = \frac{1}{2}(n/(n+1))c_1^2$.

THEOREM 2. Let M be an n-dimensional compact Kaehlerian manifold ($n \geq 2$). If

(a) $c_2 = \frac{1}{2}(n/(n+1))c_1^2$, (b) M is Einstein,

then M is a complex space form.

THEOREM 3. Let M be an n-dimensional compact Kaehlerian manifold ($n \geq 2$). If

(a) $c_2 = \frac{1}{2}(n/(n+1))c_1^2$, (b) the Bochner curvature is zero,

then M is a complex space form.

F. NEARLY KAEHLERIAN MANIFOLDS: Let M be a real n-dimensional nearly Kaehlerian manifold. Then we have

(1) If $n = 4$, then M is Kaehlerian (A. Gray [1], Takamatsu [1]).

(2) If $n = 6$, then M is Einstein.

(3) If $n = 8$, and if M is complete and simply connected, then M is $M_1 \times M_2$, where M_1 is a 6-dimensional Einstein nearly Kaehlerian manifold and M_2 is a 2-dimensional Kaehlerian manifold (A. Gray [ ]).

When $n = 10$, Matsumoto [3] proved the following

THEOREM. Let M be a 10-dimensional, connected, non-Kaehlerian, nearly Kaehlerian manifold. Then M is one of the following manifolds:

(a) M, which is complete and simply connected, is $M_1 \times M_2$, where M_1 is a 6-dimensional Einstein nearly Kaehlerian manifold and M_2 is a 4-dimensional Kaehlerian manifold;

(b) M is an Einstein manifold;

(c) M satisfies $5Q - Q^* = 8\,I$, which is not Einstein.

See also, Matsumoto [2].

G. ALMOST QUATERNION MANIFOLDS: It is well known for two given tensor fields F and G of type (1,1) in a differentiable manifold the expression

$$[F,G](X,Y) = [FX,GY] - F[X,GY] - G[FX,Y] + [GX,FY]$$
$$- G[X,FY] - F[GX,Y] + (FG+GF)[X,Y]$$

defines a tensor field [F,G] of type (1,2), and that the tensor field [F,F] plays a very important role in the discussion of integrability conditions of an almost complex structure defined by F. We call [F,G] the Nijenhuis tensor of F and G.

Let M be an almost quaternion manifold with almost quaternion structure (F,G,H). If there exists a coordinate system in which the components of F, G and H are all constant, the almost quaternion structure is said to be integrable. Yano-Ako [1] obtained

THEOREM. A necessary and sufficient condition that an almost quaternion structure (F,G,H) on M be integrable is that [F,G] = 0 and R = 0, where R is the curvature tensor of an affine connection on M.

For an affine connection on M see also Yano-Ako [1].

CHAPTER IV

SUBMANIFOLDS OF KAEHLERIAN MANIFOLDS

In this chapter, we study submanifolds of Kaehlerian manifolds, especially those of complex space forms. §1 is devoted to the study of Kaehlerian submanifolds (invariant submanifolds) of Kaehlerian manifolds. We compute the Laplacian for the second fundamental form and the Ricci tensor of a Kaehlerian submanifold of a complex space form, and give some integral formulas. As applications of these integral formulas we prove the classification theorem of Kaehlerian hypersurfaces with parallel Ricci tensor of complex space forms, and we also study Kaehlerian hypersurfaces with constant scalar curvature. In §2, we discuss anti-invariant submanifolds of Kaehlerian manifolds (see Yano-Kon [1]). We give the basic formulas for anti-invariant submanifolds of complex space forms and some examples of anti-invariant submanifolds of complex space forms. We then prove some theorems which characterize these examples. In the last §3, we discuss CR submanifolds of Kaehlerian manifolds (see Yano-Kon [18]). We construct some examples of CR submanifolds of a complex Euclidean space and complex projective space, which have parallel mean curvature vector, and flat normal connection or semi-flat normal connection. We also prove some theorems which characterize the above examples.

1. KAEHLERIAN SUBMANIFOLDS

Let $\bar{M}$ be a Kaehlerian manifold of complex dimension m (of real dimension 2m) with almost complex structure J and with Kaehlerian metric g. Let M be a complex n-dimensional analytic submanifold of $\bar{M}$, that is, the immersion $f: M \longrightarrow \bar{M}$ is holomorphic, i.e., $J \cdot f_* = f_* \cdot J$, where f_* is the differential of the immersion f and we denote by the same J the induced complex structure on M. Then the Riemannian metric g, which will be denoted by the same letter of $\bar{M}$, induced on M is Hermitian. It is easy to see that the fundamental 2-form with this Hermitian metric g is the restriction of the fudamental 2-form of $\bar{M}$ and hence is closed. This shows that every complex analytic submanifold M of a Kaehlerian manifold $\bar{M}$ is also a Kaehlerian manifold with respect to the induced structure. We call such a submanifold M of a Kaehlerian manifold $\bar{M}$ a *Kaehlerian submanifold*. In other words, a Kaehlerian submanifold M of a Kaehlerian manifold $\bar{M}$ is an *invariant submanifold* under the action of the complex structure J of $\bar{M}$, i.e., $JT_x(M) \subset T_x(M)$ for every point x of M. Then we see that $JT_x(M)^\perp \subset T_x(M)^\perp$ for every point x of M. We denote by $\bar{\nabla}$ (resp. ∇) the operator of covariant differentiation with respect to the connection in $\bar{M}$ (resp. M). For any vector fields X and Y on M we have

$$\bar{\nabla}_X JY = \nabla_X JY + B(X,JY) \quad \text{and} \quad \bar{\nabla}_X JY = J\bar{\nabla}_X Y = J\nabla_X Y + JB(X,Y).$$

Both the tangent space $T_x(M)$ and the normal space $T_x(M)^\perp$ being invariant by J, we obtain

$$\nabla_X JY = J\nabla_X Y \qquad \text{and} \qquad B(X,JY) = JB(X,Y).$$

The first identity shows once more that ∇ is Kaehlerian. From the second identity and from symmetry of B(X,Y) in X and Y, we have

LEMMA 1.1. The second fundamental form B of a Kaehlerian submanifold M satisfies

$$B(JX,Y) = B(X,JY) = JB(X,Y),$$

or equivalently

$$JA_V X = -A_V JX = A_{JV} X.$$

PROPOSITION 1.1. Any Kaehlerian submanifold M is a minimal submanifold.

Proof. For each $T_x(M)$ we can choose an orthonormal basis $e_1,\dots,e_n,Je_1,\dots,Je_n$. Then Lemma 1.1 implies

$$\mathrm{Tr}B = \sum_i [B(e_i,e_i) + B(Je_i,Je_i)] = 0,$$

which shows that M is a minimal submanifold. QED.

From equations of Gauss and Lemma 1.1 we have

PROPOSITION 1.2. Let M be a Kaehlerian submanifold of a Kaehlerian manifold $\bar{M}$ and let R and $\bar{R}$ be the Riemannian curvature tensors of M and $\bar{M}$ respectively. Then

$$g(R(X,JX)JX,X) = g(\bar{R}(X,JX)JX,X) - 2g(B(X,X),B(X,X))$$

for all vector field X on M.

As a direct consequence of Proposition 1.2, we obtain

PROPOSITION 1.3. Any Kaehlerian submanifold M of a complex space form $\bar{M}(c)$ of constant holomorphic sectional curvature c is totally geodesic if and only if M is of constant holomorphic sectional curvature c.

We now suppose that M is a complex n-dimensional Kaehlerian submanifold of a complex m-dimensional space form $\bar{M}^m(c)$ of constant holomorphic sectional curvature c. Then we have equations of Gauss and Codazzi respectively:

$$(1.1)\qquad R(X,Y)Z = \tfrac{1}{4}c[g(Y,Z)X-g(X,Z)Y+g(JY,Z)JX-g(JX,Z)JY+2g(X,JY)JZ] + A_{B(Y,Z)}X - A_{B(X,Z)}Y,$$

$$(1.2)\qquad (\nabla_X B)(Y,Z) = (\nabla_Y B)(X,Z).$$

Moreover, equation of Ricci is given by

$$(1.3)\qquad g(R(X,Y)U,V) + g([A_V,A_U]X,Y) = \tfrac{1}{2}cg(X,JY)g(JU,V).$$

From (1.1) the Ricci tensor S and the scalar curvature r of M are respectively given by

$$(1.4)\qquad S(X,Y) = \tfrac{1}{2}(n+1)cg(X,Y) - \sum_i g(B(X,e_i),B(Y,e_i)),$$

$$(1.5)\qquad r = n(n+1)c - \sum_{i,j} g(B(e_i,e_j),B(e_i,e_j)).$$

From (1.4) and (1.5) we have

PROPOSITION 1.4. Let M be a complex n-dimensional Kaehlerian submanifold of a complex space form $\bar{M}^m(c)$. Then

(1) $S - \frac{1}{2}(n+1)cg$ is negative semi-definite;

(2) $r \leq n(n+1)c$.

In the next place, we consider the Ricci tensor S of a Kaehlerian submanifold M of a complex space form $\bar{M}^m(c)$. We have already seen that S satisfies (see Proposition 4.2 of Chapter III)

$$(\nabla_Z S)(X,Y) = (\nabla_X S)(Y,Z) + (\nabla_{JY} S)(JX,Z).$$

LEMMA 1.2. If a Kaehlerian manifold M has the constant scalar curvature, then

$$(\nabla^2 S)(X,Y) = 2\sum_{i=1}^{2n} (R(e_i,X)S)(e_i,Y).$$

Proof. Since the scalar curvature r of M is constant, the Ricci tensor S of M satisfies $\Sigma(\nabla_{e_i} S)(e_i, X) = 0$. Thus we have

$$\begin{aligned}
(\nabla^2 S)(X,Y) &= \sum_i (\nabla_{e_i}\nabla_{e_i} S)(X,Y) = \sum_i \nabla_{e_i}((\nabla_{e_i} S)(X,Y)) \\
&= \sum_i [\nabla_{e_i}((\nabla_X S)(e_i,Y) + \nabla_{e_i}((\nabla_{JY} S)(e_i,JX))] \\
&= \sum_i [(R(e_i,X)S)(e_i,Y) + (R(e_i,JY)S)(e_i,JX)] \\
&= 2\sum_i (R(e_i,X)S)(e_i,Y).
\end{aligned}$$

QED.

Take an orthonoemal basis $e_1,\dots,e_{2n}$ in $T_x(M)$ such that $e_{n+t} = Je_t$ $(t = 1,\dots,n)$ and an orthonormal basis $v_1,\dots,v_{2p}$ for $T_x(M)^\perp$ such that $v_{p+s} = Jv_s$ $(s = 1,\dots,p)$, where we have put $p = m-n$. We calculate $(\nabla^2 S)(X,Y)$ for a Kaehlerian submanifold M of $\bar{M}^m(c)$ in the following way. Since M is minimal, we obtain

$$\begin{aligned}
2\sum_i (R(e_i,X)S)(e_i,Y) &= -2\sum_i [S(\bar{R}(e_i,X)e_i,Y) + S(\bar{R}(e_i,X)Y,e_i) \\
&\quad + S(A_{B(X,e_i)}e_i,Y) - S(A_{B(e_i,Y)}X,e_i) + S(A_{B(X,Y)}e_i,e_i)].
\end{aligned}$$

Moreover, we have

$$-2\sum_i [S(\bar{R}(e_i,X)e_i,Y) + S(\bar{R}(e_i,X)Y,e_i) = nc[S(X,Y) - \frac{1}{2n} r g(X,Y)].$$

The Ricci tensor S has the property $S(JX,JY) = S(X,Y)$, and hence Lemma 1.1 implies that $\Sigma S(A_{B(X,Y)}e_i,e_i) = 0$. We also have

$$\begin{aligned}
&-2\sum_i [S(A_{B(X,e_i)}e_i,Y) - S(A_{B(e_i,Y)}X,e_i)] \\
&\quad = -2\sum_{i,a} [g(A_a e_i,QY)g(A_a X,e_i) - g(A_a X,Qe_i)g(A_a Y,e_i)] \\
&\quad = -2\sum_a [g(QA_a A_a X,Y) - g(A_a Q A_a X,Y)],
\end{aligned}$$

where Q is the Ricci operator of M given by $g(QX,Y) = S(X,Y)$. Consequently, we have

$$2\sum_i (R(e_i,X)S)(e_i,Y) = c[nS(X,Y) - \tfrac{1}{2}rg(X,Y)] - 2\sum_a [g(QA_aA_aX,Y) - g(A_aQA_aX,Y)].$$

Therefore, Lemma 1.2 implies the following (Kon [7])

THEOREM 1.1. Let M be a complex n-dimensional Kaehlerian submanifold of a complex space form $\bar{M}^m(c)$. If the scalar curvature r of M is constant, then

$$g(\nabla^2 Q,Q) = c[n|Q|^2 - \tfrac{1}{2}r^2] - \sum_a |[Q,A_a]|^2.$$

We give some applications of Theorem 1.1. First we obtain

PROPOSITION 1.5. Let $\bar{M}$ be a complex space form of constant holomorphic sectional curvature $c < 0$, and let M be a Kaehlerian submanifold of $\bar{M}$. If the Ricci tensor S of M is parallel, then M is an Einstein manifold.

Proof. From the assumption we have $|Q|^2 = r^2/2n$. This is equivalent to the fact that M is Einstein. QED.

PROPOSITION 1.6. Let M be a complex n-dimensional compact Kaehlerian submanifold with constant scalar curvature of a complex space form $\bar{M}^m(c)$ $(c > 0)$. If $QA_a = A_aQ$ $(a = 1,\dots,p)$, then M is an Einstein manifold.

Proof. If $QA_a = A_aQ$ for $a = 1,\dots,p$, then we easily see that $QA_a = A_aQ$ for $a = 1,\dots,2p$. By the assumption and Theorem 1.1 we have

$$0 \leq \int_M |\nabla Q|^2 *1 = -\int_M g(\nabla^2 Q,Q)*1 = -c\int_M [n|Q|^2 - \tfrac{1}{2}r^2]*1.$$

But we have always $r^2 \leq 2n|Q|^2$, hence we obtain $\nabla Q = 0$. Therefore, we have $r^2 = 2n|Q|^2$, which shows that M is an Einstein manifold. QED.

THEOREM 1.2. Let M be a compact Kaehlerian hypersurface of a complex space form $\bar{M}^{n+1}(c)$ $(c > 0)$. If the scalar curvature of M is constant, then M is an Einstein manifold.

Proof. Since the real codimension of M is 2, we can take an orthonormal basis v, Jv for $T_x(M)^{\perp}$. Then we have

$$Q = \tfrac{1}{2}(n+1)cI - 2A_v^2.$$

Therefore we obtain $QA_v = A_vQ$, and hence M is Einstein by Proposition 1.6. QED.

We now compute the Laplacian for the square of the length of the second fundamental form of a complex n-dimensional Kaehlerian submanifold M of a complex space form $\bar{M}^m(c)$. From (3.10) of Chapter II, by a straightforward computation, we have

$$g(\nabla^2 A,A) = \tfrac{1}{2}(n+2)c|A|^2 - \sum_{a,b}|[A_a,A_b]|^2 - \sum_{a,b}(\mathrm{Tr}A_aA_b)^2. \tag{1.6}$$

LEMMA 1.3. Let M be a complex n-dimensional Kaehlerian submanifold of a complex m-dimensional Kaehlerian manifold $\bar{M}$. Then

$$\frac{1}{n}|A|^4 \leq \sum_{a,b}|[A_a,A_b]|^2 \leq |A|^4, \tag{1.7}$$

$$\frac{1}{2p}|A|^4 \leq \sum_{a,b}(\mathrm{Tr}A_aA_b)^2 \leq \tfrac{1}{2}|A|^4, \quad (p = m-n). \tag{1.8}$$

If $\bar{M}$ is of constant holomprphic sectional curvature c, then M is Einstein if and only if $\Sigma|[A_a,A_b]|^2 = |A|^4/n$.

Proof. We put

$$A^* = \sum_{a=1}^{2p} A_a^2.$$

Clearly A^* is a symmetric, positive semi-definite operator. So we have $\mathrm{Tr}A^* = |A|^2$. From Lemma 1.1 we see that

$$\sum_{a,b}|[A_a,A_b]|^2 = 2\sum_{a,b}\mathrm{Tr}A_a^2A_b^2 = 2\mathrm{Tr}(A^*)^2.$$

We easily see that $JA^* = A^*J$. Thus, for a suitable basis, A^* is represented by a matrix form

$$A^* = \begin{pmatrix} \lambda_1 & & & & \\ & \cdot & & & \\ & & \cdot & & \\ & & & \cdot & \\ & & & & \lambda_{2n} \end{pmatrix}, \quad \lambda_{n+t} = \lambda_t, \quad \lambda_t \geq 0 \ (t=1,\ldots,n).$$

Then we have

$$\begin{aligned} \mathrm{Tr}(A^*)^2 &= \sum_{i=1}^{2n}\lambda_i^2 = [(\sum_{i=1}^{2n}\lambda_i)^2 - \sum_{i\neq j}^{2n}\lambda_i\lambda_j] \\ &= (\sum_{i=1}^{2n}\lambda_i)^2 - 4\sum_{i\neq j}^{n}\lambda_i\lambda_j - \mathrm{Tr}(A^*)^2, \end{aligned}$$

from which

$$(1.9) \qquad 2\mathrm{Tr}(A^*)^2 = |A|^4 - 4\sum_{i\neq j}^{n}\lambda_i\lambda_j \leq |A|^4.$$

On the other hand, we obtain

$$(1.10) \qquad 2\mathrm{Tr}(A^*)^2 = 2\sum_{i=1}^{2n}\lambda_i^2 = \frac{1}{n}(\sum_{i=1}^{2n}\lambda_i)^2 + \frac{1}{n}\sum_{i>j}^{2n}(\lambda_i - \lambda_j)^2 \geq \frac{1}{n}|A|^4.$$

From (1.9) and (1.10) we have (1.7). If $\bar{M}$ is of constant holomorphic sectional curvature c, then the Ricci operator Q of M is giben by $Q = \frac{1}{2}(n+1)c - A^*$. If $\Sigma|[A_a,A_b]|^2 = |A|^4/n$, $\lambda_i = \lambda_j$ for all i, j. Thus Q is proportional to I and hence M is Einstein. The converse is also true.

We notice here that the following equation is satisfied:

$$(1.11) \qquad \sum|[A_a,A_b]|^2 = \frac{1}{n}|A|^4 + 2|Q|^2 - \frac{1}{n}r^2.$$

In the next place, we take a basis $v_1,\dots,v_{2p}$ of $T_x(M)^\perp$ such that

$$\sum_{a,b}(\mathrm{Tr}A_aA_b)^2 = \sum_a(\mathrm{Tr}A_a^2)^2.$$

Then we have

$$(1.12)\qquad \sum_a(\mathrm{Tr}A_a^2)^2 = \tfrac{1}{2}|A|^4 - 2\sum_{a\neq b}^{p}(\mathrm{Tr}A_a^2)(\mathrm{Tr}A_b^2) \le \tfrac{1}{2}|A|^4,$$

$$(1.13)\qquad \sum_a(\mathrm{Tr}A_a^2)^2 = \frac{1}{2p}|A|^4 + \frac{1}{2p}\sum_{a>b}^{2p}(\mathrm{Tr}A_a^2 - \mathrm{Tr}A_b^2)^2 \ge \frac{1}{2p}|A|^4.$$

From (1.12) and (1.13) we have (1.8). QED.

By Lemma 1.3 and (1.6) we have the following (Tanno [6])

THEOREM 1.3. Let M be a complex n-dimensional Kaehlerian submanifold of a complex space form $\bar{M}^m(c)$. Then

$$-g(\nabla^2 A, A) \le \tfrac{1}{2}[3|A|^2 - (n+2)c]|A|^2.$$

THEOREM 1.4. Let M be a complex n-dimensional compact Kaehlerian submanifold of a complex space form $\bar{M}^m(c)$. Then either M is totally geodesic, or $|A|^2 = (n+2)c/3$, or at some point x of M, $|A|^2(x) > (n+2)c/3$.

Proof. Since M is compact, we have

$$(1.14)\qquad 0 \le \int_M|\nabla A|^2 *1 = -\int_M g(\nabla^2 A,A)*1 \le \tfrac{1}{2}\int_M[3|A|^2 - (n+2)c]|A|^2*1.$$

Suppose $|A|^2 \le (n+2)c/3$ everywhere on M. Then $\nabla A = 0$ and $|A|^2 = (n+2)c/3$ or $|A|^2 = 0$. Except for these possibilities, we get $|A|^2(x) > (n+2)c/3$ at some point x of M. QED.

PROPOSITION 1.7. Let M be a complex n-dimensional Kaehlerian submanifold of a complex space form $\bar{M}^m(c)$. If M is Einstein, then

$$0 \le |\nabla A|^2 \le \tfrac{1}{2}(n+2)[\tfrac{1}{n}|A|^2 - c]|A|^2.$$

Proof. Since M is Einstein, $|A|^2$ is constant and hence $-g(\nabla^2 A,A) = |\nabla A|^2$. Thus, from (1.6) and Lemma 1.3, we have our inequality. QED.

PROPOSITION 1.8. Under the same assumption as that of Proposition 1.7, either M is totally geodesic or $|A|^2 \geq nc$.

THEOREM 1.5. Let M be a complex n-dimensional Kaehlerian submanifold of a complex space form $\bar{M}^m(c)$ $(c > 0)$. If $|A|^2 = (n+2)c/3$, then M is an Einstein manifold of complex dimension 1.

Proof. Since $|A|^2$ is constant, we have $-g(\nabla^2 A,A) = |\nabla A|^2$. From this and the assumption we see that the second fundamental form of M is parallel. Thus, from (1.9) and (1.12), we can assume that

$$\lambda_i = 0 \ (i = 2,\ldots,n), \qquad A_a = 0 \ (a = 2,\ldots,p).$$

Therefore we see that $A^* = 2(A_1)^2$. Since $Q = \frac{1}{2}(n+1)cI - A^*$, we obtain $QA_a = A_aQ$ for $a = 1,\ldots,p$. Therefore, by Proposition 1.6, M is an Einstein manifold. Thus we have $|A|^2 \geq nc$, which implies that $n = 1$. Thus we have our assertion. QED.

THEOREM 1.6. Let M be a Kaehlerian hypersurface of a complex space form $\bar{M}^{n+1}(c)$. Then the following conditions are equivalent:

(1) The Ricci tensor S of M is parallel;

(2) The second fundamental form of M is parallel;

(3) M is an Einstein manifold.

Proof. First of all we prove that (1) is equivalent to (2). If the second fundamental form of M is parallel, then the Ricci tensor S of M is obviousely parallel. We suppose that the Ricci tensor S of M is parallel. We take a normal basis v, Jv of $T_x(M)^{\perp}$. There exists a 1-form s such that $D_X v = s(X)Jv$ for any vector field X tangent to M. Thus we have, by (1.4),

$$(\nabla_X A)_v A_v Y + s(X)A_{Jv}A_v Y + A_v(\nabla_X A)_v Y + s(X)A_v A_{Jv} Y = 0.$$

From this and Lemma 1.1 we obtain

$$(\nabla_X A)_V A_V Y + A_V (\nabla_X A)_V Y = 0. \tag{1.15}$$

Let we take any two λ, μ of characteristic roots of A_V at a point x of M. Now we define spaces by setting

$$T_\lambda = \{X \in T_x(M) : A_V X = \lambda X\}, \quad T_\mu = \{X \in T_x(M) : A_V X = \mu X\}.$$

Then $T_\lambda \cap T_\mu = \{0\}$ when $\lambda \neq \mu$. Let Y be in T_λ. Then (1.15) implies

$$A_V(\nabla_X A)_V Y = -(\nabla_X A)_V A_V Y = -\lambda(\nabla_X A)_V Y$$

for any vector X. Thus, if $Y \in T_\lambda$, then $(\nabla_X A)_V Y \in T_{-\lambda}$ for any X. Let $X \in T_\mu$, $Y \in T_\lambda$ and $\lambda \neq \mu$. Then, using the Codazzi equation, we obtain $(\nabla_X A)_V Y \in T_{-\lambda} \cap T_{-\mu}$, and hence $(\nabla_X A)_V Y = 0$.

On the other hand, if $\lambda = \mu \neq 0$, X, $Y \in T_\lambda$, then we have $(\nabla_X A)_V Y \in T_{-\lambda}$ and hence $(\nabla_X A)_V (\nabla_X A)_V Y \in T_\lambda$. By the Codazzi equation we also have

$$(\nabla_X A)_V (\nabla_X A)_V Y = (\nabla_{(\nabla_X A)_V Y} A)_V X \in T_{-\lambda}.$$

Therefore, we have $(\nabla_X A)_V (\nabla_X A)_V Y = 0$. Thus we obtain $(\nabla_X A)_V Y = 0$.

Let $\lambda = \mu = 0$. We take X, $Y \in T_0$ at $x \in M$, and extend these to local vector fields on M which are covariant constant with respect to ∇ at x. Then, by (1.15) we have

$$g((\nabla_X A)_V (\nabla_X A)_V Y, Y) = 0,$$

from which $(\nabla_X A)_V Y = 0$.

From these considerations we have our assertion.

Next we prove that (2) is equivalent to (3). If M is Einstein, then the Ricci tensor of M is parallel and hence the second fundamental form of M is parallel.

Conversely, we suppose that the second fundamental form of M is parallel. Then Theorem 1.1 implies

$$c[n|Q|^2 - \tfrac{1}{2}r^2] = 0.$$

Thus, if $c \neq 0$, then M is Einstein. Suppose that $c = 0$. Then (1.6) shows that M is totally geodesic. Consequently, M is an Einstein manifold. QED.

We now prove the following theorem (Chern [1], Nomizu-Smyth [1], Tsunero Takahashi [2]).

THEOREM 1.7. Let M be a Kaehlerian hypersurface of a complex space form $\bar{M}^{n+1}(c)$ with parallel Ricci tensor. If $c \leq 0$, then M is totally geodesic. If $c > 0$, then either M is totally geodesic, or an Einstein manifold with $|A|^2 = nc$ and hence $r = n^2c$.

Proof. From Theorem 1.6 the second fundamental form of M is parallel. Thus, if $c \leq 0$, M is totally geodesic by (1.6). Let $c > 0$. Then, by (1.6), Lemma 1.3 and Theorem 1.6, we have

$$\tfrac{1}{2}(n+2)[c - \frac{1}{n}|A|^2]|A|^2 = 0.$$

Therefore, $|A|^2 = 0$ or $|A|^2 = nc$. From these and (1.5) we have our assertion. QED.

In the next place we give a global version of Theorem 1.7. For this purpose we prepare some lemmas.

Let M be a Kaehlerian hypersurface of a complex space form $\bar{M}^m(c)$. We take a local basis v, Jv in the normal bundle for which the second fundamental tensors are given by A_v, $A_{Jv} = JA_v$. To simplify the notation we write A_v by A. Moreover, we take a 1-form s such that $D_Xv = s(X)Jv$. We put

$$D(X,Y) = AX \wedge AY + JAX \wedge JAY,$$

where $X \wedge Y$ denotes the skew-symmetric endomorphism which maps Z upon

$g(Y,Z)X - g(X,Z)Y$. Then we have

$$D(X,Y) = -\bar{R}(X,Y) + R(X,Y)$$

$$= -\tfrac{1}{4}c[X \wedge Y + JX \wedge JY + 2g(X,JY)J] + R(X,Y).$$

LEMMA 1.4. At each point x of M we have

$$\text{Ker } A = \{X \in T_x(M) : D(X,Y) = 0 \text{ for all } Y \in T_x(M)\}$$

$$= \{X \in T_x(M) : (R - \bar{R})(X,Y) = 0 \text{ for all } Y \in T_x(M)\}.$$

Proof. Clearly Ker A is contained in the subspace defined by D. On the other hand, if $X \notin \text{Ker } A$, then $D(X,JX) = 2JAX \wedge AX \neq 0$, from which we have the first equality. The second equality is clear by the definition of D. QED.

The rank of A at $x \in M$ will be called the *rank* of M at x since Lemma 1.4 shows that it is intrinsic, that is, depends only on M.

Let $f, \bar{f} : M \longrightarrow \bar{M}$ be two Kaehlerian immersions. For each $x_0 \in M$ there is a neighborhood $U(x_0)$ of x_0 in M on which we can choose of a unit normal vector field v (resp. $\bar{v}$) for the immersion f (resp. $\bar{f}$), there by giving rise to tensor fields A and s (resp. $\bar{A}$ and $\bar{s}$) on $U(x_0)$.

LEMMA 1.5. At each point x of $U(x_0)$ we have

(1) $A = 0$ if and only if $\bar{A} = 0$;

(2) If $A = \bar{A} \neq 0$ and $\nabla A = \nabla\bar{A}$, then $s = \bar{s}$.

Proof. (1) is immediate consequence from Lemma 1.4. We next prove (2). From the Codazzi equation (1.2) we obtain

$$(\nabla_X A)Y - s(X)JAY - (\nabla_Y A)X + s(Y)JAX = 0.$$

From the assumption we have

$$-s(X)JAY + s(Y)JAX = -\bar{s}(X)JAY + \bar{s}(Y)JAX.$$

Thus we have

$$(\bar{s}(X) - s(X))JAY = (\bar{s}(Y) - s(Y))JAX.$$

Suppose that $X \notin \operatorname{Ker} A$. Then, by setting $Y = JX$, we get

$$(\bar{s}(X) - s(X))AX = (\bar{s}(JX) - s(JX))JAY.$$

Since AX ($\neq 0$) and JAX are linearly independent, we conclude that $s(X) = \bar{s}(X)$. If $X \in \operatorname{Ker} A$, then we choose $Y \notin \operatorname{Ker} A$ and get $(\bar{s}(X) - s(X))JAY = 0$, that is, $s(X) = \bar{s}(X)$. QED.

LEMMA 1.6. Assuming $R \neq \bar{R}$ (that is, $A \neq 0$) at some point of M, let x_0 be a point where the rank of M is maximal. There exists a neighborhood $U(x_0)$ of x_0 on which we may choose unit normal vector fields v and $\bar{v}$, with respect to the immersions f and $\bar{f}$, respectively, such that $A = \bar{A}$ and $s = \bar{s}$ on $U(x_0)$.

Proof. On $U(x_0)$ on which the rank of M is constant and equal to k, say, we choose unit normal vector fields v and $\bar{v}$, with respect to the immersions f and $\bar{f}$, respectively. At each point x of $U(x_0)$ we can choose an orthonormal basis $e_1,\ldots,e_n,Je_1,\ldots,Je_n$ of $T_x(M)$ for which A is

$$Ae_i = \lambda_i e_i \ (i = 1,\ldots,k), \quad Ae_j = 0 \ (j = k+1,\ldots,n).$$

Then $AJe_i = -\lambda_i Je_i$ $(i = 1,\ldots,k)$ and $AJe_j = 0$ $(j = k+1,\ldots,n)$. Since we have

$$(R - \bar{R})(e_i,Je_i) = -2Ae_i \wedge JAe_i = -2\bar{A}e_i \wedge J\bar{A}e_i,$$

and the middle form of this identity being nonzero when $i \leq k$, we see that $\bar{A}e_i$ is a linear combination of Ae_i and JAe_i, say, $\bar{A}e_i = a_iAe_i + b_iJAe_i$. Then $a_i^2 + b_i^2 = 1$. From

$$(R - \bar{R})(e_i,e_j) = Ae_i \wedge Ae_j + JAe_i \wedge JAe_j$$
$$= \bar{A}e_i \wedge \bar{A}e_j + J\bar{A}e_i \wedge J\bar{A}e_j$$

we easily see that $a_i = a_j = a$, say, and $b_i = b_j = b$, say, for $i,j = 1,\ldots,k$. However, $\mathrm{Ker}\, A = \mathrm{Ker}\, \bar{A}$, by Lemma 1.4, and therefore $\bar{A} = aA + bJA$ with $a^2 + b^2 = 1$ at each point of $U(x_0)$. From the assumption on the rank of M at x_0 we can find a differentiable vector field X on $U(x_0)$ such that $AX \neq 0$ and since $a = g(\bar{A}X,AX)/g(AX,AX)$, it follows that a (and hence b) is a differentiable function on $U(x_0)$ such that $a = \cos\theta$, $b = \sin\theta$. Then $v' = \cos\theta v + \sin\theta Jv$ is a unit normal vector field on $U(x_0)$ with respect to the immersion f and clearly $A' = \bar{A}$. By Lemma 1.5, it follows that $s' = \bar{s}$. QED.

We now prove the rigidity theorem of Kaehlerian hypersurfaces (Nomizu-Smyth [1]).

THEOREM 1.8. A connected Kaehlerian hypersurface M of a simply connected complex space form $\bar{M}^{n+1}(c)$ is rigid in $\bar{M}^{n+1}(c)$.

Proof. If $R = \bar{R}$, then M is totally geodesic and hence M is rigid. If $R \neq \bar{R}$ at some point of M, let x_0 be a point where the rank of M is maximal. Let $f, \bar{f} : M \longrightarrow \bar{M}$ be two Kaehlerian immersions. From Lemma 1.6, there exists a neighborhood $U(x_0)$ of x_0 and suitably choosen unit normal vector fields v and $\bar{v}$ on $U(x_0)$ with respect to f and $\bar{f}$ respectively such that $A = \bar{A}$ and $s = \bar{s}$ on $U(x_0)$. We now resort to local coordinates to show that $\bar{f} = \phi\cdot f$ on $U(x_0)$, ϕ being a holomorphic motion of $\bar{M}$. In fact, since the group of holomorphic isometries of $\bar{M}$ is transitive on the set of unitary frames, we may assume that without loss of generality that

$$f(x_0) = \bar{f}(x_0), \quad f_*(x_0) = \bar{f}_*(x_0), \quad v(x_0) = \bar{v}(x_0)$$

and prove that $f = \bar{f}$ in a neighborhood of x_0. Let $(x^1,\ldots,x^{2n})$ be a system of local coordinates on $U(x_0)$ and let $(u^1,\ldots,u^{2n+2})$ be a system of local coordinates on a neighborhood of $f(x_0)$ in $\bar{M}$ derived from a system of complex coordinates. We use the following ranges for indices:

$$i,j,k,l = 1,\dots,2n; \quad p,q,r,s = 1,\dots,2n+2.$$

Our notations (in the summation convention) will be

$$f^p(x) = u^p(f(x)), \quad f^p_i = \frac{\partial f^p}{\partial x^i}, \quad f^p_{ij} = \frac{\partial^2 f^p}{\partial x^i \partial x^j}, \quad \text{etc.},$$

$$f_k\left(\frac{\partial}{\partial x^i}\right) = f^p_i\left(\frac{\partial}{\partial u^p}\right), \quad v = v^r\frac{\partial}{\partial u^r}, \quad Jv = (Jv)^r\frac{\partial}{\partial u^r},$$

$$v^r_i = \frac{\partial v^r}{\partial x^i}, \quad v^r_{ij} = \frac{\partial^2 v^r}{\partial x^i \partial x^j}, \quad \text{etc., } \dots$$

The corresponding notation for $\bar{f}$ is then self-explanatory. We also use

$$h_{ij} = g\left(A\frac{\partial}{\partial x^i}, \frac{\partial}{\partial x^j}\right), \quad k_{ij} = g\left(JA\frac{\partial}{\partial x^i}, \frac{\partial}{\partial x^j}\right),$$

$$A\frac{\partial}{\partial x^i} = a^j_i\frac{\partial}{\partial x^j}, \quad s\left(\frac{\partial}{\partial x^i}\right) = s_i.$$

Note that we have $A = \bar{A}$ and $s = \bar{s}$ so that we do not need the corresponding notation for $\bar{f}$ here. The Christoffel symbols are denoted by Γ^i_{jk} for (x^i) and by Γ^p_{qr} for (u^p). We note that $(Jv)^r = -v^{r+n+1}$ and $(Jv)^{r+n+1} = v^r$. Then the Gauss and Weingarten formulas imply

$$\text{(I)} \qquad f^r_{ij} = -f^p_i f^q_j \Gamma^r_{pq} + f^r_k \Gamma^k_{ij} + h_{ij}v^r + k_{ij}(Jv)^r,$$

$$\text{(II)} \qquad v^r_i = -f^p_i v^q \Gamma^r_{pq} - a^j_i f^r_j + s_i(Jv)^r.$$

We denote the corresponding equations for the immersion $\bar{f}$ by $(\bar{\text{I}})$ and $(\overline{\text{II}})$. At x_0 we have

$$\text{(1.16)} \qquad \begin{aligned} &f^p(x_0) = \bar{f}^p(x_0), \quad f^p_i(x_0) = \bar{f}^p_i(x_0), \quad v^r(x_0) = \bar{v}^r(x_0), \\ &(Jv)^r(x_0) = (J\bar{v})^r(x_0). \end{aligned}$$

We wish to show that $f = \bar{f}$ in a neighborhood of x_0; since f^p and $\bar{f}^p$ are real analytic it sufficies to prove

(1.17) $$f^p_{ij}(x_0) = \bar{f}^p_{ij}(x_0),$$

(1.18) $$f^p_{ijk}(x_0) = \bar{f}^p_{ijk}(x_0),$$

and so on for all higher-order derivatives at x_0. (1.17) follows from (I), ($\bar{\text{I}}$), (1.16) and the equation $A = \bar{A}$ on $U(x_0)$, while

(1.19) $$v^r_i(x_0) = \bar{v}^r_i(x_0)$$

follows from (II), ($\overline{\text{II}}$), (1.16) and the equations $A = \bar{A}$ ans $s = \bar{s}$ on $U(x_0)$. Now f^r_{ijk} and $\bar{f}^r_{ijk}$ are obtained by differentiating (I) and ($\bar{\text{I}}$) and we deduce (1.18) from (1.16), (1.17), (1.19) and $A = \bar{A}$ on $U(x_0)$. In the same manner v^r_{ij} and $\bar{v}^r_{ij}$ are obtained by differentiating (II) and ($\overline{\text{II}}$). Using the previous equations together with the equations $A = \bar{A}$ and $s = \bar{s}$ on $U(x_0)$, we infer

$$v^r_{ij}(x_0) = \bar{v}^r_{ij}(x_0).$$

We can then easily obtain

$$f^p_{ijkl}(x_0) = \bar{f}^p_{ijkl}(x_0).$$

The equations for higher-order derivatives are obtained in the same fashion. Thus $f = \bar{f}$ in a neighborhood of x_0 and this completes the proof. QED.

Moreover, we have the following (Smyth [1])

THEOREM 1.9. If M_1 and M_2 are Kaehlerian hypersurfaces of $\bar{M}^{n+1}(c)$ which are simply connected complete Einstein manifolds with the same Ricci curvature, then they are holomorphically isometric.

Proof. Let f_i denote the immersion of M_i into $\bar{M}$ ($i = 1$ or 2).

Take any $x_i \in M_i$ and let v_i be any unit normal vector field of M_i with respect to f_i in a neighborhood $U(x_i)$ of x_i. Let A_i be the second fundamental tensor for f_i. We may choose an orthonormal basis $e^i_1,\ldots,e^i_n,Je^i_1,\ldots,Je^i_n$ of $T_{x_i}(M_i)$ with respect to which

$$A_i = \begin{pmatrix} \lambda I_n & 0 \\ 0 & -\lambda I_n \end{pmatrix}.$$

We consider the holomorphic linear isometry $F: T_{x_1}(M_1) \longrightarrow T_{x_2}(M_2)$ given by $F(e^1_j) = e^2_j$ and $F(Je^1_j) = Je^2_j$, $j = 1,\ldots,n$. Clearly $F \cdot A_1 = A_2 \cdot F$. On the other hand, from the assumption, the second fundamental form is parallel, and it follows from the Gauss equation that F maps the Riemannian curvature tensor of M_1 at x_1 into the Riemannian curvature tensor of M_2 at x_2. Since M_i is locally symmetric, the following lemma completes the proof. QED.

LEMMA 1.7. Let M_1 and M_2 be arbitrary simply connected complete symmetric Kaehlerian manifolds. Any holomorphic linear isometry $F: T_{x_1}(M_1) \longrightarrow T_{x_2}(M_2)$ which preserves the Riemannian curvature tensors extends uniquely to a holomorphic isometry of M_1 onto M_2.

Proof. F extends uniquely to an isometry $f: M_1 \longrightarrow M_2$ (cf. Kobayashi-Nomizu [;p.265]). Let y_1 be any point of M_1 and set $y_2 = f(y_1)$. Let c_1 be any differentiable curve joining x_1 to y_1, and set $c_2 = f \cdot c_1$. Since parallel displacement along c_2 corresponds, under f, to parallel displacement along c_1, and since the almost complex structures of M_1 and M_2 are parallel, f maps the almost complex structure of M_1 into that of M_2. Thus f is holomorphic. QED.

We now prove a global version of Theorem 1.7 (Nomizu-Smyth [1]).

THEOREM 1.10. (1) CP^n and the complex quadric are the only complete Kaehlerian hypersurfaces in CP^{n+1} which have parallel Ricci tensors.

(2) D^n (resp. C^n) is the only complete Kaehlerian hypersurface in D^{n+1} (resp. C^{n+1}) which has parallel Ricci tensor.

Proof. (1) Let $f: M \longrightarrow CP^{n+1}$ be the Kaehlerian immersion.

If M has parallel Ricci tensor, then M is Einstein by Theorem 1.6. Let $\hat{M}$ be the universal covering manifold of M and let π be the covering map. On $\hat{M}$ we take the Kaehlerian structure which makes π a holomorphic immersion. $\hat{M}$ is then simply connected and complete Einstein manifold. Moreover, $f\cdot\pi$ is a holomorphic isometric immersion of $\hat{M}$ in CP^{n+1}. From Theorems 1.7, 1.8 and 1.9 $\hat{M}$ immerses either onto a projective hyperplane or onto complex quadric in CP^{n+1}. In either case $f\cdot\pi(\hat{M})$ is a simply connected and since $f\cdot\pi$ is a covering map (see Theorem 4.6 in Kobayashi-Nomizu [1]), it is one-to-one. Hence π is one-to-one and therefore M is holomorphically isometric either to CP^n or to Q^n.

The same type of argument can be applied to (2). QED.

From Theorem 1.2 and Theorem 1.10 we have (Kon [7])

THEOREM 1.11. Let M be a compact Kaehlerian hypersurface of CP^{n+1}. If the scalar curvature of M is constant, then either M is CP^n or Q^n.

We next consider a complex n-dimensional Kaehlerian submanifold M with constant holomorphic sectional curvature k immersed in a complex space form $\bar{M}^m(c)$. First of all, the Gauss equation implies

$$\begin{aligned}(k-c)|A|^2 &= \sum_{b,i,j} g(A_b e_i, \bar{R}(e_i,e_j)A_b e_j - R(e_i,e_j)A_b e_j)\\ &= -\sum_{a,b,i,j}[g(A_b e_i, A_a e_i)g(A_a e_j, A_b e_j)\\ &\qquad - g(A_b e_i, A_a e_j)g(A_b e_j, A_a e_i)]\\ &= -\sum_{a,b}[(\mathrm{Tr}A_aA_b)^2 - \mathrm{Tr}(A_aA_b)^2] = -\sum_{a,b}(\mathrm{Tr}A_aA_b)^2,\end{aligned}$$

from which

$$\sum_{a,b}(\mathrm{Tr}A_aA_b)^2 = (c-k)|A|^2 = \frac{1}{n(n+1)}|A|^4. \tag{1.20}$$

From (1.8) and (1.20) we obtain

$$\frac{1}{2p}|A|^4 \leq \frac{1}{n(n+1)}|A|^4, \quad (p = m-n).$$

Therefore we have (O'Neill [1])

THEOREM 1.12. Let M be a complex n-dimensional Kaehlerian submanifold of a complex space form $\bar{M}^{n+p}(c)$. If M is of constant holomorphic sectional curvature k, and if $p < n(n+1)/2$, then M is totally geodesic and hence $c = k$.

From (1.6), (1.20) and Lemma 1.3 we have (Ogiue [3], [4])

THEOREM 1.13. Let M be a complex n-dimensional Kaehlerian submanifold of a complex space form $\bar{M}^{m}(c)$. If M is of constant holomorphic sectional curvature k, then

$$|\nabla A|^2 = n(n+1)(n+2)(c-k)(\tfrac{1}{2}c-k).$$

THEOREM 1.14. Let M be a complex n-dimensional Kaehlerian submanifold of a complex space form $\bar{M}^{m}(c)$ $(c > 0)$. If M is of constant holomorphic sectional curvature k, then either $c = k$, that is, M is totally geodesic, or $c \geq 2k$.

THEOREM 1.15. Under the same assumption as that of Theorem 1.13, ii the second fundamental form of M is parallel, then $c = k$ or $c = 2k$, the latter case arising only when $c > 0$.

2. ANTI-INVARIANT SUBMANIFOLDS OF KAEHLERIAN MANIFOLDS

Let $\bar{M}$ be a complex m-dimensional (real 2m-dimensional) almost Hermitian manifold with almost complex structure J and with Hermitian metric g. An n-dimensional Riemannian manifold M isometrically immersed in $\bar{M}$ is called an *anti-invariant submanifold* of $\bar{M}$ (or *totally real submanifold* of $\bar{M}$) if $JT_x(M) \subset T_x(M)^{\perp}$ for each point x of M. Then we have $m \geq n$.

In this section we study an n-dimensional anti-invariant submanifold M of a complex m-dimensional Kaehlerian manifold $\bar{M}$. We choose a local field of orthonormal frames $e_1,\dots,e_n;e_{n+1},\dots,e_m;e_{1*}=Je_1,\dots,$ $e_{n*}=Je_n;e_{(n+1)*}=Je_{n+1},\dots,e_{m*}=Je_m$ in $\bar{M}$ in such a way that, restricted to M, $e_1,\dots,e_n$ are tangent to M and hence the remaining vectors are normal to M. With respect to this frame field of $\bar{M}$, let $\omega^1,\dots,\omega^n;$ $\omega^{n+1},\dots,\omega^m;\omega^{1*},\dots,\omega^{n*};\omega^{(n+1)*},\dots,\omega^{m*}$ be the field of dual frames. Unless otherwise stated, we use the conventions that the ranges of indices are respectively:

$$i,j,k,l = 1,\dots,n; \quad a,b,c,d = n+1,\dots,m,1*,\dots,m*;$$
$$\alpha,\beta,\gamma = n+1,\dots,m; \quad \lambda,\mu,\nu = n+1,\dots,m,(n+1)*,\dots,m*.$$

Then we have

$$\omega^i_j + \omega^j_i = 0, \quad \omega^i_j = \omega^{i*}_{j*}, \quad \omega^{i*}_j = \omega^{j*}_i,$$

$$\omega^\alpha_\beta + \omega^\beta_\alpha = 0, \quad \omega^\alpha_\beta = \omega^{\alpha*}_{\beta*}, \quad \omega^{\alpha*}_\beta = \omega^{\beta*}_\alpha,$$

$$\omega^i_\alpha + \omega^\alpha_i = 0, \quad \omega^i_\alpha = \omega^{i*}_{\alpha*}, \quad \omega^{i*}_\alpha = \omega^{\alpha*}_i.$$

From $\omega^a_i = h^a_{ij}\omega^j$ we easily see that

$$h^i_{jk} = h^j_{ik} = h^k_{ij}, \tag{2.1}$$

where we write h^{i*}_{jk} as h^{i}_{jk}. That is the second fundamental form A of M satisfies

$$A_{JX}Y = A_{JY}X$$

for any vector fields X and Y on M.

We assume that the ambient manifold $\bar{M}$ is of constant holomorphic sectional curvature c. Then the Gauss equation of M is given by

$$(2.2) \qquad R^{i}_{jkl} = \tfrac{1}{4}c(\delta_{ik}\delta_{jl} - \delta_{il}\delta_{jk}) + \sum_{a}(h^{a}_{ik}h^{a}_{jl} - h^{a}_{il}h^{a}_{jk}).$$

From (2.2) we see that the Ricci tensor R_{ij} and the scalar curvature r of M are given respectively by

$$(2.3) \qquad R_{ij} = \tfrac{1}{4}(n-1)c\delta_{ij} + \sum_{a,k}(h^{a}_{kk}h^{a}_{ij} - h^{a}_{ik}h^{a}_{jk}),$$

$$(2.4) \qquad r = \tfrac{1}{4}n(n-1)c + \sum_{a,i,j}(h^{a}_{ii}h^{a}_{jj} - h^{a}_{ij}h^{a}_{ij}).$$

From (2.2), (2.3) and (2.4) we have

PROPOSITION 2.1. Let M be an n-dimensional anti-invariant minimal submanifold of a complex space form $\bar{M}^{m}(c)$. Then M is totally geodesic if and only if M satisfies one of the following conditions:

(1) M is of constant curvature $\frac{1}{4}c$;

(2) $S = \frac{1}{4}(n-1)cg$;

(3) $r = \frac{1}{4}n(n-1)c$.

Let M be an n-dimensional anti-invariant submanifold of a complex m-dimensional Kaehlerian manifold $\bar{M}$. Then we have the decomposition:

$$T_x(M)^{\perp} = JT_x(M) \oplus N_x(M),$$

where $N_x(M)$ is the orthogonal complement of $JT_x(M)$ in $T_x(M)^{\perp}$. We see that the space $N_x(M)$ is invariant under the action of J, that is, $JN_x(M) = N_x(M)$. For any vector field V normal to M we put

$$JV = tV + fV,$$

where tV is the tangential part of JV and fV the normal part of JV. We then have

$$tfV = 0, \quad f^2V = -V - JtV, \quad tJX = -X, \quad fJX = 0.$$

Therefore we have

$$f^3 + f = 0,$$

which shows that, if f does not vanish, it defines an f-structure in the normal bundle (see Chapter VII). Using the Gauss and Weingarten formulas, we obtain

$$(\nabla_X f)V = -B(X,tV) - JA_V X,$$

where we have put $(\nabla_X f)V = D_X(fV) - f(D_X V)$. If $\nabla_X f = 0$ for all tangent vector field X, then the f-structure f in the normal bundle is said to be *parallel*.

LEMMA 2.1. Let M be an n-dimensional anti-invariant submanifold of a complex m-dimensional Kaehlerian manifold $\bar{M}$. If the f-structure in the normal bundle is parallel, then

$$A_V = 0 \quad \text{for } V \in N_x(M).$$

Proof. If $V \in N_x(M)$, then $tV = 0$. Thus we have $JA_V X = 0$ and hence $A_V = 0$. QED.

From (3.9) of Chapter II and the fact that $h^a_{ijk} = h^a_{ikj}$ when the ambient manifold is of constant holomorphic sectional curvature, we have

LEMMA 2.2. Let M be an n-dimensional anti-invariant submanifold of a complex space form $\bar{M}^m(c)$. Then

$$\sum_{a,i,j} h^a_{ij}\Delta h^a_{ij} = \sum_{a,i,j,k} h^a_{ij}h^a_{kkij} + \tfrac{1}{4}c\sum_a[\mathrm{Tr}A_a^2 - (\mathrm{Tr}A_a)^2]$$
$$+ \tfrac{1}{4}c\sum_t[\mathrm{Tr}A_t^2 - (\mathrm{Tr}A_t)^2] + \sum_{a,b}[\mathrm{Tr}(A_aA_b - A_bA_a)^2$$
$$- (\mathrm{Tr}A_aA_b)^2 - \mathrm{Tr}A_b\mathrm{Tr}A_a^2A_b],$$

where we have put $A_t = A_{t*}$.

We now put

$$S_{ab} = \mathrm{Tr}A_aA_b, \quad S_a = S_{aa}, \quad S = \sum_a S_a,$$

so that S_{ab} is a symmetric $(2m-n, 2m-n)$-matrix and can be assumed to be diagonal for a suitable frame. S is the square of the length of the second fundamental form of M.

We prove the following theorem (Ludden-Okumura-Yano [2]).

THEOREM 2.1. Let M be an n-dimensional compact anti-invariant minimal submanifold of a complex space form $\bar{M}^m(c)$ $(c > 0)$. If $S \leq nqc/4(2q-1)$, where $q = 2m-n$, then M is totally geodesic.

Proof. From Lemma 2.2 and Lemma 5.1 of Chapter II we find

$$|\nabla A|^2 - \tfrac{1}{2}\Delta S = -\sum_{a,b}\mathrm{Tr}(A_aA_b - A_bA_a)^2 + \sum_a S_a^2 - \tfrac{1}{4}ncS - \tfrac{1}{4}c\sum_t \mathrm{Tr}A_t^2$$
$$\leq [(2 - 1/q)S - \tfrac{1}{4}nc]S - \frac{1}{q}\sum_{a\,b}(S_a - S_b)^2 - \tfrac{1}{4}c\sum_t \mathrm{Tr}A_t^2.$$

From the assumption we have

$$\sum_{a\,b}(S_a - S_b)^2 = 0, \qquad \sum_t \mathrm{Tr}A_t^2 = 0.$$

Therefore we see that $S_a = S_b$ for all a, b, and $A_t = 0$, which means that $S_t = 0$. Consequently, we obtain $S_a = 0$ for all a, and hence M is totally geodesic. QED.

THEOREM 2.2. Let M be an n-dimensional compact anti-invariant submanifold of a complex space form $\bar{M}^n(c)$. If M is minimal, then

$$0 \leq \int_M |\nabla A|^2 * 1 \leq \int_M [(2-1/n)S - \tfrac{1}{4}(n+1)c] * 1.$$

Proof. From Lemma 5.1 of Chapter II we have

$$\begin{aligned} (2.5) \qquad & -\sum_{t,s} \mathrm{Tr}(A_t A_s - A_s A_t)^2 + \sum_t S_t^2 - \tfrac{1}{4}(n+1)cS \\ & \leq 2\sum_{t \neq s} S_t S_s + \sum_t S_t^2 - \tfrac{1}{4}(n+1)cS \\ & = [(2-1/n)S - \tfrac{1}{4}(n+1)c]S - \frac{1}{n}\sum_{t>s}(S_t - S_s)^2. \end{aligned}$$

On the other hand, Lemma 2.2 implies

$$|\nabla A|^2 - \tfrac{1}{2}\Delta S = -\sum_{t,s} \mathrm{Tr}(A_t A_s - A_s A_t)^2 + \sum_t S_t^2 - \tfrac{1}{4}(n+1)cS.$$

Therefore, (2.5) gives our inequality. QED.

Theorem 2.2 was given by Chen-Ogiue [2]. As an application of Theorem 2.2 we have

THEOREM 2.3. Let M be an n-dimensional compact anti-invariant minimal submanifold of a complex space form $\bar{M}^n(c)$. If S satisfies $S < n(n+1)c/4(2n-1)$, then M is totally geodesic.

In Theorem 2.2 and 2.3 we study the case of $m = n$. When $m > n$, by Lemma 2.1, if we assume that the f-structure in the normal bundle is parallel, then we have the similar results as those of Theorem 2.2 and Theorem 2.3 (see Yano-Kon [3]).

In the next place we consider the case of $S = n(n+1)c/4(2n-1)$. Without loss of generality, we assume that $c = 4$. First of all we have

THEOREM 2.4. Let M be an n-dimensional ($n > 1$) anti-invariant minimal submanifold of a complex space form $\bar{M}^n(4)$. If $S = n(n+1)/(2n-1)$, then $n = 2$ and M is a flat surface of $\bar{M}^2(4)$. Moreover, with respect to an adapted dual orthonormal frame field $\omega^1, \omega^2, \omega^{1*}, \omega^{2*}$ in $\bar{M}^2(4)$, the connection form (ω_B^A) of $\bar{M}^2(4)$, restricted to M, is given by

$$\begin{pmatrix} 0 & 0 & -\lambda\omega^2 & -\lambda\omega^1 \\ 0 & 0 & -\lambda\omega^1 & \lambda\omega^2 \\ \lambda\omega^2 & \lambda\omega^1 & 0 & 0 \\ \lambda\omega^1 & -\lambda\omega^2 & 0 & 0 \end{pmatrix}, \qquad \lambda = 1/\sqrt{2} .$$

Proof. From the assumption the second fundamental form of M is parallel and (2.5) implies

$$\sum_{t,s} (S_t - S_s)^2 = 0, \qquad -\mathrm{Tr}(A_t A_s - A_s A_t)^2 = 2\mathrm{Tr}A_t^2 \mathrm{Tr}A_s^2.$$

In view of Lemma 5.1 of Chapter II we may assume that $A_t = 0$ for $t = 3,\ldots,n$, which means that $S_t = 0$ for $t = 3,\ldots,n$. On the other hand, we have $S_t = S_s$ for all t, s. Therefore we must have $n = 2$. Thus Lemma 5.1 of Chapter II implies

$$A_1 = \lambda \begin{pmatrix} 0 & 1 \\ 1 & 0 \end{pmatrix}, \qquad A_2 = \lambda \begin{pmatrix} 1 & 0 \\ 0 & -1 \end{pmatrix},$$

because of $h_{12}^1 = \lambda = h_{11}^2$. Since $S = 2$, we have $2\lambda^2 = 1$ and we may assume that $\lambda = (1/2)^{1/2}$. Then, from the Gauss equation, we see that M is flat. Moreover, we easily see the following:

$$\omega_1^{1*} = \lambda\omega^2, \qquad \omega_2^{1*} = \omega_1^{2*} = \lambda\omega^1,$$

$$\omega_2^{2*} = -\lambda\omega^2, \qquad \omega_1^2 = \omega_{1*}^{2*} = 0.$$

These prove our assertion. QED.

Example 2.1. Let S^5 be a 5-dimensional unit sphere with standard Sasakian structure (ϕ,ξ,η,g). The integral curves of the structure vector field ξ are great circles S^1 in S^5 which are the fibres of the standard fibration $\pi : S^5 \longrightarrow CP^2$ onto complex projective space CP^2 of complex dimension 2 and of constant holomorphic sectional curvature 4. Let $T = S^1 \times S^1$ be maximal torus which is imbedded in S^5 as an anti-invariant submanifold normal to the structure vector field ξ. The imbedding $X : T \longrightarrow S^5$ is given by

$$X = \frac{1}{\sqrt{3}}(\cos u^1, \sin u^1, \cos u^2, \sin u^2, \cos u^3, \sin u^3),$$

where $u^3 = -u^1 - u^2$ in C^3. We consider the following diagram:

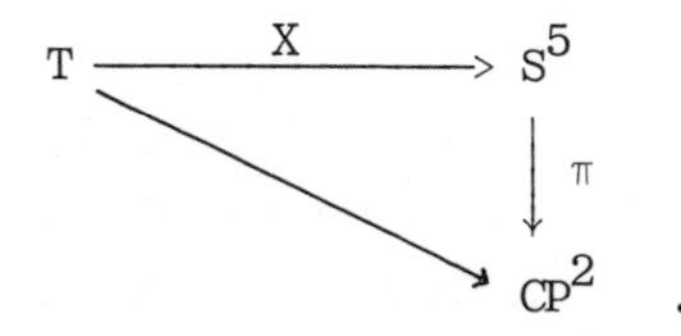

We easily see that $\pi|X(T)$ is one-to-one. Consequently T is imbedded in CP^2 by $\pi\cdot X$. Thus $S = 2$. (For the detail, see §3 of Chapter VI.)

From this example and Theorem 2.4 we obtain (Ludden-Okumura-Yano [1])

THEOREM 2.5. Let M be an n-dimensional $(n>1)$ compact anti-invariant minimal submanifold of CP^n. If $S = n(n+1)/(2n-1)$, then M is $S^1 \times S^1$ in CP^2.

LEMMA 2.3. Let M be an n-dimensional anti-invariant submanifold of a complex n-dimensional Kaehlerian manifold $\bar{M}$. Then M is flat if and only if the normal connection of M is flat.

Proof. We have

$$\Omega^{i*}_{j*} = d\omega^{i*}_{j*} + \omega^{i*}_{k*} \wedge \omega^{k*}_{j*} = d\omega^i_j + \omega^i_k \wedge \omega^k_j = \Omega^i_j,$$

which shows that $R^{i*}_{j*kl} = R^i_{jkl}$. Thus we have our assertion. QED.

LEMMA 2.4. Let M be an n-dimensional anti-invariant submanifold of a complex n-dimensional Kaehlerian manifold $\bar{M}$. If the second fundamental forms of M are commutative, i.e., $A_aA_b = A_bA_a$ for all a and b, then we can choose an orthonormal basis $\{e_t\}$ for which A_t satisfies $A_te_t = \lambda_te_t$, $A_te_s = 0$ $(t \neq s)$ for each $t = 1,\ldots,n$, that is, $h^t_{ij} = 0$ unless $t = i = j$.

Proof. If $A_aA_b = A_bA_a$, we can choose an orthonormal basis $e_1,\ldots,e_n$ for which all A_a are simultaneously diagonal, i.e., $h^a_{ij} = 0$ when $i \neq j$, that is, $h^t_{ij} = 0$ when $i \neq j$. Thus we have $h^t_{ij} = 0$ unless $t = i = j$ by $h^t_{ij} = h^i_{tj}$. QED.

PROPOSITION 2.2. Let M be an n-dimensional $(n > 1)$ anti-invariant, totally umbilical submanifold of a complex n-dimensional Kaehlerian manifold $\bar{M}$. Then M is totally geodesic.

Proof. From the assumption we have $h^t_{ij} = \delta_{ij}(\mathrm{Tr}A_t)/n$. Therefore the second fundamental forms of M are commutative. Thus Lemma 2.4 implies that $h^t_{ij} = 0$ unless $t = i = j$. On the other hand, we have $h^t_{ij} = \lambda_t\delta_{ij}/n$. Putting $t \neq i = j$, we find $\lambda_t = 0$ and hence M is totally geodesic. QED.

LEMMA 2.5. Let M be an n-dimensional anti-invariant submanifold of a complex space form $\bar{M}^n(c)$. Then M is of constant curvature $\frac{1}{4}c$ if and only if the second fundamental forms of M are commutative.

Proof. From (2.2) we have

$$R^i_{jkl} = \tfrac{1}{4}c(\delta_{ik}\delta_{jl} - \delta_{il}\delta_{jk}) + \sum_t(h^t_{ik}h^t_{jl} - h^t_{il}h^t_{jk})$$
$$= \tfrac{1}{4}c(\delta_{ik}\delta_{jl} - \delta_{il}\delta_{jk}) + \sum_t(h^i_{tk}h^j_{tl} - h^i_{tl}h^j_{tk}),$$

which proves our assertion. QED.

LEMMA 2.6. Let M be an n-dimensional anti-invariant submanifold of a complex m-dimensional Kaehlerian manifold $\bar{M}$. Then

$$\sum_{t,s} \mathrm{Tr}A_t^2A_s^2 = \sum_{t,s} (\mathrm{Tr}A_tA_s)^2.$$

Proof. $$\sum_{t,s} \mathrm{Tr}A_t^2A_s^2 = \sum_{t,s,i,j,k,l} h_{kl}^t h_{li}^t h_{ij}^s h_{jk}^s$$
$$= \sum_{t,s,i,j,k,l} h_{tl}^k h_{lt}^i h_{sj}^i h_{js}^k = \sum_{k,i} (\mathrm{Tr}A_kA_i)^2.$$

LEMMA 2.7. Let M be an n-dimensional anti-unvariant submanifold of constant curvature k of a complex space form $\bar{M}^n(c)$. Then

$$(2.6)\quad (\tfrac{1}{4}c-k)\sum_t[\mathrm{Tr}A_t^2 - (\mathrm{Tr}A_t)^2] = \sum_{t,s}[\mathrm{Tr}A_t^2A_s^2 - \mathrm{Tr}(A_tA_s)^2],$$

$$(2.7)\quad (n-1)(\tfrac{1}{4}c-k)|A|^2 = \sum_{t,s}[\mathrm{Tr}A_t^2A_s^2 - \mathrm{Tr}A_s\mathrm{Tr}A_t^2A_s].$$

Proof. From the assumption we have

$$(2.8)\quad (\tfrac{1}{4}c-k)(\delta_{ik}\delta_{jl} - \delta_{il}\delta_{jk}) = \sum_t(h_{il}^t h_{jk}^t - h_{ik}^t h_{jl}^t).$$

Multiplying the both sides of this equation by $\Sigma_s h_{il}^t h_{jk}^t$, summing up with respect to i, j, k and l and using Lemma 2.6, we have (2.6).

Putting i = k in (2.8) and summing up with respect to i, we have

$$(2.9)\quad (n-1)(\tfrac{1}{4}c-k)\delta_{ji} = \sum_{t,i}(h_{il}^t h_{ij}^t - h_{ii}^t h_{jl}^t).$$

Multiplying the both sides of (2.9) by $\Sigma_s h_{jk}^s h_{kl}^s$ and summing up with respect to j, k and l we obtain (2.7). QED.

THEOREM 2.6. Let M be an n-dimensional anti-invariant submanifold with parallel mean curvature vector of a complex space form $\bar{M}^n(c)$. If M is of constant curvature k, then

$$|\nabla A|^2 = -k\sum_t[(n+1)\mathrm{Tr}A_t^2 - 2(\mathrm{Tr}A_t)^2].$$

Proof. By the assumption we easily see that the square of the length of the second fundamental form of M is constant. Thus Lemma 2.2 implies

$$(2.10) \quad |\nabla A|^2 = -\sum_t [\tfrac{1}{4}(n+1)c\mathrm{Tr}A_t^2 - \tfrac{1}{2}(\mathrm{Tr}A_t)^2]$$

$$- \sum_{t,s} [\mathrm{Tr}(A_tA_s - \mathrm{Tr}A_sA_t)^2 - (\mathrm{Tr}A_tA_s)^2 + \mathrm{Tr}A_a\mathrm{Tr}A_t^2A_s].$$

Substituting (2.6) and (2.7) into (2.10) and using Lemma 2.6, we have our equation. QED.

THEOREM 2.7. Let M be an n-dimensional ($n > 1$) anti-invariant submanifold with parallel mean curvature vector of a complex space form $\bar{M}^n(c)$. If M is of constant curvature k, and if $c \geq 4k$, then $k \leq 0$ or M is totally geodesic and $c = 4k$.

Proof. From (2.9) we have

$$n(n-1)(\tfrac{1}{4}c-k) = \sum_t [\mathrm{Tr}A_t^2 - (\mathrm{Tr}A_t)^2].$$

Since $c \geq 4k$, we have $\Sigma_t \mathrm{Tr}A_t^2 \geq \Sigma_t (\mathrm{Tr}A_t)^2$. If $k > 0$, then Theorem 2.6 implies

$$0 = (n-1)\sum_t \mathrm{Tr}A_t^2 + 2\sum_t [\mathrm{Tr}A_t^2 - (\mathrm{Tr}A_t)^2],$$

from which we have $|A|^2 = 0$ and hence M is totally geodesic. In this case $c = 4k$. Except for this possibility we have $k \leq 0$. QED.

THEOREM 2.8. Let M be an n-dimensional ($n > 1$) anti-invariant submanifold of a complex space form $\bar{M}^n(c)$ and M be with parallel second fundamental form and of constant curvature k. If $c \geq 4k$, then either M is totally geodesic ($c = 4k$) or M is flat ($k = 0$).

Theorems 2.6, 2.7 and 2.8 are given by Yano-Kon [2]. In case of $\dim \bar{M} = m > n$, see Yano-Kon [3].

We now give some examples of anti-invariant submanifolds of complex space forms.

Example 2.2. Let RP^n be an n-dimensional real projective space. By taking real and complex homogeneous coordinates properly in RP^n

and CP^m, respectively, the imbedding of RP^n into CP^m is given as follows. Let $(x^0,\ldots,x^n)$ and $(z^0,\ldots,z^m)$ be the homogeneous coordinates of RP^n and CP^m, respectively. Our imbedding is given by natural imbeding of R^{n+1} into C^{m+1}, i.e., $(x^0,\ldots,x^n) \longrightarrow (x^0,\ldots,x^n,0,\ldots,0)$. In this case RP^n is of constant curvature $\frac{1}{4}c$.

Example 2.3. Let $S^1(r_i) = \{z_i \in C : |z_i|^2 = r_i^2\}$, $i = 1,\ldots,n$, be circles of radius r_i. We consider

$$M^n = S^1(r_1) \times \ldots \times S^1(r_n)$$

in C^n, which is obviously flat. The position vector X of M^n in C^n has components given by

$$X = (r_1\cos u^1, r_1\cos u^1,\ldots,r_n\cos u^n, r_n\cos u^n).$$

Putting $X_i = \partial_i X = \partial X/\partial u^i$, we have

$$X_i = r_i(0,\ldots,0,-\sin u^i,\cos u^i,0,\ldots,0), \quad i = 1,\ldots,n.$$

On the other hand, we can take

$$V_i = -(0,\ldots,0,\cos u^i,\sin u^i,0,\ldots,o), \quad i = 1,\ldots,n,$$

as orthonormal vectors normal to M^n. Then we obtain

$$JX_i = r_iV_i, \quad i = 1,\ldots,n.$$

Consequently M^n is a flat anti-invariant submanifold in C^n and it has parallel mean curvature vector and flat normal connection. Moreover, we see that M^n has parallel and commutative second fundamental forms. On the other hand, C^n is totally geodesic in C^m $(m > n)$. Thus M^n is an anti-invariant submanifold of C^m with parallel f-structure in the normal bundle.

Example 2.4. As an example similar to Example 2.3, we consider

$$M^n = S^1(r_1) \times \ldots \times S^1(r_p) \times R^{n-p}, \quad 1 \leq p < n.$$

Obviously R^{n-p} is a totally geodesic anti-invariant submanifold of CP^{n-p}. Thus M^n is a flat anti-invariant submanifold of C^n. Moreover, M^n is an anti-invariant submanifold with parallel and commutative second fundamental forms and with parallel f-structure of C^m $(m > n)$.

THEOREM 2.9. Let M be an n-dimensional $(n > 1)$ complete anti-invariant submanifold of C^n and M be with parallel mean curvature vector and with commutative second fundamental forms. Then either M is R^n, or a pythagorean product of the form

$$S^1(r_1) \times \ldots \times S^1(r_n)$$

or a pythagorean product of the form

$$S^1(r_1) \times \ldots \times S^1(r_p) \times R^{n-p}, \quad 1 \leq p < n.$$

Proof. By the assumption and Lemma 2.5 we see that M is flat and hence the normal connection of M is flat. Moreover, Theorem 2.6 implies that the second fundamental form of M is parallel. For a frame field, chosen as in Lemma 2.4, we define the following distributions:

$$T_t(x) = \{X \varepsilon T_x(M) : A_t X = \lambda_t X\} \quad \text{for} \quad _t \neq 0,$$

$$T_0(x) = \{X \varepsilon T_x(M) : A_t = 0, 1 \leq t \leq n\}.$$

Then each T_t is of dimension 1 and we have the decomposition :

$$T_x(M) = T_1(x) \oplus \ldots\ldots \oplus T_p(x) \oplus T_0(M) \text{ (direct sum)}.$$

Moreover, since the normal connection of M is flat and the second fundamental form of M is parallel, we see that T_t is involutive and totally geodesic in M. Thus we conclude that

$$M = M_1 \times \ldots \times M_p \times M_0 \quad \text{(Riemannian product)},$$

where M_t and M_0 are the maximal integral submanifolds of T_t and T_0 respectively. Since M is complete, we have our assertion (see Theorem 3 of Yano-Ishihara [3]). QED.

THEOREM 2.10. Let M be an n-dimensional $(n > 1)$ complete anti-invariant submanifold of a simply connected complete complex space form $\bar{M}^n(c)$ and M be with parallel and commutative second fundamental form. If M is not totally geodesic, then M is a pythagorean product of the form

$$S^1(r_1) \times \ldots \times S^1(r_n) \quad \text{in } C^n$$

or a pythagorean product of the form

$$S^1(r_1) \times \ldots \times S^1(r_p) \times R^{n-p} \quad \text{in } C^n, \quad 1 \leq p < n.$$

Proof. From Lemma 2.4 and Theorem 2.6 we see that M is flat and hence $c = 0$ by Lemma 2.5. Thus $\bar{M}^n(c)$ is C^n. Therefore our theorem follows from Theorem 2.9. QED.

In Theorem 2.6, if M is minimal, then

$$|\nabla A|^2 = -(n+1)k|A|^2.$$

In the following we prove that an n-dimensional anti-invariant minimal submanifold M of constant curvature k of $\bar{M}^n(c)$ is totally geodesic or flat, which was proved by Ejiri [1].

We need the following algebraic lemma.

LEMMA 2.8. Let T be a symmetric 3-linear map of $R^n \times R^n \times R^n$ into R such that

$$A[g(X,Z)g(Y,W) - g(X,W)g(Y,Z)] + \sum_{m=1}^{n} T(X,Z,f_m)T(Y,W,f_m) - \sum_{m=1}^{n} T(X,W,f_m)T(Y,Z,f_m) = 0 \quad \text{and } A \geq 0,$$

$$\sum_{m=1}^{n} T(X,f_m,f_m) = 0,$$

where g is the Euclidean metric of R^n and $f_1,\ldots,f_n$ is an orthonormal basis. If we choose an orthonormal basis $e_1,\ldots,e_n$, such that each e_i is a maximum point of the cubic function $T(X,X,X)$, restricted to $\{X \in R^n : |X| = 1, \text{ and } X \text{ is orthogonal to } e_1,\ldots,e_{i-1}\}$, then T has the following expression:

$$T(e_a,e_a,e_a) = (n-a)\left[\frac{A}{(n-a+1)} + \cdots + \frac{A}{(n-a+1)\ldots n}\right]^{1/2},$$

$$T(e_a,e_{i_a},e_{j_a}) = -\left[\frac{A}{(n-a+1)} + \cdots + \frac{A}{(n-a+1)\ldots n}\right]\delta_{i_a j_a},$$

where $1 \leq a \leq n$ and $a \leq i_a, j_a \leq n$ unless $i_a = j_a = a$.

THEOREM 2.11. Let M be an n-dimensional anti-invariant minimal submanifold of constant curvature k of a complex n-dimensional space form $\bar{M}^n(c)$. Then M is totally geodesic or flat (k = 0).

Proof. We put $T = -JB$. Then T is a symmetric tensor field of type (1,2) on M and satisfies

$$g(T(X,Y),Z) = g(T(X,Z),Y).$$

Moreover, equations of Gauss and Codazzi are given respectively by

$$(c-k)[g(X,Z)g(Y,W) - g(X,W)g(Y,Z)] + g(T(X,Z),T(Y,W)) - g(T(X,W),T(Y,Z)) = 0,$$

$$(\nabla_X T)(Y,Z) = (\nabla_Y T)(X,Z).$$

Since M is minimal, T satisfies the condition of Lemma 2.8 for A = c - k. We may assume that M is not totally geodesic, i.e., $c \neq k$ and hence $A \neq 0$. We easily obtain a local field of orthonormal frames $e_1,\ldots,e_n$ such that the Lemma 2.8 holds. We denote by ω^i_j the Levi-Civita connection with respect to $e_1,\ldots,e_n$. Using the Codazzi equation, we have

$$-(A/n)^{1/2}\sum_{i=1}^n \omega^i_a(e_1)e_i - \sum_{i=1}^n \omega^i_a(e_1)T(e_i,e_1) - \sum_{i=1}^n \omega^i_a(e_1)T(e_a,e_i)$$
$$-(n-1)(A/n)^{1/2}\sum_{i=1}^n \omega^i_1(e_a)e_i + 2\sum_{i=1}^n \omega^i_1(e_a)T(e_i,e_1) = 0,$$

for all $a \neq 1$. Taking the innerproduct of it and e_1, we obtain $\omega^1_a(e_1) = 0$. This, together with the innerproduct of the above and e_b $(b \neq 1)$, implies $\omega^a_1(e_b) = 0$. As a result, e_1 is a parallel vector field on M. Thus M is flat. QED.

3. CR SUBMANIFOLDS OF KAEHLERIAN MANIFOLDS

Let $\bar{M}$ be a complex m-dimensional (real 2m-dimensional) Kaehlerian manifold with almost complex structure J and with Kaehlerian metric g. Let M be a real n-dimensional Riemannian manifold isometrically immersed in $\bar{M}$. We denote by the same g the Riemannian metric tensor field induced on M. from that of $\bar{M}$.

For any vector field X tangent to M, we put

$$JX = Px + FX, \tag{3.1}$$

where PX is the tangential part of JX and FX the normal part of JX. Then P is an endomorphism on the tangent bundle T(M) and F is a normal bundle valued 1-form on the tangent bundle T(M).

For any vector field V normal to M, we put

$$JV = tV + fV, \tag{3.2}$$

where tV is the tangential part of JV and fV the normal part of JV. Clearly, P is skew-symmetric on T(M) and f is skew-symmetric on $T(M)^{\perp}$. From (3.1) and (3.2) we have

$$g(FX,V) + g(X,tV) = 0, \tag{3.3}$$

which gives the relation between F and t. We also have

$$P^2 = -I - tF, \qquad FP + fF = 0, \tag{3.4}$$

$$Pt + tf = 0, \qquad f^2 = -I - Ft. \tag{3.5}$$

We define the covariant derivative $\nabla_X P$ of P by $(\nabla_X P)Y = \nabla_X(PY) - P\nabla_X Y$ and the covariant derivative $\nabla_X F$ of F by $(\nabla_X F)Y = D_X(FY) - F\nabla_X Y$.

Similarly, we define the covariant derivatives $\nabla_X t$ of t and $\nabla_X f$ of f by $(\nabla_X t)V = \nabla_X(tV) - tD_X V$ and $(\nabla_X f)V = D_X(fV) - fD_X V$, respectively. Then, from the Gauss and Weingarten formulas we have

$$tB(X,Y) = fB(X,Y) = (\nabla_X P)Y - A_{FY}X + B(X,PY) + (\nabla_X F)Y.$$

Comparing the tangential and normal parts of the both sides of this equation, we find

$$(\nabla_X P)Y = A_{FY}X + tB(X,Y), \tag{3.6}$$

$$(\nabla_X F)Y = -B(X,PY) + fB(X,Y). \tag{3.7}$$

Similarly, we have

$$-PA_V X - FA_V X = (\nabla_X t)V - A_{fV}X + B(X,tV) + (\nabla_X f)V,$$

from which

$$(\nabla_X t)V = A_{fV}X - PA_V X, \tag{3.8}$$

$$(\nabla_X f)V = -FA_V X - B(X,tV). \tag{3.9}$$

Suppose now that the ambient manifold $\bar{M}$ is of constant holomorphic sectional curvature c. Then the curvature tensor R of M is given by

$$R(X,Y)Z = \tfrac{1}{4}c[g(Y,Z)X-g(X,Z)Y+g(JY,Z)JX-g(JX,Z)JY+2g(X,JY)JZ]$$
$$+ A_{B(Y,Z)}X - A_{B(X,Z)}Y + (\nabla_Y B)(X,Z) - (\nabla_X B)(Y,Z).$$

Comparing the tangential and normal parts of the both sides of this equation, we have the following equations of Gauss and Codazzi respectively:

$$(3.10)\quad R(X,Y)Z = \tfrac{1}{4}c[g(Y,Z)X-g(X,Z)Y+g(PY,Z)PX-g(PX,Z)PY+2g(X,PY)PZ] + A_{B(Y,Z)}X - A_{B(X,Z)}Y,$$

$$(3.11)\quad (\nabla_X B)(Y,Z) - (\nabla_Y B)(X,Z) = \tfrac{1}{4}c[g(PY,Z)FX - g(PX,Z)FY + 2g(X,PY)FZ].$$

Moreover, equation of Ricci is given by

$$(3.12)\quad g(R^{\perp}(X,Y)U,V) + g([A_V,A_U]X,Y) = \tfrac{1}{4}c[g(FY,U)g(FX,V)-g(FX,U)g(FY,V)+2g(X,PY)g(fU,V)].$$

Definition. Let $\bar{M}$ be a Kaehlerian manifold with almost complex structure J. A submanifold M of $\bar{M}$ is called a CR *submanifold* of $\bar{M}$ if there exists a differentiable distribution $D : x \longrightarrow D_x \in T_x(M)$ on M satisfying the following conditions:

(1) is invariant, i.e., $JD_x = D_x$ for each $x \in M$, and

(2) the complementary orthogonal distribution $D^{\perp}: x \longrightarrow D_x^{\perp} \in T_x(M)$ is anti-invariant, i.e., $JD_x^{\perp} \in T_x(M)^{\perp}$ for each $x \in M$.

In the sequel, we put $\dim \bar{M} = 2m$, $\dim M = n$, $\dim D = h$, $\dim D^{\perp} = q$ and codim $M = 2m-n = p$. If $q = 0$, then a CR submanifold M is a Kaehlerian submanifold of $\bar{M}$, and if $h = 0$, then M is an anti-invariant submanifold of $\bar{M}$. If $p = q$, then a CR submanifold M is called a *generic submanifold* of $\bar{M}$. If $h > 0$ and $q > 0$, then a CR submanifold M is said to be *non-trivial* (or *proper*).

Remark. Sometimes, the definitions of CR submanifolds, generic submanifolds and anti-invariant submanifolds (totally real submanifolds) are respectively given as follows (cf. Wells [1], [2]).

Let M be a real n-dimensional submanifold of a complex m-dimensional complex manifold $\bar{M}$. We define

$$H_x(M) = T_x(M) \cap JT_x(M),$$

as holomorphic tangent space to M at x. If $\dim_C H_x(M)$ is constant on M, then M is called a CR submanifold. It is well known that

$$\max(n-m,0) \leq \dim_c H_x(M) \leq \tfrac{1}{2}n.$$

If $\dim_c H_x(M) = \max(n-m,0)$ at each point of M, then M is called a generic submanifold. Moreover, if $\dim_c H_x(M) = 0$, then M is called a totally real submanifold.

Remark. We now state the result which justifies the name of CR submanifold. Let M be a differentiable manifold and $T(M)^C$ its complexified tangent bundle. A CR *structure* on M is a complex subbundle H of $T(M)^C$ such that $H_x \cap \bar{H}_x = \{0\}$ and H is involutive, i.e., for complex vector fields X and Y in H, [X,Y] is in H. It is well known that on a CR manifold there exists a (real) distribution D and a field of endomorphism $p : D \longrightarrow D$ such that $p^2 = -I_D$. D is just $\mathrm{Re}(H \oplus \bar{H})$ and $H_x = \{X - \sqrt{-1}pX : X \in D_x\}$. Blair-Chen [1] prove the following:

Let M be a CR submanifold of a Haermitian manifold $\bar{M}$. If M is non-trivial, then M is a CR manifold.

We next give some characterizations of CR submanifolds of Kaehlerian manifolds. First of all, we prove the following (Yano-Kon [12])

THEOREM 3.1. In order for a submanifold M of a Kaehlerian manifold $\bar{M}$ to be a CR submanifold, it is necessary and sufficient that $FP = 0$.

Proof. Suppose that M is a CR submanifold of $\bar{M}$. We denote by l and $l^\perp$ the projection operators on D_x and $D_x^\perp$ respectively. Then

$$l + l^\perp = I, \quad l^2 = l, \quad l^{\perp 2} = l^\perp, \quad l l^\perp = l^\perp l = 0.$$

From (3.1) we have $l^\perp P l = 0$, $Fl = 0$ and $Pl = P$, from which and the second equation of (3.4), we find

(3.13) $$FP = 0.$$

Thus we have

(3.14) $$fF = 0.$$

From (3.3) and (3.14) we obtain

(3.15) $$tf = 0,$$

and hence, from the first equation of (3.5)

(3.16) $$Pt = 0.$$

Thus, the first equation of (3.4) implies

(3.17) $$P^3 + P = 0.$$

From the second equation of (3.5) we also have

(3.18) $$f^3 + f = 0.$$

Conversely, for a submanifold M of a Kaehlerian manifold $\bar{M}$, assume that we have (3.13), that is, FP = 0. Then we have (3.14), (3.15), (3.16), (3.17) and (3.18). We put

$$l = -P^2, \qquad l^{\perp} = I - l.$$

Then we see

$$l + l^{\perp} = I, \quad l^2 = l, \quad l^{\perp 2} = l \ , \quad l l^{\perp} = l^{\perp} l = 0,$$

which show that l and $l^{\perp}$ are complementary projection operators and consequently define complementary orthogonal distributions $\mathcal{D}$ and $\mathcal{D}^\perp$ respectively. From equation $l = -P^2$ we have $Pl = P$. This equation can be written as $Pl^\perp = 0$. But $g(PX,Y)$ is skew-symmetric and $g(l^\perp X,Y)$ is symmetric and consequently $l^\perp P = 0$. Thus we have $l^\perp Pl = 0$. Moreover, by $l = -P^2$ we have $Fl = 0$. These equations show that the distribution $\mathcal{D}$ is invariant and the distribution $\mathcal{D}^\perp$ is anti-invariant. QED.

From (3.17) and (3.18) we have (Yano-Kon [12])

THEOREM 3.2. Let M be a CR submanifold of a Kaehlerian manifold $\bar{M}$. Then P is an f-structure in M and f is an f-structure in the normal bundle of M.

Let $\bar{R}$ be the curvature tensor of a complex space form $\bar{M}(c)$. Then we have (Blair-Chen [1])

THEOREM 3.3. Let M be a submanifold of a complex space form $\bar{M}(c)$ with $c \neq 0$. Then M is a CR submanifold of $\bar{M}$ if and only if the maximal holomorphic subspace $\mathcal{D}_x = T_x(M) \cap JT_x(M)$ defines a non-trivial differentiable distribution $\mathcal{D}$ on M such that

$$g(\bar{R}(X,Y)Z,W) = 0 \tag{3.19}$$

for all $X, Y \in \mathcal{D}$ and $Z, W \in \mathcal{D}^\perp$, where $\mathcal{D}^\perp$ denoting the orthogonal complementary distribution of $\mathcal{D}$ in M.

Proof. If M is a CR submanifold of $\bar{M}(c)$, then we have $\bar{R}(X,Y)Z = \frac{1}{2}cg(X,JY)JZ$. From this equation we have (3.19).

Conversely, from (3.19) we have $g(\bar{R}(JX,X)Z,W) = \frac{1}{2}cg(X,X)g(JZ,W) = 0$ for all $X \in \mathcal{D}$ and $Z, W \in \mathcal{D}^\perp$. From this we see that $J\mathcal{D}_x^\perp$ is perpendicular to D_x. Since $\mathcal{D}$ is holomorphic, $J\mathcal{D}_x^\perp$ is also perpendicular to $\mathcal{D}_x$. Therefore, $J\mathcal{D}_x \in T_x(M)^\perp$. This shows that M is a CR submanifold.

QED.

Let us now suppose that $\bar{M}$ is a Hermitian manifold and let Ω be the fundamental 2-form of $\bar{M}$, i.e., $\Omega(X,Y) = g(X,JY)$. $\bar{M}$ is a Kaehlerian manifold if and only if $d\Omega = 0$. However we consider a class of Hermitian manifolds slightly larger than that of Kaehlerian manifolds for which $d\Omega = \Omega \wedge \omega$, ω being a 1-form called the Lee form. When ω is closed we call these manifolds locally conformal symplectic manifolds. They include the well known Hopf manifolds. We prove the following theorem (Blair-Chen [1]).

THEOREM 3.4. Let $\bar{M}$ be a Hermitian manifold with $d\Omega = \Omega \wedge \omega$. Then in order for M to be a CR submanifold of $\bar{M}$ it is necessary that $D^{\perp}$ be integrable.

Proof. Let X be vector field in D and Z, W vector fields in D . Then $\Omega(X,Z) = 0$ and $\Omega(Z,W) = 0$. Therefore $\Omega \wedge \omega(X,Z,W) = 0$ and hence

$$0 = 3d\Omega(X,Z,W) = -g([Z,W],JX),$$

but X and hence JX is arbitrary in D and [Z,W] is tangent to M, therefore [Z,W] is in $D^{\perp}$. QED.

From Theorem 3.4 we see that the distribution $D^{\perp}$ of a CR submanifold of a Kaehlerian manifold is integrable. We next consider the condition that the distribution D is integrable. If the distribution is integrable and moreover if the almost complex structure P induced on each integral submanifold of D is integrable, then we say that the f-structure P is *partially integrable*.

THEOREM 3.5. Let M be a CR submanifold of a Kaehlerian manifold $\bar{M}$. Then the f-structure P is partially integrable if and only if

$$B(PX,Y) = B(X,PY)$$

for any vector fields X and Y in D.

Proof. Let X and Y be vector fields in D. Then (3.7) implies

$$\begin{aligned} F[X,Y] &= F\nabla_X Y - F\nabla_Y X = -(\nabla_X F)Y + (\nabla_Y F)X \\ &= B(X,PY) - B(PX,Y). \end{aligned}$$

Thus D is integrable if and only if $B(X,PY) = B(PX,Y)$. In this case the integral submanifold of $\mathcal{D}$ is invariant in $\bar{M}$ and hence it is also a Kaehlerian manifold. Thus the almost complex structure induced from P on the integral submanifold of $\mathcal{D}$ is integrable. QED.

In the next place we give some examples of generic submanifolds and CR submanifolds of complex space forms.

Example 3.1. Let C^m be the complex number space of complex dimension m. Let M be a product Riemannian manifold of the form $C^p \times M^q$, where M^q is a real q-dimensional anti-invariant submanifold of C^q C^{p+q}. Then M is a generic submanifold of C^{p+q}, and M is moreover a CR submanifold of C^m with $m > p+q$.

Example 3.2. Let $S^m(r)$ be an m-dimensional sphere of radius r. We consider an immersion:

$$S^{m_1}(r_1) \times \ldots \times S^{m_k}(r_k) \longrightarrow C^{(n+k)/2}, \quad n = \sum_{i=1}^{k} m_i,$$

where $m_1,\ldots,m_k$ are odd numbers. Then n+k is evev. We now consider

$$S^{m_i}(r_i) \subset C^{(m_i+1)/2} \quad (i = 1,\ldots,k)$$

and

$$C^{(n+k)/2} = C^{(m_1+1)/2} \times \ldots \times C^{(m_k+1)/2}.$$

Then each $S^{m_i}(r_i)$ is a real hypersurface of $C^{(m_i+1)/2}$. We denote by v_i the unit normal of $S^{m_i}(r_i)$ in $C^{(m_i+1)/2}$. Then Jv_i is tangent to $S^{m_i}(r_i)$. Therefore $S^{m_1}(r_1) \times \ldots \times S^{m_k}(r_k)$ is a generic submanifold of $CP^{(n+k)/2}$, and hence CR submanifold of C^m $(2m > n+k)$ with parallel mean curvature vector and flat normal connection. Similarly, we can consider an immersion:

$$S^{m_1}(r_1) \times \ldots \times S^{m_k}(r_k) \times R^p \longrightarrow C^{(n+k)/2} \times C^p \subset C^m,$$

where $(n+k)+2p < 2m$. Then the submanifold is a generic submanifold of $C^{(n+k+2p)/2}$ and CR submanifold of C^m with parallel mean curvature vector and flat normal connection.

In the following we need some results of Riemannian fibre bundles, which will be proved in Chapter IX.

Let S^{2m+1} be a $(2m+1)$-dimensional unit sphere and CP^m be a complex m-dimensional projective space with constant holomorphic sectional curvature 4. Then there is a fibering

$$\bar{\pi} : S^{2m+1} \longrightarrow CP^m.$$

Let N be an $(n+1)$-dimensional submanifold immersed in S^{2m+1} and M be an n-dimensional submanifold immersed in CP^m. We assume that N is tangent to the vertical vector field ξ of S^{2m+1} and there exists a fibration $\pi : N \longrightarrow M$ such that the diagram

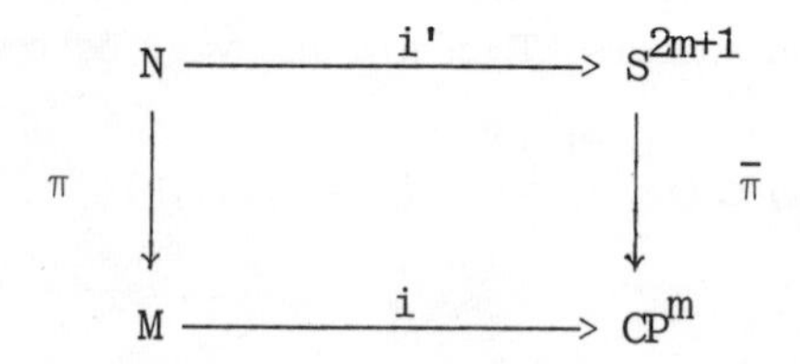

commutes and the immersion i' is a diffeomorphism on the fibres. We denote by α the second fundamental form of the immersion i'. Then we have the following lemmas.

LEMMA 3.1. The second fundamental form α of N is parallel if and only if the second fundamental form B of M satisfies

$$(\nabla_X B)(Y,Z) = g(X,PY)FZ + g(X,PZ)FY \tag{3.20}$$

and

$$fB(X,Y) = B(X,PY) + B(Y,PX). \tag{3.21}$$

LEMMA 3.2. The normal connection of N is flat if and only if M satisfies the conditions

$$R^{\perp}(X,Y)V = 2g(X,PY)fV \tag{3.22}$$

and

$$(\nabla_X f)V = 0. \tag{3.23}$$

Example 3.3. Let $S^m(r)$ be an m-dimensional sphere with radius r. We consider the following immersion:

$$N_{m_1,\ldots,m_k} \longrightarrow S^{n+k}, \qquad n+1 = \sum_{i=1}^{k} m_i,$$

where $m_1,\ldots,m_k$ are odd numbers and $r_1^2 + \ldots + r_k^2 = 1$ and we have put

$$N_{m_1,\ldots,m_k} = S^{m_1}(r_1) \times \ldots \times S^{m_k}(r_k).$$

Then we have the following commutative diagram:

$$\begin{array}{ccc} N_{m_1,\ldots,m_k} & \xrightarrow{\ i'\ } & S^{n+k} \\ \pi \big\downarrow & & \big\downarrow \bar{\pi} \\ M_{m_1,\ldots,m_k} & \xrightarrow{\ i\ } & CP^{(n+k-1)/2}, \end{array}$$

where we have put $M_{m_1,\ldots,m_k} = \pi(N_{m_1,\ldots,m_k})$. Then we see that $M_{m_1,\ldots,m_k}$ is a generic submanifold of $CP^{(n+k-1)/2}$ and is a CR submanifold of CP^m $(2m+1 > n+k)$. Since the normal connection of $N_{m_1,\ldots,m_k}$ is flat, we see that the normal connection of $M_{m_1,\ldots,m_k}$ satisfies $R^{\perp}(X,Y) = 2g(X,PY)f$ and $\nabla f = 0$ by Lemma 3.2. Moreover, the second fundamental form B of $M_{m_1,\ldots,m_k}$ satisfies (3.20) and (3.21) because of Lemma 3.1. (See Example 4.1 of Chapter VI.)

Example 3.4. In Example 3.3, if $r_i = (m_i/(n+1))^{1/2}$ $(i = 1,\ldots,k)$, then $N_{m_1,\ldots,m_k}$ is minimal and hence $M_{m_1,\ldots,m_k}$ is also minimal.

Let M be an n-dimensional CR submanifold of a complex projective space $\bar{M}^m(c)$. If the curvature tensor $R^\perp$ of the normal bundle of M satisfies

$$R^\perp(X,Y) = \tfrac{1}{2}cg(X,PY)fV \tag{3.24}$$

for any vector fields X, Y tangent to M and any vector field V normal to M, then the normal connection of M is said to be *semi-flat*. The justification of this definition is given by Lemma 3.2. We notice that, if M is a generic submanifold of $\bar{M}^m(c)$, then f vanishes identically and hence $R^\perp = 0$.

LEMMA 3.3. Let M be an n-dimensional CR submanifold of CP^m with semi-flat normal connection. Suppose that dim $\mathcal{D} = h \geq 4$ and the f-structure in the normal bundle of M is parallel. If B satisfies (3.20), then the second fundamental form α of N of S^{2m+1} is parallel.

The proof of this lemma will be given in § of Chapter IX.

LEMMA 3.4. Let M be an n-dimensional CR submanifold of CP^m. Then we have

$$g(\nabla B,\nabla B) \geq 2hq,$$

where $h = \dim \mathcal{D}$ and $q = \dim \mathcal{D}^\perp$, and the equality holds if and only if (3.20) holds.

Proof. We put

$$T(X,Y,Z) = (\nabla_X B)(Y,Z) + g(Y,PX)FZ + g(Z,PX)FY.$$

Then $T = 0$ if and only if (3.20) holds. Let $\{e_i\}$ be an orthonormal basis of $T_x(M)$. Then

$$|T|^2 = |\nabla B|^2 + 2hq + 4\sum_{i,j} g((\nabla_{e_i} B)(e_i, Pe_i), Fe_j).$$

On the other hand, equation of Codazzi implies

$$\sum_{i,j} g((\nabla_{e_i} B)(Pe_i, e_j), Fe_j) = \sum_{i,j} g((\nabla_{e_j} B)(e_i, Pe_i), Fe_j) - hq.$$

Since B is symmetric and P is skew-symmetric, the first term in the right hand side of the equation above vanishes. Consequently, we have

$$|T|^2 = |\nabla B|^2 - 2hq,$$

which proves our assertion. QED.

LEMMA 3.5. Let M be a CR submanifold of a Kaehlerian manifold $\bar{M}$. Then the f-structure in the normal bundle of M is parallel if and only if

$$A_V tU = A_U tV \tag{3.25}$$

for any vector fields U and V normal to M.

Proof. From (3.9) we have

$$\begin{aligned} g((\nabla_X f)V, U) &= -g(FA_V X, U) - g(B(X, tV), U) \\ &= g(A_V tU, X) - g(A_U tV, X), \end{aligned}$$

which proves our lemma. QED.

LEMMA 3.6. Let M be a CR submanifold of a complex space form $\bar{M}^m(c)$ with semi-flat normal connection. If the f-structure in the normal bundle of M is parallel, then $A_{fV} = 0$ for any vector field V normal to M.

Proof. From equation of Ricci we find

$$g([A_V,A_U]X,Y) = \tfrac{1}{4}c[g(FY,U)g(FX,V) - g(FX,U)g(FY,V)].$$

Thus we see that $A_{fV}A_U = A_UA_{fV}$ for any vector fields U and V normal to M. On the other hand, (3.7) implies

$$0 = g((\nabla_X f)fV,FY) = -g(f^2V,(\nabla_X F)Y)$$

$$= g(A_{f^2V}X,PY) - g(A_{fV}X,Y),$$

from which

$$A^2_{fV} = -A_{fV}PA_{f^2V}.$$

Therefore we have

$$\mathrm{Tr}A^2_{fV} = -\mathrm{Tr}A_{fV}PA_{f^2V} = \mathrm{Tr}A_{f^2V}PA_{fV} = \mathrm{Tr}A_{fV}A_{f^2V}P$$

$$= \mathrm{Tr}A_{f^2V}A_{fV}P = \mathrm{Tr}A_{fV}PA_{f^2V} = -\mathrm{Tr}A^2_{fV}.$$

Consequently, we have $\mathrm{Tr}A^2_{fV} = 0$ and hence $A_{fV} = 0$. QED.

THEOREM 3.6. Let M be an n-dimensional complete CR submanifold of CP^m with semi-flat normal connection and $h \geq 4$. If the f-structure f in the normal bundle of M is parallel and if $|\nabla B|^2 = 2hq$, then M is totally geodesic invariant submanifold $CP^{n/2}$ of CP^m or M is a generic submanifold of $CP^{(n+q)/2}$ in CP^m and is

$$\pi(S^{m_1}(r_1) \times \ldots \times S^{m_k}(r_k)), \quad \Sigma m_i = n+1, \quad \Sigma r_i^2 = 1,$$

where $2 \leq k \leq n-3$ and $m_1,\ldots,m_k$ are odd numbers and $q = k-1$.

Proof. We assume that $q = 0$. Then M is a Kaehlerian submanifold of CP^m. From equation of Ricci we have $A_VA_U - A_UA_V = 0$ for any vector fields U and V normal to M. Thus Lemma 1.1 implies $0 = A_{JU}A_U - A_UA_{JU}$

$= 2JA_U^2$, and hence $A_U = 0$. Therefore, M is totally geodesic in CP^m and is $CP^{n/2}$.

Let us next assume that $q \geq 1$. From Lemmas 3.3 and 3.4 we see that the second fundamental form α of N in S^{2m+1} is parallel. From Lemma 3.6 we also have $A_{fV} = 0$. If $A_{FX} = 0$ for some vector field X tangent to M, (3.12) implies that $c = 0$ when $q \geq 2$. This is a contradiction. When $q = 1$, if $A_{FX} = 0$, then M is totally geodesic in CP^m and hence M is invariant or anti-invariant in CP^m. This is also a contradiction. Thus the first normal space of M is of dimension q. Moreover, the first normal space of M is parallel. Indeed, from (3.7) and (3.21) we have $(\nabla_X F)Y = B(Y,PX)$, and hence $g(D_X FY, fV) = g((\nabla_X F)Y, fV) = g(B(Y,PX), fV) = 0$. Then we can see that the first normal space of N in S^{2m+1} is also parallel. Therefore, there is a totally geodesic $(n+1+q)$-dimensional submanifold S^{n+1+q} of S^{2m+1}, and hence $CP^{(n+q)/2}$ of CP^m containing N and M respectively. Then, from the definition, M is a generic submanifold of $CP^{(n+q)/2}$. Now, using Lemma 3.2, Example 3.3, Theorem 5.7 of Chapter II and Example 4.3 of Chapter II, we have our assertion. QED.

THEOREM 3.7. Let M be an n-dimensional complete generic submanifold of CP^m with flat normal connection. If $h \geq 4$ and $|\nabla B|^2 = 2hp$ ($p = \text{codim } M$), then M is

$$\pi(S^{m_1}(r_1) \times \ldots \times S^{m_k}(r_k)), \qquad \Sigma m_i = n+1, \quad \Sigma r_i^2 = 1,$$

where $2 \leq k \leq n-3$, $m = (n+k-1)/2$ and $m_1, \ldots, m_k$ are odd numbers.

Theorem 3.6 was proved by Yano-Kon [12] and Theorem 3.7 was proved by Yano-Kon [11].

Let M be an n-dimensional CR submanifold of a complex m-dimensional Kaehlerian manifold $\bar{M}$. Then we have the decomposition of the tangent space $T_x(M)$ at each point x of M:

$$T_x(M) = \mathcal{D}_x \oplus \mathcal{D}_x^{\perp} .$$

Similarly, we have

$$T_x(M)^{\perp} = F\mathcal{D}_x^{\perp} \oplus N_x,$$

where N_x is the orthogonal complement of $F\mathcal{D}_x^{\perp}$ in $T_x(M)^{\perp}$. Thus $JN_x = fN_x = N_x$. We take an orthonormal frame $\{e_1,\ldots,e_{2m}\}$ of $\bar{M}$ such that, restricted to M, $e_1,\ldots,e_n$ are tangent to M. Then $e_1,\ldots,e_n$ form an orthonormal frame of M. We can take $\{e_1,\ldots,e_n\}$ in such a way that $e_1,\ldots,e_{n-q}$ form an orthonormal frame of $\mathcal{D}_x$ and $e_{n-q+1},\ldots,e_n$ form an orthonormal frame of $\mathcal{D}_x^{\perp}$, where $q = \dim \mathcal{D}_x^{\perp}$ and $n-q = h = \dim \mathcal{D}_x$. Moreover, we can take $\{e_{n+1},\ldots,e_{2m}\}$ in such a way that $e_{n+1},\ldots,e_{n+q}$ form an orthonormal frame of F_x and $e_{n+q+1},\ldots,e_{2m}$ form an orthonormal frame of N_x. Unless otherwise stated, we use the conventions that the ranges of indices are respectively:

$$i,j,k = 1,\ldots,n; \quad r,s,t = 1,\ldots,n-q; \quad a,b,c = n-q+1,\ldots,n;$$

$$x,y,z = n+1,\ldots,n+q; \quad \lambda,\mu,\nu = n+q+1,\ldots,2m.$$

LEMMA 3.7. Let M be a CR submanifold of CP^m with semi-flat normal connection and parallel f-structure f. If $PA_V = A_VP$ for any vector field V normal to M, then

$$g(A_UX,A_VY) = g(X,Y)g(tU,tV) - g(FX,U)g(FY,V)$$

$$- \sum_i g(A_UtV,e_i)g(A_{Fe_i}X,Y).$$

Proof. From the assumption we have $g(A_UPX,tV) = 0$, from which

$$g((\nabla_YA)_UPX,tV) + g(A_U(\nabla_YP)X,tV) + g(A_UPX,(\nabla_Yt)V) = 0.$$

Changing Y to PY in this equation, and using (3.6), (3.8) and Lemma 3.6, we have

$$g((\nabla_{PY}A)_UPX,tV) + g(A_UtB(PY,X),tV) - g(PA_UX,P^2A_VY) = 0,$$

from which

$$g((\nabla_{PY}A)_U PX,tV) - \sum_i g(A_U tV,e_i)g(A_{Fe_i}X,PY) = g(PA_U X,A_V Y) = 0.$$

From this and the Codazzi equation (3.11) we have

$$g(PX,PY)g(tU,tV) - \sum_i g(A_U tV,e_i)g(A_{Fe_i}PX,PY) + g(P^2A_U X,A_V Y) = 0.$$

On the other hand, we have

$$g(PX,PY)g(tU,tV) = g(X,Y)g(tU,tV) - g(FX,FY)g(tU,tV),$$

$$-\sum_i g(A_U tV,e_i)g(A_{Fe_i}PX,PY) = -\sum_i g(A_U tV,e_i)g(A_{Fe_i}X,Y) + g(A_U tV,A_{FY}X),$$

$$g(P^2A_U X,A_V Y) = -g(A_U X,A_V Y) - g(A_U X,A_{FY}tV).$$

Moreover, from (3.12), we obtain

$$g(A_U tV,A_{FY}X) - g(A_U X,A_{FY}tV) = g(tU,tV)g(FX,FY) - g(FX,U)g(FY,V).$$

From these equations we have our assertion. QED.

LEMMA 3.8. Let M be a CR submanifold of CP^m with semi-flat normal connection and parallel f-structure f. If the mean curvature vector of M is parallel and if $PA_V = A_V P$ for any vector field V normal to M, then the square of the length of the second fundamental form of M is constant.

Proof. From Lemmas 3.6 and 3.7 the square of the length of the second fundamental form of M is given by

$$|A|^2 = \sum_x \mathrm{Tr}A_x^2 = (n-1)q + \sum_{x,y} g(A_x te_x,te_y)\mathrm{Tr}A_y.$$

On the other hand, for any vector field $V \in FD^\perp$, we have $D_X V \in FD^\perp$, because of $\nabla f = 0$. From (3.7), we also have, for any $V \in N$, $D_X V \in N$.

Therefore, since $R^{\perp}(X,Y)V = 0$ for any $V \in F\mathcal{D}^{\perp}$, we can choose an orthonormal frame $\{e_x\}$ of $F\mathcal{D}^{\perp}$ such that $De_x = 0$ for all x (see Proposition 3.3 of Chapter II). Then we see that $\nabla_X(te_x) = -PA_xX$. Since the mean curvature vector of M is parallel and $PA_V = A_VP$, from the Codazzi equation (3.11) and (3.25), we find

$$\nabla_X|A|^2 = \sum_{x,y} g((\nabla_{te_x}A)_y te_x, X)\mathrm{Tr}A_y.$$

On the other hand, using $PA_V = A_VP$ and (3.6), we have

$$\sum_i g((\nabla_{Pe_i}A)_x Pe_i, tV) = 0 \text{ and } \sum_i g((\nabla_{Pe_i}A)_x Pe_i, PX) = 0.$$

Consequently, we have

$$\sum_s (\nabla_{e_s}A)_x e_s = \sum_i (\nabla_{Pe_i}A)_x Pe_i = 0$$

for each x. Since the mean curvature vector of M is parallel, using the Codazzi equation (3.11), we obtain

$$0 = \sum_i (\nabla_{e_i}A)_x e_i = \sum_s (\nabla_{e_s}A)_x e_s + \sum_a (\nabla_{e_a}A)_x e_a$$

$$= \sum_a (\nabla_{e_a}A)_x e_a = \sum_y (\nabla_{te_y}A)_x te_y$$

for each x, and hence $|A|^2$ is constant. QED.

LEMMA 3.9. Let M be an n-dimensional CR submanifold of CP^m with semi-flat normal connection, parallel f-structure f and parallel mean curvature vector. If $PA_V = A_VP$ for any vector field V normal to M, then we have $|\nabla A|^2 = 2(n-q)q$.

Proof. From Lemma 3.6 and Proposition 3.1 of Chapter II we have

$$g(\nabla^2 A, A) = \sum_{x,i,j} g((\nabla_{e_i}\nabla_{e_i}A)_x e_j, A_x e_j)$$

$$= (n-3)\sum_x \mathrm{Tr}A_x^2 - \sum_x (\mathrm{Tr}A_x)^2 + 6\sum_x [\mathrm{Tr}(A_xP)^2 - \mathrm{Tr}A_x^2P^2]$$

$$+ 3 \sum_{x,y} [g(A_x te_y, A_x te_y) - g(A_x te_x, A_y te_y)]$$

$$- \tfrac{1}{2} \sum_{x,y,i} g([A_x,A_y]e_i,[A_x,A_y]e_i) + \sum_{x,y} [3g(A_x te_x, te_y)TrA_y$$

$$- (TrA_xA_y)^2 + (TrA_y)(TrA_x^2A_y)],$$

where we take $\{e_x\}$ such that $De_x = 0$ for all x and used the fact that $(\nabla_X A)_V Y = 0$ for any $V \varepsilon N_x$. On the other hand, the Ricci equation (3.12) implies

$$\sum_{x,y,i} g([A_x,A_y]e_i,[A_x,A_y]e_i) = 2q(q-1),$$

$$\sum_{x,y,i} [g(A_x te_y, A_x te_y) - g(A_x te_x, A_y te_y)] = q(q-1).$$

From these equations we get

$$g(\nabla^2 A, A) = (n-3)\sum_x TrA_x^2 - \sum_x (TrA_x)^2 + 3\sum_x |[P,A_x]|^2 + 2q(q-1)$$

$$+ \sum_{x,y} [3g(A_x te_x, te_y)TrA_y - (TrA_xA_y)^2 + TrA_y TrA_x^2A_y].$$

Moreover, Lemma 3.7 implies

$$\sum_{x,y} TrA_y g(A_x te_x, te_y) = \sum_x TrA_x^2 - (n-1)q,$$

and Lemma 3.8 shows that $|\nabla A|^2 = -g(\nabla^2 A, A)$. Consequemtly, we have

$$|\nabla A|^2 - 2(n-q)q = \sum_{x,y} (TrA_xA_y)^2 - n\sum_x TrA_x^2 + \sum_x (TrA_x)^2$$

$$- \sum_{x,y} TrA_x TrA_y^2 A_x + (n-1)q.$$

On the other hand, Lemma 3.7 implies

$$\sum_{x,y}(\mathrm{Tr}A_xA_y)^2 = (n-1)\sum_x \mathrm{Tr}A_x^2 + \sum_{x,y,z}\mathrm{Tr}A_z\mathrm{Tr}A_xA_y g(A_x te_y, te_z),$$

$$-\sum_{x,y}\mathrm{Tr}A_x\mathrm{Tr}A_y^2A_x = -\sum_x(\mathrm{Tr}A_x)^2 + \sum_{x,y}\mathrm{Tr}A_x g(A_x te_y, te_y).$$

Therefore, we have

$$|\nabla A|^2 - 2(n-q)q = \sum_{x,y}\mathrm{Tr}A_x g(A_x te_y, te_y) - \sum_x \mathrm{Tr}A_x^2 + (n-1)q = 0.$$

Thus we obtain $|\nabla A|^2 = 2(n-q)q$. QED.

Here, we notice that $|\nabla B|^2 = |\nabla A|^2$ in general. Moreover, if $PA_V = A_VP$, then we can prove Theorem 3.6 without the assumption $h \geq 4$. Indeed, we see that

$$g(fB(X,Y),V) - g(B(X,PY),V) - g(B(Y,PX),V)$$

$$= -g(A_VX,PY) - g(A_VPX,Y) = 0.$$

Thus B satisfies (3.21). Therefore, Theorem 3.6 and Lemma 3.9 prove the following (Yano-Kon [13])

THEOREM 3.8. Let M be an n-dimensional complete CR submanifold of CP^m with semi-flat normal connection, parallel f-structure f and parallel mean curvature vector. If $PA_V = A_VP$ for any vector field V normal to M, then M is a totally geodesic Kaehlerian submanifold $CP^{n/2}$ of CP^m or M is a generic submanifold of $CP^{(n+q)/2}$ in CP^m and is

$$\pi(S^{m_1}(r_1) \times \dots \times S^{m_k}(r_k)), \qquad \Sigma m_i = n+1, \quad \Sigma r_i^2 = 1,$$

where $q = k-1$ and $m_1,\dots,m_k$ are odd numbers.

From Theorem 3.8 we also have the following (Ki-Pak-Kim [1])

THEOREM 3.9. Let M be an n-dimensional complete generic submanifold of CP^m with flat normal connection and parallel mean curvature vector. If $PA_V = A_V P$ for any vector field V normal to M, then M is

$$\pi(S^{m_1}(r_1) \times \ldots \times S^{m_k}(r_k)), \quad \Sigma m_i = n+1, \quad \Sigma r_i^2 = 1,$$

where $2m-n = k-1$ and $m_1,\ldots,m_k$ are odd numbers.

We give the meaning of the condition $PA_V = A_V P$ in terms of the f-structure.

Let M be a CR submanifold of a Kaehlerian manifold $\bar{M}$. The Nijenhuis tensor of the f-structure P on M is given by

$$N_P(X,Y) = P^2[X,Y] + [PX,PY] - P[PX,Y] - P[X,PY].$$

We put

$$S(X,Y) = N_P(X,Y) - t[(\nabla_X F)Y - (\nabla_Y F)X].$$

The f-structure P is said to be *normal* if S vanishes identically.

THEOREM 3.10. Let M be a CR submanifold of a Kaehlerian manifold $\bar{M}$. Then the f-structure P on M is normal if and only if $PA_{FX} = A_{FX}P$ for any vector field X tangent to M.

Proof. From the definition we have

$$\begin{aligned} S(X,Y) &= (\nabla_{PX}P)Y - (\nabla_{PY}P)X + P[(\nabla_Y P)X - (\nabla_X P)Y] \\ &\quad - t[(\nabla_X F)Y - (\nabla_Y F)X] \\ &= (A_{FY}P - PA_{FY})X - (A_{FX}P - PA_{FX})Y. \end{aligned}$$

If $PA_{FX} = A_{FX}P$, then $S = 0$. Conversely, we suppose that $S = 0$. Then, for any $X \in D^\perp$ and $Y \in D$, we have $A_{FX}PY = PA_{FX}Y$. Let $Z \in D^\perp$. Then

$$g(A_{FX}PY,Z) = g(PA_{FX}Y,Z) = 0,$$

from which $A_{FX}Z \in \mathcal{D}^{\perp}$ for $Z \in \mathcal{D}^{\perp}$. Thus we have $A_{FX}PZ = PA_{FX}Z$ for any $Z \in \mathcal{D}^{\perp}$. Consequently, we obtain $PA_{FX} = A_{FX}P$ for any vector field X tangent to M. QED.

Let M be an n-dimensional CR submanifold of C^m with flat normal connection and with parallel f-structure f. Then the Ricci equation (3.12) implies $A_V A_U = A_U A_V$ and $A_{fV} = 0$ by Lemma 3.6. We suppose that $PA_V = A_V P$. Then, by a method quite similar to that used in the proof of Lemma 3.7, we have

$$g(A_U X, A_V Y) = -\sum_i g(A_U tV, e_i) g(A_{Fe_i} X, Y).$$

From this we see that $|A|^2$ is constant when the mean curvature vector of M is parallel. We can also prove the following

LEMMA 3.10. Let M be a CR submanifold of C^m with flat normal connection and with parallel f-structure f. If the mean curvature vector of M is parallel and if $PA_V = A_V P$ for any vector field V normal to M, then the second fundamental form of M is parallel.

From Lemma 3.10 and Theorem 5.5 of Chapter II we see that the sectional curvature of M is non-negative, and hence Theorem 4.1 of Chapter II implies the following (Yano-Kon [13])

THEOREM 3.11. Let M be an n-dimensional complete CR submanifold of C^m with flat normal connection and with parallel f-structure f. If the mean curvature vector of M is parallel and if $PA_V = A_V P$ for any vector field V normal to M, then M is a sphere $S^n(r)$, a plane R^n, a pythagorean product of the form

$$S^{p_1}(r_1) \times \dots \times S^{p_N}(r_N), \quad p_1,\dots,p_N \geq 1, \ \Sigma p_i = n, \ 1 < N \leq m-n,$$

or

$$S^{p_1}(r_1) \times \dots \times S^{p_N}(r_N) \times R^p, \ p_1,\dots,p_N,p \geq 1, \ \Sigma p_i + p = n, \ 1 < N \leq m-n.$$

From Theorem 3.11 we have (Ki-Pak [1])

THEOREM 3.12. Let M be a complete n-dimensional generic submanifold of C^m with flat normal connection and with parallel mean curvature vector. If $PA_V = A_V P$ for any vector field V normal to M, then we have the same conclusion as that of Theorem 3.11.

LEMMA 3.11. Let M be an n-dimensional CR submanifold of CP^m with semi-flat normal connection and with parallel f-structure f. If U is a parallel section in the normal bundle of M, then

$$\mathrm{div}(\nabla_{tU}tU) = (n-1)g(tU,tU) + \sum_x \mathrm{Tr}A_x g(A_x tU,tU) - \mathrm{Tr}A_U^2 + \tfrac{1}{2}|L(tU)g|^2.$$

Proof. From equation of Ricci (3.12) and Lemma 3.6 we have $fU = 0$ and hence $U \in FD^\perp$. We also have $\nabla_X tU = -PA_U X$, from which

$$\mathrm{div}\ tU = \sum_i g(\nabla_{e_i} tU,e_i) = -\mathrm{Tr}PA_U = 0$$

because of A_U is symmetric and P is skew-symmetric. Thus, from Theorem 4.3 of Chapter I, we find

$$\mathrm{div}(\nabla_{tU}tU) = S(tU,tU) + \tfrac{1}{2}|L(tU)g|^2 - |\nabla tU|^2,$$

where S is the Ricci tensor of M. On the other hand, from (3.10) and Lemma 3.6 we have

$$S(tU,tU) = (n-1)g(tU,tU) + \sum_x \mathrm{Tr}A_x g(A_x tU,tU) - \sum_x g(A_x^2 tU,tU).$$

We also have

$$|\nabla tU|^2 = \mathrm{Tr}A_U^2 - \sum_x g(A_x^2 tU,tU).$$

From these equations we have our equation. QED.

We notice here that

$$|L(tU)g|^2 = |[P,A_U]|^2,$$

which means that tU is an infinitesimal isometry of M if and only if $PA_U = A_UP$.

We now prove the following (Yano-Kon [13])

THEOREM 3.13. Let M be an n-dimensional compact minimal CR submanifold of CP^m with semi-flat normal connection and parallel f-structure f. Then

$$\int_M[(n-1)q - |A|^2]*1 = -\tfrac{1}{2}\int_M\sum_x|[P,A_x]|^2*1.$$

Proof. We can take an orthonormal frame $\{e_x\}$ such that $De_x = 0$ for each x. From this and Lemma 3.11 we have our equation. QED.

As an application of Theorem 3.13 we have

THEOREM 3.14. Let M be an n-dimensional compact minimal CR submanifold of CP^m with semi-flat normal connection and with parallel f-structure f. If $|A|^2 \leq (n-1)q$, then M is $CP^{n/2}$, or M is a generic minimal submanifold of $CP^{(n+q)/2}$ in CP^m and is

$$\pi(S^{m_1}(r_1) \times \ldots \times S^{m_k}(r_k)), \quad r_i = (m_i/(n+1))^{1/2} \ (i=1,\ldots,k),$$

where $\Sigma m_i = n+1$ and $m_1,\ldots,m_k$ are odd numbers.

Proof. From Theorem 3.13 we obtain $PA_x = A_xP$ for all x. Therefore our theorem follows from Theorem 3.8. QED.

THEOREM 3.15. Let M be an n-dimensional compact minimal generic submanifold of CP^m with flat normal connection. If $|A|^2 \leq (n-1)q$, then M is

$$\pi(S^{m_1}(r_1) \times \ldots \times S^{m_k}(r_k), \quad r_i = (m_i/(n+1))^{1/2} \ (i=1,\ldots,k),$$

where $\Sigma m_i = n+1$ and $m_1,\ldots,m_k$ are odd numbers.

THEOREM 3.16. Let M be an n-dimensional complete CR submanifold of CP^m with flat normal connection and parallel mean curvature vector. If $PA_V = A_V P$ for any vector field V normal to M, then M is

$$\pi(S^1(r_1) \times \ldots \times S^1(r_{n+1})), \qquad \Sigma r_i^2 = 1$$

in CP^n in CP^m, or M is

$$\pi(S^{m_1}(r_1) \times \ldots \times S^{m_k}(r_k)), \quad \Sigma m_i = n+1, \quad \Sigma r_i^2 = 1,$$

where $m_1, \ldots, m_k$ are odd numbers and $q = k-1$, $2m = n+q$.

Proof. From the assumption and the Ricci equation (3.12) we have

$$g([A_{fU}, A_U]PX, X) = 2g(PX,PX)g(fU,fU),$$

from which

$$\mathrm{Tr}A_{fU}A_U P - \mathrm{Tr}A_U A_{fU} P = 2(n-q)g(fU,fU) = 0.$$

Thus we have $n = q$, that is, $P = 0$ and M is anti-invariant submanifold of CP^m, or $f = 0$, that is, M is a generic submanifold of CP^m. Therefore our theorem follows from Theorems 3.8 and 3.9. QED.

We next study the pinching problem of the sectional curvature of a generic minimal submanifold M of CP^m (Yano-Kon [14], [17]).

THEOREM 3.17. Let M be an n-dimensional compact generic minimal submanifold of $CP^{(n+p)/2}$ with flat normal connection. If the minimum of the sectional curvature of M is $(n-p)/n(n-1)$, then $p = 1$ and M is the geodesic minimal hypersphere

$$\pi(S^n(r_1) \times S^1(r_2)), \quad r_1 = (n/(n+1))^{1/2}, \ r_2 = (1/(n+1))^{1/2},$$

or $p = n$ and M is the flat anti-invariant submanifold

$$\pi(S^1(r_1) \times \ldots \times S^1(r_{n+1})), \ r_1 = \ldots = r_{n+1} = (1/(n+1))^{1/2}.$$

Proof. From Lemma 3.11 we have

$$\operatorname{div}\Big(\sum_a \nabla_{Jv_a} Jv_a\Big) = (n-1)p - \sum_a \operatorname{Tr} A_a^2 + \tfrac{1}{2}\sum_a |[P,A_a]|^2,$$

where $\{v_a\}$ denotes an orthonormal frame for $T(M)^{\perp}$ such that $Dv_a = 0$ for each a. On the other hand, using (3.1) of Chapter II, we obtain

$$g(\nabla^2 A, A) = \sum_{a,i,j} g((R(e_j,e_i)A_a)e_j, A_a e_i) + \frac{3}{2}\sum_a |[P,A_a]|^2 - 3\sum_a \operatorname{Tr} A_a^2 + 3(p-1)p,$$

where $\{e_i\}$ is an orthonormal frame of M. Since M is compact, these equations imply

$$0 \leq \int_M [|\nabla A|^2 - 2(n-p)p]*1$$

$$= \int_M [(n-p)p - \sum_{a,i,j} g((R(e_j,e_i)A_a)e_j, A_a e_i)]*1.$$

We can choose $\{e_i\}$ such that $A_a e_i = \lambda_i^a e_i$ $(i = 1,\ldots,n)$. Then

$$\sum_{i,j} g((R(e_j,e_i)A_a)e_j, A_a e_i) = \tfrac{1}{2}\sum_{i,j} (\lambda_j^a - \lambda_i^a)^2 K_{ij},$$

where K_{ij} denotes the sectional curvature of M with respect to the section spanned by e_i, e_j. By the assumption, we have $K_{ij} \geq (n-p)/n(n-1)$ and hence

$$\sum_{i,j} g((R(e_j,e_i)A_a)e_j, A_a e_i) \geq \frac{(n-p)}{2n(n-1)}\sum_{i,j} (\lambda_j^a - \lambda_i^a)^2 = \frac{n-p}{n-1}\operatorname{Tr} A_a^2.$$

Therefore, we have

$$0 \leq \int_M [|\nabla A|^2 - 2(n-p)p]*1 \leq \frac{n-p}{n-1}\int_M [(n-1)p - \sum_a \operatorname{Tr} A_a^2]*1$$

$$= -\frac{(n-p)}{2(n-1)} \int_M \sum_a |[P,A_a]|^2 *1.$$

From this we have $|\nabla A|^2 = 2(n-p)p$. We also have $n = p$ or $|A|^2 = (n-1)p$ and $PA_a = A_aP$ for all a. If $n = p$, then M is an anti-invariant submanifold with flat normal connection and hence M is flat and obviously $P = 0$. Consequently, Theorem 3.9 implies that M is

$$\pi(S^1(r_1) \times \ldots \times S^1(r_{n+1})), \quad r_1=\ldots=r_{n+1}=(1/(n+1))^{1/2},$$

or

$$\pi(S^{m_1}(r_1) \times \ldots \times S^{m_k}(r_k), \quad r_i = (m_i/(n+1))^{1/2} \ (i=1,\ldots,k).$$

But, by the assumption $K_{ij} \geq (n-p)/n(n-1) > 0$ when $n \neq p$. Thus, we conclude that M is a geodesic hypersphere. QED.

Any real hypersurface of a Kaehlerian manifold is obviously a generic submanifold. Thus Theorem 3.17 implies (Kon [14])

THEOREM 3.18. Let M be a compact real minimal hypersurface of $CP^{(n+1)/2}$. If the minimum of the sectional curvature of M is $1/n$, then M is the geodesic hypersphere

$$\pi(S^n(r_1) \times S^1(r_2)), \quad r_1 = (n/(n+1))^{1/2}, \ r_2 = (1/(n+1))^{1/2}.$$

A CR submanifold M of a Kaehlerian manifold $\bar{M}$ is called a CR *product* if it is locally a Riemannian product of a holomorphic submanifold M^T and an anti-invariant submanifold M .

First we give a characterization of CR product (Chen [2]).

THEOREM 3.19. A CR submanifold M of a Kaehlerian manifold $\bar{M}$ is a CR product submanifold if and only if the f-structure P on M is parallel.

Proof. Suppose that P is parallel. Then (3.6) gives

$$A_{FY}X = -tB(X,Y),$$

from which $tB(X,PY) = 0$ for any vector fields X and Y tangent to M.

Let Y be in $\mathcal{D}$. Then, for any vector fields X, Z tangent to M, we have

$$g(F\nabla_X Y, FZ) = -g((\nabla_X F)Y, FZ) = g(B(X,PY),FZ) - g(fB(X,Y),FZ)$$
$$= -g(tB(X,PY),Z) + g(B(X,Y),fFZ) = 0.$$

Thus we have $F\nabla_X Y = 0$ and consequently the distribution $\mathcal{D}$ is parallel.

Let Y be in $\mathcal{D}^\perp$. Then we have $P\nabla_X Y = -(\nabla_X P)Y = 0$ and hence $\mathcal{D}^\perp$ is also parallel. Therefore, M is a CR product submanifold.

Conversely, if M is a CR product submanifold, we easily see that P is parallel. QED.

Example 3.5. We define a mapping

$$f_{hp} : CP^h \times CP^p \longrightarrow CP^{h+p+hp}$$

by

$$(z_0,\ldots,z_h;w_0,\ldots,w_p) \longrightarrow (z_0w_0,\ldots,z_iw_j,\ldots,z_hw_p),$$

where $(z_0,\ldots,z_h)$ and $(w_0,\ldots,w_p)$ are the homogeneous coordinates of CP^h and CP^p respectively. It is easy to see that f_{hp} is a Kaehlerian imbedding. Let $M^\perp$ be a p-dimensional anti-invariant submanifold of CP^p. Then $CP^h \times M^\perp$ induces a natural CR product in CP^{h+p+hp}.

A CR product submanifold $M = M^T \times M^\perp$ in CP^m is called a *standard* CR product if

(1) $m = h+p+hp$, and

(2) M^T is a totally geodesic Kaehlerian submanifold of CP^m,

where $h = \dim_C \mathcal{D}_x$ and $p = \dim_R \mathcal{D}_x^\perp$.

We shall prove that $m = h+p+hp$ is in fact the smallest dimension of CP^m for admitting a CR product.

LEMMA 3.12. Let M be a CR product of CP^m. Then $\{B(X_i,Z_x)\}$, $i = 1,\ldots,2h$; $x = 1,\ldots,p$, are orthonormal vectors in N_x, where $\{X_i\}$ and $\{Z_x\}$ are orthonormal basis for $\mathcal{D}_x$ and $\mathcal{D}_x^\perp$ respectively.

Proof. Since $tB(X,PY) = 0$, we obtain $A_{FZ}PY = 0$ for any vector fields X and Y tangent to M. From this, (3.7) and (3.11) we have

$$g(B(PX,Z),fB(X,Z)) = g(X,X)g(Z,Z)$$

for any $X \in \mathcal{D}_x$ and $Z \in \mathcal{D}_x^\perp$. On the other hand, (3.7) implies

$$0 = (\nabla_Z F)X = -B(Z,PX) + fB(X,Z).$$

Thus we have

$$|B(X,Z)|^2 = g(X,X)g(Z,Z).$$

We suppose that $|X| = |Z| = 1$. Then $|B(X,Z)| = 1$. Therefore we obtain by linearlity that

$$g(B(X_i,Z),B(X_j,Z)) = 0, \quad i \neq j.$$

Moreover, we see that $B(X,Z) \in N_x$ by $A_{FZ}PY = 0$. Hence, if $\dim \mathcal{D}_x^\perp = 1$, the lemma is proved. If $\dim \mathcal{D}_x^\perp = p \geq 2$, then

$$g(B(X_i,Z_x),B(X_j,Z_y)) + g(B(X_i,Z_y),B(X_j,Z_x)) = 0$$

for $i \neq j$, $x \neq y$. Since M is a CR product, we see that $g(R(X_i,X_j)Z_x,Z_y) = 0$ and hence, by (3.10) we obtain

$$g(B(X_i,Z_x),B(X_j,Z_y)) - g(B(X_i,Z_y),B(X_j,Z_x)) = 0.$$

Therefore we obtain $g(B(X_i,Z_x),B(X_j,Z_y)) = 0$ and hence we have our assertion. QED.

As an immediate consequence of Lemma 3.12 we have

THEOREM 3.20. Let M be a CR product submanifold of CP^m. Then $m \geq h+p+hp$.

THEOREM 3.21. Every CR product M of CP^m with $m = h+p+hp$ is a standard CR product.

Proof. For any X, Y, Z in D and $W \in D^\perp$, the Gauss equation (3.10) implies

$$g(B(X,W),B(Y,Z)) = g(B(X,X),B(Y,W)).$$

In particular, if $Y = PX$, then

$$g(B(X,X),B(PX,W)) = g(B(PX,Z),B(X,W))$$

$$= g(fB(X,Z),B(X,W)) = -g(B(X,Z),B(PX,W)),$$

from which

$$g(B(X,Z),B(PX,W)) = g(B(X,PZ),B(X,W)) = 0.$$

Therefore we obtain

$$g(B(X,Z),B(X,W)) = 0$$

for any X, Z in D and W in $D^\perp$. Then by linearlity we have

$$g(B(X,Z),B(Y,W)) + g(B(Y,Z),B(X,W)) = 0.$$

Thus we have

$$g(B(X,Z),B(Y,W)) = 0.$$

On the other hand, Lemma 3.12 implies that B(X,Z) lies in $FD^\perp$ for any X, Z in D, because of $m = h+p+hp$. But we have $A_{FX}PY = 0$, and hence we must have $B(X,Z) \in N$. Consequently, we have $B(X,Z) = 0$ for any X, Z in D. Therefore, M^T must be totally geodesic in CP^m. QED.

THEOREM 3.22. Let M be a CR product of CP^m. Then we have

$$|A|^2 \geq 4hp.$$

If the equality holds, then M^T and $M^\perp$ are both totally geodesic in CP^m.

Proof. First of all, we have $|B(X,Z)| = 1$ for any unit vectors X in $\mathcal{D}$ and Z in $\mathcal{D}^\perp$. Thus we have

$$|A|^2 = 4hp + \sum_{i,j=1}^{2h} |B(X_i,X_j)|^2 + \sum_{x,y=1}^{p} |B(Z_x,Z_y)|^2,$$

where $\{X_i\}$ and $\{Z_x\}$ are orthonormal basis of $\mathcal{D}$ and $\mathcal{D}^\perp$ respectively. From this equation we have our assertion. QED.

Example 3.6. Let RP^p be a real p-dimensional projective space. Then RP^p is a totally geodesic anti-invariant submanifold of CP^p. Then the composition of the immersions

$$CP^h \times RP^p \longrightarrow CP^h \times CP^p \longrightarrow CP^{h+p+hp} \longrightarrow CP^m$$

gives a CR product in CP^m with $|A|^2 = 4hp$.

Theorems 3.20 and 3.21 are proved by Chen [2].

EXERCISES

A. COMPLEX SPACE FORMS IMMERSED IN COMPLEX SPACE FORMS: Let $M^n(k)$ be a complex n-dimensional space form of constant holomorphic sectional curvature k immersed in a complex (n+p)-dimensional space form $\bar{M}^{n+p}(c)$. Ogiue [4] proved the following

THEOREM 1. If $p = n(n+1)/2$, then either $c = k$ or $c = 2k$, the latter case arising only when $c > 0$.

Furthermore, Nakagawa-Ogiue [1] proved the following theorems.

THEOREM 2. If $c > 0$ and the immersion is full, then $c = \nu k$ and $n+p = \binom{n+\nu}{\nu} - 1$ for some positive integer ν.

THEOREM 3. If $c \leq 0$, then $c = k$, that is, $M^n(k)$ is totally geodesic in $\bar{M}^{n+p}(c)$.

These theorems are the local version of a classification theorem of Kaehlerian imbeddings of complete and simply connected complex space forms into complete and simply connected complex space forms (Calabi [1]).

In Theorem 1, the second fundamental form of M is parallel. Nakagawa-Takagi [1] showed the classification theorem of complete Kaehlerian submanifolds imbedded in CP^m with parallel second fundamental form.

B. KAEHLERIAN SUBMANIFOLDS WITH $R(X,Y)S = 0$: Let M be a complex n-dimensional Kaehlerian submanifold of a complex space form $\bar{M}^{n+p}(c)$. We denote by S the Ricci tensor of M and by R the Riemannian curvature tensor of M. We consider the condition

(*) $\quad R(X,Y)S = 0 \quad$ for any vector fields X and Y tangnet to M.

Then we have (Kon [5])

THEOREM 1. Let M be a complex n-dimensional Kaehlerian submanifold of a complex space form $\bar{M}^m(c)$ satisfying the condition (*). If $c < 0$, then M is totally geodesic.

When $c > 0$, Nakagawa-Takagi [2] proved the following

THEOREM 2. Let M be a complex n-dimensional Kaehlerian submanifold of a complex space form $\bar{M}^{n+p}(c)$ $(c > 0)$. If M satisfies the condition (*) and the codimension p is less than n-1, then M is Einstein.

When M is a Kaehlerian hypersurface, Ryan [2] proved

THEOREM 3. The complete Kaehlerian manifolds with (*) which occur as hypersurfaces in a complex space form $\bar{M}^{n+1}(c)$ $(c \neq 0)$ are

(1) the complex projective space CP^n and the complex quadric Q^n;

(2) the disk D^n of holomorphic sectional curvature $c < 0$.

When $c = 0$, Tsunero Takahashi [3] proved the following

THEOREM 4. A complete complex hypersurface in C^{n+1} satisfying the condition (*) is cylindrical.

C. POSITIVELY CURVED KAEHLERIAN SUBMANIFOLDS: Let M be a complex n-dimensional complete Kaehlerian submanifold in a complex projective space CP^{n+p} with constant holomorphic sectional curvature 1. Then we have (Ogiue [2])

THEOREM 1. If every Ricci curvature of M is greater than n/2, then M is totally geodesic.

We denote by H the holomorphic sectional curvature of M. Then we have (Ogiue [1])

THEOREM 2. If $H > 1 - (n+2)/2(n+2p)$, then M is totally geodesic.

Moreover, we obtain the following theorem (Itoh [1]).

THEOREM 3. If $H > 3n/(4n+2)$, then M is totally geodesic.

For positively curved Kaehlerian submanifolds, see Ogiue [5].

D. NORMAL CURVATURE: Let M be a complex n-dimensional Kaehlerian submanifold of a complex space form $\bar{M}^m(c)$. We denote by R the curvature tensor in the normal bundle of M. We consider the following condition on $R^\perp$:

$$(*) \qquad R^\perp(X,Y) = fg(X,JY)J,$$

where X and Y are arbitrary vector fields tangent to M and f is a function on M. I. Ishihara [1] proved the following

THEOREM. Let M be a complex n-dimensional ($n \geq 2$) Kaehlerian submanifold of a complex space form $\bar{M}^m(c)$ and assume that M satisfies the condition (*). Then either M is totally geodesic or M is an Einstein Kaehlerian hypersurface of $\bar{M}^m(c)$ with scalar curvature n^2c. The latter case occurs only when $c > 0$.

Combining this theorem with Theorem 1.10, we can determine complete Kaehlerian submanifolds with (*) of simply connected complete complex space forms.

E. KAEHLERIAN IMMERSIONS WITH VANISHING BOCHNER CURVATURE TENSOR: Kon [9] proved the following theorems:

THEOREM 1. Let $\bar{M}$ be a Kaehlerian manifold of complex dimension n+p with vanishing Bochner curvature tensor, and let M be a Kaehlerian submanifold of $\bar{M}$ of complex dimension n with vanishing Bochner curvature tensor. If $p < (n+1)(n+2)/(4n+2)$, then M is totally geodesic in $\bar{M}$.

THEOREM 2. Under the same assumption as in Theorem 1, if $p = 1$ and $n \geq 2$, then M is totally geodesic in $\bar{M}$.

F. ANTI-INVARIANT SUBMANIFOLDS WITH FLAT NORMAL CONNECTION: Let M be an n-dimensional anti-invariant submanifold of a complex m-dimensional Kaehlerian manifold $\bar{M}$. If $n = m$, by Lemma 2.3, we see that M is flat if and only if the normal connection of M is flat. When $n < m$, Yano-Kon-Ishihara [1] proved the following theorem.

THEOREM. Let M be an n-dimensional ($n \geq 3$) anti-invariant submanifold of a complex space form $\bar{M}^{n+p}(c)$ ($c \neq 0$) with parallel mean curvature vector. If the normal connection of M is flat, then M is a flat anti-invariant submanifold of some $\bar{M}^{n}(c)$ in $\bar{M}^{n+p}(c)$, where $\bar{M}^{n}(c)$ is a totally geodesic Kaehlerian submanifold of $\bar{M}^{n+p}(c)$ of complex dimension n.

G. PARALLEL SECOND FUNDAMENTAL FORM: For an anti-invariant minimal submanifold of a complex space form with parallel second fundamental form we have (Kon [6])

THEOREM. Let M be an n-dimensional anti-invariant minimal submanifold with parallel second fundamental form of a complex space form $\bar{M}^{n}(c)$. Then either M is totally geodesic or M has non-negative scalar curvature $r \geq 0$. Moreover, if $r = 0$, then M is flat.

H. TOTALLY UMBILICAL ANTI-INVARIANT SUBMANIFOLDS: Yano [7] proved the following theorems.

THEOREM 1. Let M be an n-dimensional ($n \geq 3$) totally umbilical, anti-invariant submanifold of a complex m-dimensional Kaehlerian manifold with vanishing Bochner curvature tensor. Then M is conformally flat.

THEOREM 2. Let M be an n-dimensional ($n \geq 4$) anti-invariant submanifold of a complex n-dimensional Kaehlerian manifold $\bar{M}$ with vanishing Bochner curvature tensor. If the second fundamental forms of M are commutative, then M is conformally flat.

I. CONFORMALLY FLAT ANTI-INVARIANT SUBMANIFOLDS: Let M be an n-dimensional anti-invariant submanifold of a complex projective space CP^n with constant holomorphic sectional curvature $c > 0$. We denote by K and r the sectional curvature and the scalar curvature of M respectively. Verheyen-Verstraelen [1] proved the following theorems.

THEOREM 1. Let M be compact, conformally flat and of dimension $n \geq 4$. Then $r > ((n-1)^3(n+2)/4(n^2+n-4))c$ implies that M is totally geodesic.

THEOREM 2. Let M be complete, conformally flat and of dimension $n \geq 4$. Then $K > ((n-1)^2/4n(n^2+n-4))c$ implies that M is totally geodesic.

J. TOTALLY UMBILICAL CR SUBMANIFOLDS: Bejancu [3] proved the following theorem.

THEOREM. Let M be a totally umbilical non-trivial CR submanifold of a Kaehlerian manifold $\bar{M}$. If $\dim \mathcal{D}^{\perp} > 1$, then M is totally geodesic in $\bar{M}$.

K. REAL HYPERSURFACES: Let M be a real (2n-1)-dimensional real hypersurface of a Kaehlerian manifold $\bar{M}$ of complex dimension n (real dimension 2n). Then M is obviously a generic submanifold of $\bar{M}$. We give examples of real hypersurfaces in a complex projective space CP^n with constant holomorphic sectional curvature 4 (see R. Takagi [2]).

Let C^{n+1} be the space of (n+1)-tuples of complex numbers $(z_1,\ldots,z_{n+1})$. Put

$$S^{2n+1} = \{(z_1,\ldots,z_{n+1}) \in C^{n+1} : \sum_{j=1}^{n+1} |z_j|^2 = 1\}.$$

For a positive number r we denote by $M_0'(2n,r)$ a hypersurface of S^{2n+1} defined by

$$\sum_{j=1}^{n} |z_j|^2 = r|z_{n+1}|^2, \qquad \sum_{j=1}^{n+1} |z_j|^2 = 1.$$

For an integer m ($2 \leq m \leq n-1$) and a positive number s, a hypersurface $M'(2n,m,s)$ of S^{2n+1} is defined by

$$\sum_{j=1}^{m} |z_j|^2 = s \sum_{j=m+1}^{n+1} |z_j|^2, \qquad \sum_{j=1}^{n+1} |z_j|^2 = 1.$$

For a number t ($0 < t < 1$) we denote by $M'(2n,t)$ a hypersurface of S^{2n+1} defined by

$$\left|\sum_{j=1}^{n+1} z_j^2\right|^2 = t, \qquad \sum_{j=1}^{n+1} |z_j|^2 = 1.$$

Let be the natural projection of S^{2n+1} onto CP^n. Then $M_0(2n-1,r) = \pi(M_0'(2n,r))$ is a connected compact real hypersurface of CP^n with two constant principal curvatures. We call $M_0(2n-1,r)$ a *geodesic hypersphere* of CP^n. Moreover, $M(2n-1,m,s) = \pi(M'(2n,m,s))$ $(n \geq 3)$ and $M(2n-1,t) = \pi(M'(2n,t))$ $(n \geq 2)$ are connected compact real hypersurfaces of CP^n with three constant principal curvatures. R. Takagi [1], [2] proved the following theorems.

THEOREM 1. If M is a connected complete real hypersurface in CP^n $(n \geq 2)$ with two constant principal curvatures, then M is a geodesic hypersphere.

THEOREM 2. If M is a connected complete real hypersurface in CP^n $(n \geq 3)$ with three constant principal curvatures, then M is congruent to some $M(2n-1,m,s)$ or $M(2n-1,t)$.

We denote by C a unit normal of a real hypersurface of a Kaehlerian manifold $\bar{M}$. We put $JC = -U$. Then U is a unit vector field on M. We define a 1-form u by $u(X) = g(X,U)$. If the Ricci tensor S of M is of the form

$$S(X,Y) = ag(X,Y) + bu(X)u(Y)$$

for some constants a and b, then M is called a *pseudo-Einstein real hypersurface* of $\bar{M}$. We have (Kon [13])

THEOREM 3. If M is a connected complete pseudo-Einstein real hypersurface of CP^n $(n \geq 3)$, then M is congruent to some geodesic hypersphere $M_0(2n-1,r)$ or $M(2n-1,m,(m-1)/(n-m))$ or $M(2n-1,1/(n-1))$.

When a and b are functions, see Cecil-Ryan [1].

L. GENERIC MINIMAL SUBMANIFOLDS: Let M be a compact n-dimensional generic minimal submanifold of a real (n+p)-dimensional complex projective space $CP^{(n+p)/2}$ with constant holomorphic sectional curvature 4. Then we have (Kon [15])

THEOREM. If the Ricci tensor S of M satisfies

$$S(X,X) \geq (n-1)g(X,X) + 2g(PX,PX),$$

then M is a real projective space RP^n $(p = n)$, or M is the pseudo-Einstein real hypersurface $\pi(S^m(r) \times S^m(r))$ $(m = (n+1)/2,\ r = (1/2)^{\frac{1}{2}})$ of $CP^{(n+1)/2}$ $(p = 1)$.

M. SUBMANIFOLDS OF A QUATERNION KAEHLERIAN MANIFOLD: Let $\bar{M}$ be a 4n-dimensional quaternion Kaehlerian manifold with structure (F,G,H). Let M be a Riemannian manifold of dimension m $(m \leq n)$ immersed in $\bar{M}$ by an isometric immersion f. We call M a totally real submanifold of $\bar{M}$ if $T_x(M) \perp FT_x(M)$, $T_x(M) \perp GT_x(M)$, $T_x(M) \perp HT_x(M)$ for any point x of M. Then Funabashi [1] proved the following

THEOREM. Let HP^n be a quaternion projective space of dimension 4n and M a connected and complete submanifold of dimension n immersed by $f : M \longrightarrow HP^n$. Assume M is a compact, totally real and minimal submanifold satisfying the inequality $|A|^2 < (n+1)/2(3n-1)$ for the square of the length of the second fundamental form A of M. Then M is an n-dimensional real projective space RP^n, and the immersion f being congruent to the standard immersion $i : RP^n \longrightarrow HP^n$ or, M is the unit sphere S^n, f being congruent to the standard immersion $i \cdot \pi : S^n \longrightarrow HP^n$, where $\pi : S^n \longrightarrow RP^n$ is the natural projection.

CHAPTER V

CONTACT MANIFOLDS

In this chapter, we study the various almost contact manifolds and contact manifolds. In §1, we give the definitions almost contact manifolds and almost contact metric manifolds. §2 is devoted to the study of contact manifolds. We give some fundamental properties and examples of contact manifolds and contact metric manifolds. In §3, we define the torsion tensor field of a almost contact manifold and study the normality of the manifolds. Moreover, we define a K-contact Riemannian manifold and give some conditions for the manifold to be K-contact. In §4, we consider the contact distribution on a contact manifold. §5 is devoted to the study of Sasakian manifolds. We define a Sasakian manifold and give some examples of Sasakian manifolds. Moreover, we define the ϕ-sectional curvature of a Sasakian manifold and give the typical examples of Sasakian space forms, that is, Sasakian manifolds of constant ϕ-sectional curvature. In §6, we discuss regular contact manifolds. We prove theorems of Boothby-Wang [1], and consider a principal fibre bundle, which is called the Boothby-Wang fibration. Furthermore, we consider the relation of the Boothby-Wang fibration and Sasakian structures and prove a theorem of Hatakeyama [1]. In the last §7, we consider the Brieskorn manifold. We give a contact structure on a Brieskorn manifold. Moreover, we show that there exists a Sasakian structure on a Brieskorn manifold by using the deformation theorey of the standard Sasakian structure on a unit sphere.

For contact manifolds, we refered to Blair [1] and Sasaki [2].

1. ALMOST CONTACT MANIFOLDS

Let M be a (2n+1)-dimensional manifold and ϕ, ξ, η be a tensor field of type (1,1), a vector field, a 1-form on M respectively. If ϕ, ξ and η satisfy the conditions

$$\eta(\xi) = 1, \tag{1.1}$$

$$\phi^2 X = -X + \eta(X)\xi \tag{1.2}$$

for any vector field X on M, then M is said to have an *almost contact structure* (ϕ,ξ,η) and is called an *almost contact manifold*.

By the definition every almost contact manifold must have a non-singular vector field ξ over M. However, the Euler characteristic of any compact manifold is equal to zero and then there exists at least one non-singular vector field over the manifold. Therefore, the condition for an almost contact structure and that for an almost complex structure may be considered to impose almost the same degree of restrictions for odd and even dimensional manifolds respectively.

PROPOSITION 1.1. For almost contact structure (ϕ,ξ,η) we have

$$\phi\xi = 0, \tag{1.3}$$

$$\eta(\phi X) = 0, \tag{1.4}$$

$$\operatorname{rank}\phi = 2n. \tag{1.5}$$

Proof. From (1.1) and (1.2) we have $\phi^2\xi = 0$, and hence $\phi\xi = 0$ or $\phi\xi$ is a non-trivial eigenvector of ϕ corresponding to the eigenvalue 0. Suppose that $\phi\xi \neq 0$. Then we have $0 = \phi^2(\phi\xi) = -\phi\xi + \eta(\phi\xi)\xi$, from which $\phi\xi = \eta(\phi\xi)\xi$, and hence $\eta(\phi\xi) \neq 0$. But we have $0 = \phi^2\xi = \eta(\phi\xi)\phi\xi$. This contradicts to the fact that $\phi\xi \neq 0$ and $\eta(\phi\xi) \neq 0$.

Therefore, we have $\phi\xi = 0$. From this and (1.2) we easily see that $\eta(\phi X) = 0$.

In the next place, since $\phi\xi = 0$, rank$\phi \leq 2n$. If X is another vector of M such that $\phi X = 0$, then (1.2) implies that $X = \eta(X)\xi$, that is, X is proportional to ξ. Therefore, we have rank $= 2n$. QED.

We now prove that every almost contact manifold admits a Riemannian metric tensor field which plays an anologous role to an almost Hermitian metric tensor field. We first prove the following lemma.

LEMMA 1.1. Every almost contact manifold M admits a Riemannian metric tensor field h such that

$$h(X,\xi) = \eta(X)$$

for any vector field X on M.

Proof. Since M admits a Riemannian metric tensor field f (which exists provided M is paracompact), we obtain h by settibg

$$h(X,Y) = f(X - \eta(X)\xi, Y - \eta(Y)\xi) + \eta(X)\eta(Y). \qquad \text{QED.}$$

PROPOSITION 1.2. Every almost contact manifold M admits a Riemannian metric tensor field g such that

$$\eta(X) = g(X,\xi), \tag{1.6}$$

$$g(\phi X,\phi Y) = g(X,Y) - \eta(X)\eta(Y). \tag{1.7}$$

Proof. We put

$$g(X,Y) = \tfrac{1}{2}(h(X,Y) + h(\phi X,\phi Y) + \eta(X)\eta(Y)).$$

Then we can easily verify that this satisfies (1.6) and (1.7). QED.

From (1.2), (1.6) and (1.7) we have

$$g(\phi X,Y) + g(X,\phi Y) = 0. \tag{1.8}$$

This means that ϕ is a skew-symmetric tensor field with respect to g.

We call the metric tensor field g, appearing in Proposition 1.2, an associated Riemannian metric tensor field to the given almost contact structure (ϕ,ξ,η). If M admits tensor field (ϕ,ξ,η,g), g being an associated Riemannian metric tensor field of an almost contact structure (ϕ,ξ,η), then M is said to have an *almost contact metric structure* (ϕ,ξ,η,g) and is called an *almost contact metric manifold*.

PROPOSITION 1.3. Let M be a (2m+1)-dimensional manifold with almost contact structure (ϕ,ξ,η). Then the structure group of its tangent bundle reduces to $U(n) \times 1$. The converse is also true.

Proof. First of all, we can choose 2n+1 mutually orthogonal unit vectors $e_1,\ldots,e_n,\phi e_1,\ldots,\phi e_n,\xi$, which form an orthonormal frame of M, and is called an adapted frame. Then with respect to this frame, we see

$$g = \begin{pmatrix} I_n & 0 & 0 \\ 0 & I_n & 0 \\ 0 & 0 & 1 \end{pmatrix}, \quad \phi = \begin{pmatrix} 0 & I_n & 0 \\ -I_n & 0 & 0 \\ 0 & 0 & 0 \end{pmatrix}. \tag{1.9}$$

Now take another adapted frame $\bar{e}_1,\ldots,\bar{e}_n,\phi\bar{e}_1,\ldots,\phi\bar{e}_n,\xi$ with respect to which g and ϕ have the same components as (1.9) and put

$$\bar{e}_i = re_i, \qquad \phi\bar{e}_i = r\phi e_i, \qquad \xi = r\xi,$$

then we can easily see that the orthogonal matrix

$$r = \begin{pmatrix} A_n & B_n & 0 \\ C_n & D_n & 0 \\ 0 & 0 & 1 \end{pmatrix}$$

must have the form

$$r = \begin{pmatrix} A_n & B_n & 0 \\ -B_n & A_n & 0 \\ 0 & 0 & 1 \end{pmatrix}.$$

Thus the structure group of the tangent bundle of M can be reduced to $U(n) \times 1$.

Conversely, if the structure group of the tangent bundle of M can be reduced to $U(n) \times 1$, then we can define g and ϕ as tensors having (1.9) as components with respect to the adapted frames. We can also give a 1-form η and a vector field ξ by $(0,\ldots,0,1)$ and ${}^t(0,\ldots,0,1)$ respectively. They satisfy the desired properties. QED.

Since the structure group of the tangent bundle of an almost contact manifold M reduces to $U(n) \times 1$ and the determinant of every element of $U(n) \times 1$ is positive, we have the following

PROPOSITION 1.4. Every almost contact manifold is orientable.

2. CONTACT MANIFOLDS

A (2n+1)-dimensional manifold M is said to have a *contact structure* and is called a *contact manifold* if it carries a global 1-form η such that

$$\eta \wedge (d\eta)^n \neq 0 \tag{2.1}$$

everywhere on M, where the exponent denotes the nth exterior power. We call η a *contact form* of M.

A quadratic form θ of the Grassman algebra ΛV^*, V^* being dual to a vector space V, is said to have rank 2r if the exterior product $\theta^r \neq 0$ and $\theta^{r+1} = 0$. Equivalently, $\text{rank}\theta = \dim V - \dim V_0$, where $V_0 = \{X : X \in V,\ \theta(X,V) = 0\}$. It follows that on a contact manifold M the condition (2.1) implies that the quadratic form $d\eta$ in the Grassman algebra $\Lambda T_x(M)^*$ has rank 2n. We then have that $V_0 = \{X : X \in T_x(M),$

$d\eta(X,T_x(M)) = 0\}$ is a subspace of dimension 1 on which $\eta \neq 0$, and which is thus complementary to the 2n-dimensional subspace on which $\eta = 0$. Let ξ_x be the element of V_0 on which η has value 1, then ξ is a vector field, which we call an *associated vector field* to η, defined over M by η, and which is never zero since $\eta(\xi) = 1$.

THEOREM 2.1. Let M be a (2n+1)-dimensional manifold with contact structure η. Then there exists an almost contact metric structure (ϕ,ξ,η,g) such that

$$g(X,\phi Y) = d\eta(X,Y).$$

Proof. For the contact form η there exists a vector field ξ such that $\eta(\xi) = 1$ and $d\eta(\xi,T_x(M)) = 0$ at every point x of M. We define a skew-symmetric tensor field ϕ of type (1,1) as follows.

First of all we can prove that there exists a Riemannian metric tensor field h such that $\eta(X) = h(X,\xi)$. On the other hand, $d\eta$ is a symplectic form on the orthogonal complement of ξ and hence that there exists a metric g' and an endomorphism ϕ on the orthogonal complement of ξ such that $g'(X,\phi Y) = d\eta(X,Y)$ and $\phi^2 = -I$. Extending g' to a metric g agreeing with h in the direction ξ and extending ϕ so that $\phi\xi = 0$, we have an almost contact structure (ϕ,ξ,η,g). QED.

For an almost contact metric structure (ϕ,ξ,η,g) on M we put

$$\Phi(X,Y) = g(X,\phi Y). \tag{2.2}$$

We call Φ the *fundamental 2-form* of the almost contact metric structure. Since ϕ has rank 2n, we have $\eta \wedge \Phi^n \neq 0$.

An almost contact metric structure constructed from a contact form η, appearing in Theorem 2.1, is called a *contact metric structure* associated to η and a manifold with such a structure is called a *contact metric manifold*.

An almost contact metric structure with $\Phi = d\eta$ is a contact metric structure.

In the next place, we give a definition of contact structure due to Spencer [1], which is called a *contact structure in the wider sense.* First of all, we notice that the following theorem of Darboux was obtained (see Cartan [1], Sternberg [1]).

THEOREM 2.2. Let ω be a 1-form on an n-dimensional manifold M and suppose that $\omega \wedge (d\omega)^p \neq 0$ and $(d\omega)^{p+1} = 0$ on M. Then about every point there exists a coordinate system $(x^1,\ldots,x^p,y^1,\ldots,y^{n-p})$ such that

$$\omega = dy^{p+1} - \sum_{i=1}^{p} y^i dx^i.$$

From this we see that for every point of a (2n+1)-dimensional contact manifold M, there exists coordinates (x^i,y^i,z), $i = 1,\ldots,n$, such that

$$\eta = dz - \sum_{i=1}^{n} y^i dx^i.$$

Let $(x^1,\ldots,x^n,y^1,\ldots,y^n,z)$ be cartesian coordinates in (2n+1)-dimensional Euclidean space R^{2n+1}, and let η_0 be the 1-form on R^{2n+1} defined by

$$\eta_0 = dz - \sum_{i=1}^{n} y^i dx^i. \tag{2.3}$$

Then we can easily verify that

$$\eta_0 \wedge (d\eta_0)^n \neq 0. \tag{2.4}$$

η_0 is called a contact form on R^{2n+1}.

A diffeomorphism $f : U \longrightarrow U'$, where U and U' are open subsets of R^{2n+1} is called a *contact transformation* if and only if $f^*\eta_0 = \tau\eta_0$, where τ is a non-zero, real valued function on U. We denote by Γ the set of all contact transformations. Γ is a pseudo-group in the following sense:

(i) if $f : U \longrightarrow U'$ and $g : V \longrightarrow V'$ are contact transformations and $U' \cap V \neq \phi$, then $g \cdot f : f^{-1}(U' \cap V) \longrightarrow g(U' \cap V)$ is also a contact transformation,

(ii) by the composition in (i), Γ is associative and

(iii) each element of Γ admits its inverse in Γ.

A contact transformation $f \in \Gamma$ such that

$$f^*\eta_0 = \eta_0 \tag{2.5}$$

is called a *strict contact transformation*. The set Γ_0 of all such transformations is a sub-pseudo-group of Γ.

A (2n+1)-dimensional manifold M will be called a *contact manifold in the wider sense* if there exists an open covering $\{U_i\}$ of M with homeomorphism $f_i : U_i \longrightarrow V_i \subset R^{2n+1}$ such that $f_{ij} = f_i \cdot f_j^{-1}$ for all pairs (i,j) such that f_{ij} is defined. Two such coordinate systems $\{U_i, f_i\}$ and $\{U_i', f_i'\}$ will be called equivalent if $f_i' \cdot f_j^{-1} \in \Gamma$ whenever defined. An equivalence class will be called a contact structure in the wider sense on M.

From the definition we see that there exists a non-zero function ρ_{ij} on $f_j(U_i \cap U_j)$ such that

$$(f_i \cdot f_j^{-1})^* \eta_0 = \rho_{ij}\eta_0.$$

Therefore we have

$$f_i^*\eta_0 = f_j^*(\rho_{ij})f_j^*\eta_0.$$

If we define the 1-form η_i on every U_i by setting

$$\eta_i = f_i^*\eta_0,$$

then we have

$$\eta_i = f_j^*(\rho_{ij})\eta_j,$$

on non-empty $U_i \cap U_j$. Since η_0 satisfies (2.4), we obtain

$$\eta_i \wedge (d\eta_i)^n \neq 0.$$

Let D be the subbundle of the tangent bundle T(M) whose fibre D_x is given by

$$D_x = \{X \in T_x(M): \eta_i(X) = 0\}$$

for $x \in U_i$. Recall that a vector bundle over a manifold with standard fibre R^p is said to be orientable if the structure group of its associated principal fibre bundle with group GL(p,R) can be reduced to $GL^+(p,R)$, which is a subgroup of GL(p,R) consisting of matrices with positive determinants. We put $\eta_i = \tau_{ij}\eta_j$ on $U_i \cap U_j$. Then we have

$$\eta_i \wedge (d\eta_i)^n = \tau_{ij}^{n+1}(\eta_j \wedge (d\eta_j)^n),$$

and τ_{ij}^{n+1} is just the Jacobian of the coordinate transformation. Thus, if M is orientable and n is even, τ_{ij} must be positive and hence vector bundle D is orientable.

We now prove the following theorem (G. W. Gray [1]).

THEOREM 2.3. Let M be a (2n+1)-dimensional orientable contact manifold in the wider sense. If n is even, then M is a contact manifold.

Proof. From the assumption we see that T(M) and D are orientable, and hence the quotient bundle T(M)/D admits a global cross section S without zeros. On the other hand, η_i defines a local cross section S_i over U_i by the equation $\eta_i(S_i) = 1$, and hence $S_i = h_i S$, where the h_i's are non-vanishing functions of the same sign. We define η by $\eta = h_i\eta_i$ on U_i. Then we obtain a global 1-form η such that $\eta \wedge (d\eta)^n \neq 0$. QED.

We now give some examples of contact manifolds.

Example 2.1. Let R^{2n+1} be a (2n+1)-dimensional Euclidean space. Then

$$\eta = dz - \sum_{i=1}^{n} y^i dx^i$$

is a contact form on R^{2n+1}, (x^i, y^i, z) being cartesian coordinates. Then the vector field ξ is given by $\partial/\partial z$.

Example 2.2. Let M be a (2n+1)-dimensional regular hypersurface of R^{2n+2} (i.e., C^∞ with a unique tangent plane at every point). In R^{2n+2} with cartesian coordinates $(x^1, \ldots, x^{2n+2})$, we consider a 1-form defined by

$$\alpha = x^1 dx^2 - x^2 dx^1 + \ldots + x^{2n+1} dx^{2n+2} - x^{2n+2} dx^{2n+1}.$$

Then we have

$$d\alpha = 2(dx^1 \wedge dx^2 + \ldots + dx^{2n+1} \wedge dx^{2n+2}),$$

from which

$$\alpha \wedge (d\alpha)^n = 2^{n-1} n! [\sum_{i=1}^{2n+1} (-1)^{i-1} x^i dx^1 \wedge dx^2 \wedge \ldots$$
$$\ldots \wedge dx^{i-1} \wedge dx^{i+1} \wedge \ldots \wedge dx^{2n+2}].$$

We denote by $v_1, \ldots, v_{2n+1}$ 2n+1 linearly independent vectors which span the tangent space of M at $x_0 = (x_0^1, \ldots, x_0^{2n+2})$. We put

$$w^j = {*dx^j}(v_1, \ldots, v_{2n+1}),$$

where * denotes the Hodge star operator of the Euclidean metric on R^{2n+2}. Then a vector w with components w^j is normal to the hypersurface spanned by $v_1, \ldots, v_{2n+1}$. We also have

$$(\alpha \wedge (d\alpha)^n)(v_1, \ldots, v_{2n+1}) = x_0 \cdot w,$$

where (·) denotes the ordinary scalar product in R^{2n+2}.

On the other hand, the equation of the tangent space of M at x_0 is given by $w\cdot(x-x_0) = 0$. Therefore, the tangent space of M at x_0 passes through the origin if and only if $w\cdot x_0 = 0$, that is, $\alpha \wedge (d\alpha)^n = 0$ at x_0. Moreover, we see that $\eta = i^*\alpha$, i being an immersion of M into R^{2n+2}, satisfies

$$\eta \wedge (d\eta)^n = i^*(\alpha \wedge (d\alpha)^n).$$

Therefore, we see that $\eta \wedge (d\eta)^n$ vanishes at x_0 on M if and only if the tangent space of M at x_0 passes through the origin of R^{2n+2}. Consequently, we have the following theorem (G. W. Gray [1]).

THEOREM 2.4. Let M be a smooth hypersurafce immersed in R^{2n+2}. If the tangent space of M does not pass through the origin of R^{2n+2}, then M has a contact structure.

As a special case of Theorem 2.4, we see that an odd-dimensional sphere S^{2n+1} in R^{2n+2} carries a conatact structure. Furthermore, since α is invariant under substitution $(x^1,\dots,x^{2n+2}) \longrightarrow (-x^1,\dots,-x^{2n+2})$, α also induces a contact structure on the real projective space RP^{2n+1}.

Example 2.3. In R^{2n} we put

$$\beta = \sum_{i=1}^{n} x^i dx^{n+i}.$$

Let R_1^n be the subspace of R^{2n} defined by $x^i = 0$, $i = 1,\dots,n$ and R_2^n the subspace of R^{2n} defined by $x^j = 0$, $j = n+1,\dots,2n$. Then β induces a contact form on a hypersurafce M of dimension 2n-1 immersed in R^{2n} if and only if $M \cap R_1^n = \phi$, $\dim(M \cap R_2^n) = n-1$ and no tangent space to $M \cap R_2^n$ in R_2^n containes the origin in R_2^n.

Example 2.4. Let M be an (n+1)-dimensional Riemannian manifold and $T(M)^*$ its cotangent bundle. We denote by $(x^1,\dots,x^{n+1})$ local coordinates on U and $(p^1,\dots,p^{n+1})$ fibre coordinates over U defined with respect to dx^i's. If $\pi : T(M)^* \longrightarrow M$ is the natural projection, then $(p^i, q^i = x^i\cdot\pi)$ are local coordinates on $T(M)^*$. We put

$$\beta = \sum_{i=1}^{n+1} p^i dq^i$$

on a coordinate neighborhood. We denote by $T_1(M)^*$ the bundle of unit cotangent vectors. Then $T_1(M)^*$ has empty intersection with the zero section of $T(M)^*$. Moreover, the intersection with any fibre of $T(M)^*$ is an n-dimensional sphere and no tangent space to this intersection contains the origin of the fibre. Therefore, from Example 2.3, we see that β induces a contact structure on the hypersurface $T_1(M)^*$.

Similarly, we obtain a contact structure on the bundle $T_1(M)$ of unit tangent vectors.

We denote by g_{ji} the components of the metric with respect to the coordinates $(x^1,\ldots,x^{n+1})$ and by $(v^1,\ldots,v^{n+1})$ the fibre coordinates on $T(M)$. We define a 1-form β locally by

$$\beta = \sum_{i,j} g_{ij} v^j dq^i,$$

where we have put $q^i = x^i \cdot \pi$. From this we have our assertion.

Example 2.5. We now give an example of a contact manifold in the wider sense.

Let $M = R^{n+1} \times RP^n$, where RP^n denotes n-dimensional real projective space. Let $(x^0,\ldots,x^n)$ be coordinates in R^{n+1} and $(t_0,\ldots,t_n)$ homogeneous coordinates in RP^n. The subsets $\{U_i\}$, $i = 1,\ldots,n$ defined by $t_i \neq 0$, form an open covering of M. In U_i we define a 1-form η_i by

$$\eta_i = \frac{1}{t_i} \sum_{j=0}^{n} t_j dx^j.$$

Then we have $\eta_i \wedge (d\eta_i)^n \neq 0$ and $\eta_i = (t_j/t_i)\eta_j$. Thus, M has a contact structure in the wider sense, but for n even, M is non-orientable and hence cannot carry a global contact form.

3. TORSION TENSOR OF ALMOST CONTACT MANIFOLDS

Let M be a $(2n+1)$-dimensional almost contact manifold with almost contact structure (ϕ,ξ,η). We consider a product manifold $M \times R$, where R denotes a real line. Then a vector field on $M \times R$ is given by $(X,f(d/dt))$, where X is a vector field tangent to M, t the coordinate of R and f a function on $M \times R$. We define a linear map J on the tangent space of $M \times R$ by

$$J(X,f\frac{d}{dt}) = (\phi X-f\xi,\eta(X)\frac{d}{dt}). \tag{3.1}$$

Then we have $J^2 = -I$ and hence J is an almost complex structure on $M \times R$. The almost complex structure J is said to be integrable if its Nijenhuis torsion N_J vanishes, where

$$N_J(X,Y) = J^2[X,Y] + [JX,JY] - J[JX,Y] - J[X,JY].$$

If the almost complex structure J on $M \times R$ is integrable, we say that the almost contact structure (ϕ,ξ,η) is *normal*.

In the following we seek to express the condition of normality in term of the Nienhuis torsion N_ϕ of ϕ, which is defined by

$$N_\phi(X,Y) = \phi^2[X,Y] + [\phi X,\phi Y] - \phi[\phi X,Y] - \phi[X,\phi Y].$$

Since N_J is a tensor field of type (1,2), it sufficies to compute $N_J((X,0),(Y,0))$ and $N_J((X,0),(0,d/dt))$ for any vector fields X and Y on M. From (3.1) we have

$$\begin{aligned} N_J((X,0),(Y,0)) = & -([X,Y],0) + ([\phi X,\phi Y],(\phi X\eta(Y)-\phi Y\eta(X))\frac{d}{dt}) \\ & -(\phi[\phi X,Y]+(Y\eta(X))\xi,\eta([\phi X,Y])\frac{d}{dt}) \\ & -(\phi[X,\phi Y]-(X\eta(Y))\xi,\eta([X,\phi Y])\frac{d}{dt}) \end{aligned}$$

$$= (N_\phi(X,Y)+2d\eta(X,Y)\xi, ((L_{\phi X}\eta)Y-(L_{\phi Y}\eta)X)\frac{d}{dt}),$$

$$N_J((X,0),(0,d/dt)) = (-[\phi X,\xi],(\xi\eta(X))\frac{d}{dt})+(\phi[x,\xi],\eta([X,\xi])\frac{d}{dt})$$

$$= ((L_\xi\phi)X,(L_\xi\eta)(X)\frac{d}{dt}).$$

Here we define four tensors $N^{(1)}$, $N^{(2)}$, $N^{(3)}$ and $N^{(4)}$ respectively by

$$N^{(1)}(X,Y) = N_\phi(X,Y) + 2d\eta(X,Y)\xi,$$

$$N^{(2)}(X,Y) = (L_{\phi X}\eta)Y - (L_{\phi Y}\eta)X,$$

$$N^{(3)}(X) = (L_\xi\phi)X, \qquad N^{(4)}(X) = (L_\xi\eta)X.$$

It is clear that the almost contact structure (ϕ,ξ,η) is normal if and only if these four tensors vanish.

LEMMA 3.1. If $N^{(1)} = 0$, then $N^{(2)} = N^{(3)} = N^{(4)} = 0$.

Proof. If $N^{(1)} = 0$, then we have

$$[\xi,X] + \phi[\xi,\phi X] - (\xi\eta(X))\xi = 0. \tag{3.2}$$

Thus we have

$$\eta([\xi,X]) - \xi\eta(X) = 0,$$

which shows that $N^{(4)} = 0$. From this equation we also have $\eta([\xi,\phi X]) = 0$. On the other hand, applying ϕ to (3.2), we see that

$$0 = \phi L_\xi X - L_\xi\phi X + \eta([\xi,\phi X]),$$

from which $(L_\xi\phi)X = 0$ and hence $N^{(3)} = 0$. Finally, from $N^{(1)} = 0$ we have

$$0 = N_\phi(\phi X,Y) + 2d\eta(\phi X,Y)$$

$$= -[\phi X,Y] - [X,\phi Y] - (\phi Y\eta(X))\xi - \eta(X)[\phi Y,\xi]$$

$$-\phi[-X+\eta(X)\xi,Y] - \phi[\phi X,\phi Y] + (\phi X\eta(Y))\xi.$$

Applying η to this equation and using $\eta([\xi,\phi X]) = 0$, we get

$$\phi X\eta(Y) - \eta([\phi X,Y]) - \phi Y\eta(X) + \eta([\phi Y,X]) = 0.$$

Thus we have $N^{(2)} = 0$. QED.

In view of Lemma 3.1 we have

PROPOSITION 3.1. The almost contact structure (ϕ,ξ,η) of M is normal if and only if

$$N_\phi + 2d\eta \otimes \xi = 0.$$

LEMMA 3.2. Let M be a contact metric manifold with contact metric structure (ϕ,ξ,η,g). Then $N^{(2)}$ and $N^{(4)}$ vanish. Moreover, $N^{(3)}$ vanishes if and only if ξ is a Killing vector field with respect to g.

Proof. We have

$$d\eta(\phi X,\phi Y) = \Phi(\phi X,\phi Y) = g(\phi X,\phi^2 Y)$$

$$= -g(X,\phi^3 Y) = g(X,\phi Y) = d\eta(X,Y),$$

from which

$$d\eta(\phi X,Y) + d\eta(X,\phi Y) = 0.$$

This is equivalent to $N^{(2)} = 0$. On the other hand, we have

$$0 = g(X,\phi\xi) = d\eta(X,\xi) = \tfrac{1}{2}(X\eta(\xi) - \xi\eta(X) - \eta([X,\xi])).$$

Thus we obtain

$$\xi\eta(X) - \eta([\xi,X]) = 0.$$

Therefore, we have $L_\xi\eta = 0$ and hence $N^{(4)} = 0$. Moreover, we see that

$$(L_\xi g)(X,\xi) = \xi\eta(X) - \eta([\xi,X]) = (L_\xi\eta)X = 0.$$

On the other hand, we easily see that $L_\xi d\eta = 0$ and consequently

$$(L_\xi d\eta)(X,Y) = (L_\xi\Phi)(X,Y) = 0,$$

from which

$$0 = \xi g(X,\phi Y) - g([\xi,X],\phi Y) - g(X,\phi[\xi,Y])$$

$$= (L_\xi g)(X,\phi Y) + g(X,(L_\xi\phi)Y)$$

$$= (L_\xi g)(X,\phi Y) + g(X,N^{(3)}(Y)).$$

Thus ξ is a Killing vector field if and only if $N^{(3)} = 0$. QED.

LEMMA 3.3. For an almost contact metric structure (ϕ,ξ,η,g) of M we have

$$2g((\nabla_X\phi)Y,Z) = 3d\Phi(X,\phi Y,\phi Z) - 3d\Phi(X,Y,Z) + g(N^{(1)}(Y,Z),\phi X)$$
$$+ N^{(2)}(Y,Z)\eta(X) + 2d\eta(\phi Y,X)\eta(Z) - 2d\eta(\phi Z,X)\eta(Y).$$

Proof. The Riemannian connection ∇ with respect to g is given by

$$2g(\nabla_X Y,Z) = Xg(Y,Z) + Yg(X,Z) - Zg(X,Y)$$
$$+ g([X,Y],Z) + g([Z,X],Y) - g([Y,Z],X).$$

On the other hand, $d\Phi$ is given by

$$3d\Phi(X,Y,Z) = X\Phi(Y,Z) + Y\Phi(Z,X) + Z\Phi(X,Y)$$
$$- \Phi([X,Y],Z) - \Phi([Z,X],Y) - \Phi([Y,Z],X).$$

From these equations and (2.2) we have our equation. QED.

LEMMA 3.4. For a contact metric structure (ϕ,ξ,η,g) of M with $\Phi = d\eta$ and $N^{(2)} = 0$, we have

$$2g((\nabla_X\phi)Y,Z) = g(N^{(1)}(Y,Z),\phi X) + 2d\eta(\phi Y,X)\eta(Z) - 2d\eta(\phi Z,X)\eta(Y).$$

Especially we have $\nabla_\xi\phi = 0$.

Proof. The first equation is trivial by the assumption. We prove that $\nabla_\xi\phi = 0$. From $N^{(2)} = 0$ we have $d\eta(X,\xi) = 0$. Thus the first equation implies that $\nabla_\xi\phi = 0$. QED.

In the case of Lemma 3.4 it is also easy to see that the integral curves of ξ are geodesics, that is, $\nabla_\xi\xi = 0$.

Let M be a (2n+1)-dimensional contact metric manifold with contact metric structure (ϕ,ξ,η,g). If the structure vector field ξ is a Killing vector field with respect to g, then the contact structure on M is called a K-*contact structure* and M is called a K-*contact manifold*.

From Lemma 3.2 we have the following

PROPOSITION 3.2. Let M be a contact metric manifold. Then M is a K-contact manifold if and only if $N^{(3)}$ vanishes.

Since we have

$$g(X,\phi Y) = d\eta(X,Y) = \tfrac{1}{2}(g(\nabla_X\xi,Y) - g(\nabla_Y\xi,X)) = g(\nabla_X\xi,Y),$$

for a K-contact structure, we obtain $\nabla_X\xi = -\phi X$. Conversely, if $\nabla_X\xi = -\phi X$, as ϕ is skew-symmetric, ξ is a Killing vector field. Thus we have

PROPOSITION 3.3. Let M be a contact metric manifold. Then M is a K-contact manifold if and only if

$$\nabla_X \xi = -\phi X.$$

We now give a geometric characterization of K-contact manifolds.

THEOREM 3.1. In order that a (2n+1)-dimensional Riemannian manifold M is K-contact, it is necessary and sufficient that the following two conditions are satisfied:

(1) M admits a unit Killing vector field ξ;

(2) The sectional curvature for plane sections containing ξ are equal to 1 at every point of M.

Proof. Let M be a K-contact manifold. Then

$$g(R(X,\xi)\xi,X) = g(-\phi^2 X,X) = g(X,X) = 1,$$

where X is a unit vector field orthogonal to ξ.

Conversely, we suppose that M satisfies the conditions (1) and (2). Since ξ is a Killing vector field, we have

$$R(X,\xi)Y = \nabla_X \nabla_Y \xi - \nabla_{\nabla_X Y}\xi. \tag{3.3}$$

We put $\eta(X) = g(X,\xi)$ and $\phi X = -\nabla_X \xi$. Then we easily see that $\phi\xi = 0$. From (3.3) we also have

$$1 = g(R(X,\xi)\xi,X) = -g(\phi^2 X,X),$$

where X is a unit vector field on M orthogonal to ξ. Therefore, we obtain $\phi^2 X = -X$ for every vector field X of M orthogonal to ξ and hence

$$\phi^2 Y = -Y + \eta(Y)\xi$$

for any vector field Y of M. Moreover, we see that

$$d\eta(X,Y) = \tfrac{1}{2}(g(\nabla_X\xi,Y) - g(\nabla_Y\xi,X)) = -g(\nabla_Y\xi,X) = g(X,\phi Y).$$

Consequently, (ϕ,ξ,η,g) is a K-contact structure on M. QED.

From (3.3) we also have

$$R(X,\xi)\xi = -\phi^2 X = X - \eta(X)\xi. \tag{3.4}$$

We easily see that a $(2n+1)$-dimensional Riemannian manifold M admitting a unit Killing vector field ξ which satisfies (3.4) is a K-contact manifold.

4. CONTACT DISTRIBUTION

Let M be a $(2n+1)$-dimensional contact manifold with contact form η. Then $\eta = 0$ defines a 2n-dimensional distribution D of the tangent bundle. The distribution D is called the *contact distribution* which is as far from being integrable as possible from the fact that $\eta \wedge (d\eta)^n \neq 0$.

In the following we consider an integral submanifold of the distribution D (Sasaki [1]).

THEOREM 4.1. Let M be a $(2n+1)$-dimensional contact manifold with contact form η. Then there exist integral submanifolds of the contact distribution D of dimension n but no higher dimension.

Proof. We can choose local coordinates (x^i,y^i,z) such that $\eta = dz - \Sigma y^i dx^i$ on the coordinate neighborhood. Then for a point x with coordinates (x_0^i,y_0^i,z_0) in the coordinate neighborhood, $x^i = x_0^i$, $z = z_0$ defines an n-dimensional integral submanifold and a maximal integral submanifold containing this coordinate slice is an integral submanifold of D in M.

Let N be an r-dimensional integral submanifold of D. We suppose that $r > n$. We denote by $e_1,\dots,e_r$ r linearly independent local vector fields tangent to N and extend these to a basis $e_{r+1},\ldots,e_{2n},e_{2n+1}=\xi$ of M. Then we have

$$\eta(e_i) = 0, \qquad d\eta(e_i,e_j) = 0, \quad i,j = 1,\ldots,r.$$

Thus, since $r > n$, we see that $(\eta \wedge (d\eta)^n)(e_1,\ldots,e_{2n+1}) = 0$, which is a contradiction. QED.

PROPOSITION 4.1. Let N be an r-dimensional submanifold immersed in a (2n+1)-dimensional contact manifold M. Then N is an integral submanifold of D if and only if η and $d\eta$ vanish on N. Let (ϕ,ξ,η,g) be an associated almost contact metric structure. Then N is an integral submanifold of D if and only if every tangent vector X of N belongs to D and ϕX is normal to N in M.

Proof. For any vector fields X and Y tangent to N we see that $\eta(X) = \eta(Y) = 0$ and hence $d\eta(X,Y) = 0$. Conversely, if η and $d\eta$ vanish on N, we have

$$0 = d\eta(X,Y) = -\tfrac{1}{2}\eta([X,Y])$$

for any vector fields X and Y tangent to N. Thus N is an integral submanifold of D.

The second statement follows immediately from the fact that $d\eta(X,Y) = g(X,\phi Y)$. QED.

LEMMA 4.1. Let (x^i,y^i,z) be local coordinates for $x = (x_0^i,y_0^i,z_0)$ such that $\eta = dz - \Sigma y^i dx^i$ on the coordinate neighborhood. In order that r linearly independent vectors X_t, $t = 1,\ldots,r \leq n$, at x with components (a_t^i,b_t^i,c_t) be tangent to an r-dimensional integral submanifold it is necessary and sufficient that $\eta(X_t) = 0$ and $d\eta(X_t,X_s) = 0$, that is,

$$c_t = \sum_i y_0^i a_t^i, \qquad \sum_i a_t^i b_s^i = \sum_i a_s^i b_t^i.$$

Proof. Since the necessity is clear, we prove the sufficiency.

We put $c_{ts} = \Sigma a^i_t b^i_s$ and choose a sufficiently small neighborhood V of the origin of R^r with coordinates $(u^1,\ldots,u^r)$ such that

$$x^i = x^i_0 + \sum_t a^i_t u^t, \qquad y^i = y^i_0 + \sum_t b^i_t u^t,$$

$$z = z_0 + \sum_t c_t u^t + \tfrac{1}{2}\sum_{t,s} c_{ts} u^t u^s$$

define a mapping i of V into M. Then

$$\partial x^i/\partial u^t = a^i_t, \qquad \partial y^i/\partial u^t = b^i_t,$$

$$\begin{aligned}\partial z/\partial u^t &= c_t + \sum_s c_{ts} u^s = \sum_i y^i_0(\partial x^i/\partial u^t) + \sum_{i,s}(\partial x^i/\partial u^t)(\partial y^i/\partial u^s)u^s \\ &= \sum_i y^i(\partial x^i/\partial u^t)\end{aligned}$$

and hence the mapping i defines an integral submanifold tangent to $X_1,\ldots,X_r$ at x. QED.

THEOREM 4.2. Let X be a vector at a point x of M belonging to D. Then there exists an r-dimensional integral submanifold N $(1 \le r \le n)$ of D through x such that X is tangent to N.

Proof. Let (a^i_1, b^i_1, c_1) be the components of X with respect to the local coordinates (x^i, y^i, z). Since $X \in D$ we have $c_1 = \Sigma_i y^i_0 a^i_1$. If not all the a^i_1's vanish, choose $a^i_2,\ldots,a^i_r$ such that $\mathrm{rank}(a^i_t) = r$ and define $c_2,\ldots,c_r$ by $c_t = \Sigma_i y^i_0 a^i_t$, $t = 2,\ldots,r$. We now define $b^i_2,\ldots,b^i_r$ inductively as follows. We suppose that b^i_1,s are given. We take b^i_{s+1} $(1 \le s \le r-1)$ as a set of solutions of

$$\sum_i a^i_1 f^i = \sum_i a^i_{s+1} b^i_1, \ \ldots\ldots, \sum_i a^i_s f^i = \sum_i a^i_{s+1} b^i_s$$

which exists as rank $(a^i_t) = r$. Then $\{(a^i_t, b^i_t, c_t)\}$ satisfy the conditions of the previous Lemma 4.1 and hence we have an integral submanifold N with X tangent as desired.

If on the other hand, all of the a^i_1's vanish, then c_1 also vanishes, so choosing $b^i_2,\ldots,b^i_r$ such that $\mathrm{rank}(b^i_t) = r$ we again have an r-dimensional integral submanifold with X tangent to by Lemma 4.1. QED.

A diffeomorphism $f\colon M \longrightarrow M$ is called a *contact transformation* if $f^*\eta = \tau\eta$ for some non-vanishing function τ on M. If moreover $\tau = 1$, then f is called a *strict contact transformation*.

The following lemma is trivial.

LEMMA 4.2. A diffeomorphism f on a contact manifold M is a contact transformation if and only if f_*X belongs to D for every X in D.

THEOREM 4.3. A diffeomorphism f on a contact manifold M is a contact transformation if and only if f maps n-dimensional integral submanifolds of D onto n-dimensional integral submanifolds of D.

Proof. Let f be a contact transformation and N be an n-dimensional integral submanifold. Then we have $f_*(T_x(N))$ in D. Thus f(N) is an integral submanifold. Conversely, for any vector X at x in D, we have seen that there exists an integral submanifold N through x with X as a tangent vector. Since f(N) is also an integral submanifold, f_*X is in D and f is a contact transformation by Lemma 4.2. QED.

5. SASAKIAN MANIFOLDS

Let M be a (2n+1)-dimensional contact metric manifold with contact metric structure (ϕ,ξ,η,g). If the contact metric structure of M is normal, then M is said to have a *Sasakian structure* (or *normal contact metric structure*) and M is called a *Sasakian manifold* (or *normal contact metric manifold*).

We denote by ∇ the operator of covariant differentiation with respect to g. Then we have

THEOREM 5.1. An almost contact metric structure (ϕ,ξ,η,g) on M is a Sasakian structure if and only if

$$(\nabla_X\phi)Y = g(X,Y)\xi - \eta(Y)X. \tag{5.1}$$

Proof. If the structure is normal, we have $\Phi = d\eta$ and $N^{(1)} = N^{(2)} = 0$. Thus Lemma 3.3 implies (5.1).

Conversely, we suppose that the structure satisfies (5.1). Putting $Y = \xi$ in (5.1), we have $-\phi\nabla_X\xi = \eta(X)\xi - X$, and hence, applying ϕ to this, we obtain $\nabla_X\xi = -\phi X$. Since ϕ is skew-symmetric, we see that ξ is a Killing vector field. Moreover, we obtain

$$d\eta(X,Y) = \tfrac{1}{2}((\nabla_X\eta)Y - (\nabla_Y\eta)X) = g(X,\phi Y) = \Phi(X,Y).$$

Thus the structure is a contact metric structure. Furthermore, by a straightforward computation we have $N_\phi + 2d\eta \otimes \xi = 0$. Therefore, the structure is Sasakian. QED.

If M is a Sasakian manifold, from (5.1) we have

$$R(X,Y)\xi = \eta(Y)X - \eta(X)Y, \tag{5.2}$$

where R denotes the Riemannian curvature tensor of M. From (5.2) we also have

$$R(X,\xi)Y = -g(X,Y)\xi + \eta(Y)X. \tag{5.3}$$

THEOREM 5.2. Let M be a (2n+1)-dimensional Riemannian manifold admitting a unit Killing vector field ξ. Then M is a Sasakian manifold if and only if (5.3) holds.

Proof. From Theorem 3.1 we see that M has a K-contact structure (ϕ,ξ,η,g). Then we have

$$R(X,\xi)Y = \nabla_X\nabla_Y\xi - \nabla_{\nabla_X Y}\xi = -(\nabla_X\phi)Y.$$

Thus Theorem 5.1 proves our assertion. QED.

We now give some examples of Sasakian manifolds.

Example 5.1. Let M be a real (2n+1)-dimensional hypersurface of a Kaehlerian manifold $\bar{M}$ of complex dimension n+1. We denote by J the almost complex structure tensor field of $\bar{M}$ and $\bar{\nabla}$ the operator of covariant differentiation in $\bar{M}$. The operator of covariant differentiation with respect to the induced connection on M will be denoted by ∇. We denote by C a unit normal of M in $\bar{M}$ and by A the second fundamental tensor of M. We put $JC = -\xi$. Then ξ is a unit vector field on M. For any vector field X tangent to M we put

$$JX = \phi X + \eta(X)C,$$

where η is a 1-form dual to ξ. Then we have

$$\phi^2 X = -X + \eta(X)\xi, \quad \phi\xi = 0, \quad \eta(\phi X) = 0, \quad \eta(X) = g(\xi,X).$$

Thus (ϕ,ξ,η,g) defines an almost contact metric structure on M.

Moreover, the Gauss and Weingarten formulas for M are respectively given by

$$\bar{\nabla}_X Y = \nabla_X Y + g(AX,Y)C \quad \text{and} \quad \bar{\nabla}_X C = -AX.$$

From these equations we have

$$(\nabla_X \phi)Y = \eta(Y)AX - g(AX,Y)\xi, \quad \nabla_X \xi = \phi AX.$$

We suppose that $AX = -X + \beta\eta(X)\xi$, where β is a function on M. Then we have

$$g(\nabla_X \xi,Y) + g(\nabla_Y \xi,X) = -g(\phi X,Y) - g(\phi Y,X) = 0.$$

Therefore, ξ is a Killing vector field on M. Furthermore, we have

$$(\nabla_X \phi)Y = g(X,Y)\xi - \eta(Y)X.$$

Thus Theorem 5.1 shows that M is a Sasakian manifold.

Conversely, if M is a Sasakian manifold, we obtain

$$AX = -X + \eta(AX+X)\xi,$$

from which

$$AX = -X + \beta\eta(X)\xi,$$

where we have put $\beta = \eta(A\xi) + 1$.

Consequently, a real hypersurface M of a Kaehlerian manifold $\bar{M}$ is a Sasakian manifold if and only if its second fundamental form A satisfies $A = -I + \beta\eta \otimes \xi$. We notice that such a real hypersurface is an η-umbilical hypersurface.

For a real hypersurface M of a Kaehlerian manifold $\bar{M}$ we see that

$$N_\phi(X,Y) + 2d\eta(X,Y)\xi = \eta(X)(\phi A-A\phi)Y - \eta(Y)(\phi A-A\phi)X.$$

From this we easily see that $\phi A = A\phi$ if and only if $N_\phi(X,Y)+2d\eta(X,Y)\xi = 0$. Therefore, the almost contact metric structure (ϕ,ξ,η,g) on M is normal if and only if ξ is a Killing vector field on M.

Example 5.2. Let C^{n+1} be a complex (n+1)-dimensional number space. Then C^{n+1} admits a Kaehlerian structure J. Let S^{2n+1} be a (2n+1)-dimensional unit sphere, i.e.,

$$S^{2n+1} = \{z \in C^{n+1} : \ |z| = 1\}.$$

Then S^{2n+1} is a real hypersurface of C^{n+1} and the second fundamental form A of S^{2n+1} in C^{n+1} is given by $A = -I$. Therefore, from Example 1.1, S^{2n+1} admits a Sasakian structure (ϕ,ξ,η,g) which is called a natural Sasakian structure on S^{2n+1}.

Example 5.3. Let R^{2n+1} be a (2n+1)-dimensional number space. We put

$$\eta = \tfrac{1}{2}(dz - \sum_{i=1}^{n} y^i dx^i),$$

(x^i, y^i, z) being cartesian coordinates. Then the structure vector field ξ is given by $\xi = -2\partial/\partial z$ and the Riemannian metric tensor field g is given by

$$g = \tfrac{1}{4}(\eta \otimes \eta + \sum_{i=1}^{n} ((dx^i)^2 + (dy^i)^2).$$

This gives a contact metric structure on R^{2n+1} as follows. First of all we have

$$g = \tfrac{1}{4}\begin{pmatrix} \delta_{ij}+y^i y^j & 0 & -y^i \\ 0 & \delta_{ij} & 0 \\ -y^i & 0 & 1 \end{pmatrix}.$$

We give a tensor field ϕ of type (1,1) by a matrix form

$$\phi = \begin{pmatrix} 0 & \delta_{ij} & 0 \\ -\delta_{ij} & 0 & 0 \\ 0 & y^j & 0 \end{pmatrix}.$$

The vector fields $X_i = 2\partial/\partial y^i$, $X_{n+i} = 2(\partial/\partial x^i + y^i\partial/\partial z)$, ξ form a ϕ-basis for the contact metric structure. On the other hand, we can see that $N_\phi + 2d\eta \otimes \xi = 0$ and hence the contact metric structure is normal.

Let M be a (2n+1)-dimensional Sasakian manifold with Sasakian structure (ϕ, ξ, η, g). From (5.1) we easily see

$$\text{(5.4)} \quad R(X,Y)\phi Z = \phi R(X,Y)Z + g(\phi X,Z)Y - g(Y,Z)\phi X + g(X,Z)\phi Y - g(\phi Y,Z)X.$$

From (5.4) we also have the following equations:

$$\text{(5.5)} \quad R(X,Y)Z = -\phi R(X,Y)\phi Z + g(Y,Z)X - g(X,Z)Y - g(\phi Y,Z)\phi X + g(\phi X,Z)\phi Y,$$

$$\text{(5.6)} \quad g(R(\phi X,\phi Y)\phi Z,\phi W) = g(R(X,Y)Z,W) - \eta(Y)\eta(Z)g(X,W) - \eta(X)\eta(W)g(Y,Z) + \eta(Y)\eta(W)g(X,Z) + \eta(X)\eta(Z)g(Y,W).$$

A plane section in $T_x(M)$ is called a ϕ-*section* if there exists a unit vector X in $T_x(M)$ orthogonal to ξ such that $\{X,\phi X\}$ is an orthonormal basis of the plane section. Then the sectional curvature $K(X,\phi X) = g(R(X,\phi X)\phi X,X)$ is called a ϕ-*sectional curvature*, which will be denoted by $H(X)$. We shall show that on a Sasakian manifold the ϕ-sectional curvatures determine the curvature completely.

In the following we prepare some lemmas. We put

$$P(X,Y;Z,W) = g(Y,Z)g(\phi X,W) - g(\phi X,Z)g(Y,W) + g(\phi Y,Z)g(X,W) - g(X,Z)g(\phi Y,W).$$

Then we have

$$P(X,Y;Z,W) = -P(Z,W;X,Y).$$

If $\{X,Y\}$ is an orthonormal pair orthogonal to ξ and if we put $g(X,\phi Y) = \cos\theta$, $0 \le \theta \le \pi$, then

$$P(X,Y;X,Y) = -\sin^2\theta.$$

We now put

$$B(X,Y) = g(R(X,Y)Y,X)$$

for any vectors X and Y and

$$D(X) = B(X,\phi X)$$

for any vector X orthogonal to ξ. Then we have

LEMMA 5.1. For any vectors X and Y orthogonal to ξ we obtain

$$B(X,Y) = \frac{1}{32}[D(X+\phi Y) + 3D(X-\phi Y) - D(X+Y) - D(X-Y)$$
$$- 4D(X) - 4D(Y) - 24P(X,Y;X,\phi Y)].$$

Proof. First of all we have

$$D(X+Y) + D(X-Y) = 2[D(X) + D(Y) + 2B(X, Y)$$
$$+ 2g(R(X,\phi X)\phi Y,Y) + 2g(R(X,\phi Y)\phi X,Y)].$$

Putting ϕY instead of Y in this equation we obtain

$$D(X+\phi Y) + D(X-\phi Y) = 2[D(X) + D(\phi Y) + 2B(X,Y)$$
$$+ 2g(R(X,\phi X)\phi Y,Y) + 2g(R(X,Y)\phi Y,\phi X)].$$

Since $D(\phi Y) = D(Y)$, we find

$$3D(X+\phi Y) + 3D(X-\phi Y) - D(X+Y) - D(X-Y) - 4D(X) - 4D(Y)$$
$$= 12B(X,Y) - 4B(X,\phi Y) + 8g(R(X,\phi X)\phi Y,Y)$$
$$+ 12g(R(X,Y)\phi Y,\phi X) + 4g(R(X,\phi Y)Y,\phi X).$$

On the other hand, from (5.4) and the Bianchi's identity, we have

$$8g(R(X,\phi X)\phi Y,Y) = 8[B(X,Y) + B(X,\phi Y) + 2P(X,Y;X,\phi Y)].$$

We also have the following equations:

$$12g(R(X,Y)\phi Y,\phi X) = 12[B(X,Y) + P(X,Y;X,\phi Y)],$$

$$4g(R(X,\phi Y)Y,\phi X) = 4[-B(X,\phi Y) + P(X,\phi Y;X,Y)].$$

From these equations we have our assertion. QED.

We notice that $D(X) = H(X)$ if and only if X is a unit vector and $B(X,Y) = K(X,Y)$ if and only if $\{X,Y\}$ is an orthonormal pair.

LEMMA 5.2. Let $\{X,Y\}$ be an orthonormal pair of the tangent space of a Sasakian manifold M orthogonal to ξ. If we put $g(X,\phi Y) = \cos\theta$ $(0 \leq \theta \leq \pi)$, then

$$K(X,Y) = \frac{1}{8}[3(1+\cos\theta)^2 H(\frac{X+\phi Y}{|X+\phi Y|}) + 3(1-\cos\theta)^2 H(\frac{X-\phi Y}{|X-\phi Y|})$$
$$- H(\frac{X+Y}{|X+Y|}) - H(\frac{X-Y}{|X-Y|}) - H(X) - H(Y) + 6\sin^2\theta].$$

Proof. Since $D(Z) = |Z|^4 H(Z/|Z|)$ for any Z, we see that

$$D(X+\phi Y) = |X+\phi Y|^4 H(\frac{X+\phi Y}{|X+\phi Y|}) = 4(1+\cos\theta)^2 H(\frac{X+\phi Y}{|X+\phi Y|}).$$

Similarly, we obtain

$$D(X-\phi Y) = 4(1-\cos\theta)^2 H(\frac{X-\phi Y}{|X-\phi Y|}),$$

$$D(X+Y) = 4H(\frac{X+Y}{|X+Y|}), \quad D(X-Y) = 4H(\frac{X-Y}{|X-Y|}).$$

From these equations and Lemma 5.1 we have our equation. QED.

THEOREM 5.3. The ϕ-sectional curvatures determine the curvature of a Sasakian manifold.

Proof. Since the sectional curvatures of a Riemannian manifold determine the curvature, it sufficies to show that the sectional curvatures are determined by the ϕ-sectional curvatures uniquely.

Let $\{X,Y\}$ be an orthonormal pair. We put

$$X = \eta(X)\xi + aZ, \qquad Y = \eta(Y)\xi + bW,$$

where $a = (1 - \eta(X)^2)^{1/2}$, $b = (1 - \eta(Y)^2)^{1/2}$, Z, W being orthogonal to ξ. Then Z and W are unit vectors. By a simple computation we find

$$K(X,Y) = a^2\eta(Y)^2 - 2ab\eta(X)\eta(Y)g(Z,W) + b^2\eta(X)^2 + a^2b^2g(R(Z,W)W,Z).$$

Noticing that

$$g(Z,W) = -\frac{1}{ab}\eta(X)\eta(Y),$$

$$g(R(Z,W)W,Z) = [1-g(Z,W)^2]K(Z,W) = [1 - \frac{1}{a^2b^2}\eta(X)^2\eta(Y)^2]K(Z,W),$$

we obtain

$$K(X,Y) = \eta(X)^2 + \eta(Y)^2 + [1-\eta(X)^2-\eta(Y)^2]K(Z,W).$$

On the other hand, by Lemma 5.2, $K(Z,W)$ is determined by ϕ-sectional curvatures. This completes the proof. QED.

THEOREM 5.4. If the ϕ-sectional curvature at any point of a Sasakian manifold of dimension ≥ 5 is independent of the choice of ϕ-section at the point, then it is constant on the manifold and the curvature tensor is given by

$$R(X,Y)Z = \tfrac{1}{4}(c+3)[g(Y,Z)-g(X,Z)Y] + \tfrac{1}{4}(c-1)[\eta(X)\eta(Z)Y-\eta(Y)\eta(Z)X$$
$$+g(X,Z)\eta(Y)\xi-g(Y,Z)\eta(X)\xi+g(\phi Y,Z)\phi X-g(\phi X,Z)\phi Y+2g(X,\phi Y)\phi Z],$$

where c is the constant ϕ-sectional curvature.

Proof. We notice that $R(X,\xi)X = -\xi$ and $R(\xi,X)\xi = -X$ for any vector X orthogonal to ξ. Thus we have actually prove that any vector field of type (5.3) on a Sasakian manifold satisfying the symmetries of the curvature tensor, the Bianchi's identity, (5.4) and which coincides with the values of the ϕ-sectional curvatures must be the curvature tensor. From this we see that $R(X,Y)Z$ is of the given form.

Conversely, we easily see that c is constant when dimension ≥ 5.

QED.

A Sasakian manifold M is called a *Sasakian space form* if M has constant ϕ-sectional curvature c, and will be denoted by M(c).

Example 5.4. Let S^{2n+1} be a unit sphere with natural Sasakian structure (ϕ,ξ,η,g). We consider the deformed structure:

$$\eta^* = \alpha\eta, \qquad \xi^* = \alpha^{-1}\xi, \qquad \phi^* = \phi,$$

$$g^* = \alpha g + (\alpha^2 - \alpha)\eta \otimes \eta,$$

where α is a positive constant. We call this deformation D-*homothetic deformation*. Then $(\phi^*,\xi^*,\eta^*,g^*,\alpha)$ is a Sasakian structure on S^{2n+1} with constant ϕ-sectional curvature $c = 4/\alpha - 3$. We denote by $S^{2n+1}(c)$ the Sasakian manifold with this structure.

Example 5.5. Let R^{2n+1} be a (2n+1)-dimensional number space with Sasakian structure as in Example 5.3. It is checked that R^{2n+1} is of constant ϕ-sectional curvature -3 and we denote it by $R^{2n+1}(-3)$.

Example 5.6. Let CD^n be a simply connected bounded complex domain in C^n with constant holomorphic sectional curvature $c < 0$. We denote by (J,G) a Kaehlerian structure on CD^n. Since the fundamental 2-form Ω of the Kaehlerian structure is closed, $\Omega = d\omega$ for some real analytic 1-form ω. Let t denote the coordinate on R and put $\eta = 2\omega + dt$ on a product space $R \times CD^n$. If we consider R as an additive group, then η is an infinitesimal connection form on the trivial line bundle (R,CD^n). We have $\xi = \partial/\partial t$ and $g = \pi^*G + \eta \otimes \eta$, where $\pi : (R,CD^n) \longrightarrow CD^n$ is the projection. η is also written as $\eta = 2\pi^*\omega + dt$, and we have $d\eta = 2\pi^*\omega$. Therefore, these tensors define a Sasakian structure on (R, CD^n) with constant ϕ-sectional curvature $k = c-3 < -3$. We denote this by $(R, CD^n)(k)$.

We show that three types of model spaces in these examples 5.4, 5.5 and 5.6 are unique up to isomorphisms, where an isomorphism means a diffeomorphism which maps the structure tensors into the corresponding structure tensors (Tanno [4]).

THEOREM 5.5. Let M be a (2n+1)-dimensional complete simply connected Sasakian manifold with constant ϕ-sectional curvature c.

(1) If $c > -3$, then M is isomorphic to $S^{2n+1}(c)$ or M is D-homothetic to S^{2n+1};

(2) If $c = -3$, then M is isomorphic to $R^{2n+1}(-3)$;

(3) If $c < -3$, then M is isomorphic to $(R, CD^n)(c)$.

Proof. From the assumption M admits local ϕ-holomorphic free mobility and hence M admits global ϕ-holomorphic free mobility because of M is complete and simply connected. Thus M admits an automorphism group Aut(M) such that, for any point x and y, any ϕ-section at x carried to any other ϕ-section at y by some element of Aut(M). Aut(M) is of $(n+1)^2$-dimension and M is diffeomorphic to a homogeneous space Aut(M)/(isotoropy group). Thus we can assume that M is real analytic, and also that g is real analytic. We denote by M^* one of the model spaces corresponding to $c > -3$, $= 3$ or < -3 and by $(\phi^*, \xi^*, \eta^*, g^*)$ the structure tensors of M^*. For any point x of M and x^* of M^*, let $(e_1, \ldots, e_n, \phi e_1, \ldots, \phi e_n, \xi)$ and $(e_1^*, \ldots, e_n^*, \phi e_1^*, \ldots, \phi e_n^*, \xi^*)$ be orthonormal ϕ-basis at x and x^* respectively. We define a linear isomorphism $F: T_x(M) \longrightarrow T_{x^*}(M^*)$ by $Fe_i = e_i^*$, $F\phi e_i = \phi^* e_i^*$ $(i = 1, \ldots, n)$ and $F\xi = \xi^*$. Then we have $F\phi = \phi^* F$ and F is isometric at x, that is, F is isomorphic at x. Since both ϕ- and ϕ^*-sectional curvatures are equal to c, F maps R into R^*, F being considered as a map of tensor algebra. The covariant derivatives of ϕ and ξ are written in terms of ϕ, ξ, g and hence the covariant derivative of R is expressed by ϕ, ξ and g, that is, F maps $(\nabla R)_x$ into $(\nabla^* R^*)_{x^*}$. Likewise, we see that F maps

$(\nabla^k R)_x$ into $(\nabla^{*k} R^*)_{x^*}$ for every positive integer k. Then we have an isometry f of M onto M* such that f(x) = x* and the differential of f at x is F (cf. Kobayashi-Nomizu [1; p.259-261]). We then have that $(\nabla\xi)_x$ is mapped to $(\nabla^*\xi^*)_{x^*}$. Thus we have

$$(\nabla^*(f\xi))_{x^*} = f\cdot(\nabla\xi)_x = F\cdot(\nabla\xi)_x = (\nabla^*\xi^*)_{x^*}.$$

Since f is an isometry, $f\xi$ is also a Killing vector field. By $(f\xi)_{x^*} = \xi^*_{x^*}$ and $(\nabla^*(f\xi))_{x^*} = (\nabla^*\xi^*)_{x^*}$, we get $f\xi = \xi^*$. Because ϕ and η (ϕ^* and η^*) are determined by g and ξ (g* and ξ^*), f is an isomorphism between M and M*. QED.

In the following we study the properties of the Ricci tensor of a Sasakian manifold. First, let M be a (2n+1)-dimensional K-contact manifold with structure tensors (ϕ,ξ,η,g). We denote by S and Q the Ricci tensor and the Ricci operator of M respectively. We prove

PROPOSITION 5.1. If the Ricci tensor S of a K-contact manifold M is parallel, then M is an Einstein manifold.

Proof. Since $R(X,\xi)\xi = X - \eta(X)\xi$, we obtain $S(\xi,\xi) = 2n$. Thus we have $(\nabla_X S)(\xi,\xi) = 2S(\phi X,\xi) = 0$, and hence $S(X,\xi) = 2n\eta(X)$. From this we see that $S(X,\phi Y) = 2ng(X,\phi Y)$. Therefore, we obtain $S(X,Y) = 2ng(X,Y)$ for any vector fields X and Y, which means that M is an Einstein manifold. QED.

PROPOSITION 5.2. Let M be a K-contact manifold. If M is locally symmetric, then M is a Sasakian manifold with constant curvature 1.

Proof. By the assumption we have

$$-R(X,\phi Y)\xi - R(X,\xi)\phi Y = g(\phi Y,X)\xi + \eta(X)\phi Y.$$

Replacing Y by ϕY in this equation, we see that

$$R(X,Y)\xi + R(X,\xi)Y = 2\eta(Y)X - \eta(X)\xi - g(X,Y)\xi.$$

From this we find

$$R(X,Y)Z + R(X,Z)Y = 2g(Z,Y)X - g(X,Y)Z - g(Z,X)Y.$$

Thus, for any orthonormal pair $\{X,Y\}$, we obtain

$$K(X,Y) = g(R(X,Y)Y,X) = 1,$$

which shows that the sectional curvature of M is 1 and hence M is a Sasakian manifold. QED.

From Propositions 5.1 and 5.2 we see that the notion of parallel Ricci tensor and locally symmetric are not so essential on a K-contact manifold and consequently on a Sasakian manifold.

From (5.6) and the Bianchi's identity we have

LEMMA 5.3. The Ricci tensor S of a (2n+1)-dimensional Sasakian manifold M is given by

$$S(X,Y) = \tfrac{1}{2}\sum_{i=1}^{2n+1} g(\phi R(X,\phi Y)e_i,e_i) + (2n-1)g(X,Y) + \eta(X)\eta(Y).$$

Moreover, by (5.6), we easily see that

LEMMA 5.4. The Ricci tensor S of a (2n+1)-dimensional Sasakian manifold M satisfies the following equations:

$$S(X,\xi) = 2n\eta(X), \qquad S(\phi X,\phi Y) = S(X,Y) - 2n\eta(X)\eta(Y).$$

From Lemma 5.3 and the Bianchi's identity we also have

LEMMA 5.5. The Ricci tensor S of a (2n+1)-dimensional Sasakian manifold M satisfies

$$(\nabla_Z S)(X,Y) = (\nabla_X S)(Y,Z) + (\nabla_{\phi Y} S)(\phi X,Z) - \eta(X)S(\phi Y,Z)$$

$$- 2\phi(Y)S(\phi X,Z) + 2n\eta(X)g(\phi Y,Z) + 4n\eta(Y)g(\phi X,Z).$$

If the Ricci tensor S of a K-contact manifold M is of the form

$$S(X,Y) = ag(X,Y) + b\eta(X)\eta(Y),$$

a and b being constant, then M is called an *η-Einstein manifold*.

Obviously, we have

PROPOSITION 5.3. If a Sasakian manifold M is of constant ϕ-sectional curvature, then M is an η-Einstein manifold.

PROPOSITION 5.4. Let M be a (2n+1)-dimensional Sasakian manifold. If the Ricci tensor S of M satisfies $S(X,Y) = ag(X,Y) + b\eta(X)\eta(Y)$, then a and b are constant.

Proof. From the assumption on the Ricci tensor S we have $a+b = 2n$ and $r = (2n+1)a + b$, where r denotes the scalar curvature of M. Then we have $Za = -Zb$ and $Zr = (2n+1)Za + Zb = -2nZb$. On the other hand, Lemma 5.5 implies

$$Zr = 2Za + 2(\xi b)\eta(Z) = -2Zb + 2(\xi b)\eta(Z).$$

Therefore, we obtain

$$(n-1)(Zb) = -(\xi b)\eta(Z).$$

Putting $Z = \xi$ in this equation, we find $\xi b = 0$ and hence $Zb = 0$, which shows that b is constant. Then a is also constant. QED.

6. REGULAR CONTACT MANIFOLDS

Let M be a (2n+1)-dimensional contact manifold with a contact form η. We denote by ξ the associated vector field to η. We say that a contact structure η is *regular* if each point has a *regular coordinate neighborhood* , i.e., a cubical coordinate neighborhood U such that the integral curvaes of ξ passing through U pass through the neighborhood only once. Then the vector field ξ is said to be *regular*.

Hereafter we will assume that the manifold M to be compact. Then, if ξ is regular, ξ is a closed vector field, that is, the orbit of ξ through an arbitrary point is a closed curve. Let B be the set of orbits of ξ. Then we have the natural projection

$$\pi : M \longrightarrow B,$$

making correspond every point p of M to the orbit through p. Then we see that B with the quotient topology is a 2n-dimensional differentiable manifold and π is a differentiable map (see Palais [1]).

We put

$$\mu_\xi(p) = \inf\{ t : t > 0, \quad \phi_t(p) = p\}.$$

Clearly, μ_ξ is constant on every orbit. We call $\mu_\xi(p)$ a *period functions* of ξ. If ξ is regular, we see that $\mu_\xi(p)$ is constant on M.

We now give the theorems of Boothby-Wang [1].

THEOREM 6.1. Let η be a regular contact form on a compact manifold M of dimension 2n+1. Then there exists a gauge transformation $\eta' = \sigma\eta$ such that the vector field ξ' associated to η' has the following properties:

(1) The group of diffeomorphisms of M which is generated by ξ' is a 1-dimensional compact Lie group;

(2) Each element of this group except the identity does not leave any point of M invariant.

Proof. We denote the period by μ and put $\eta' = (1/\mu)\eta$. Then we have $\xi' = \mu\xi$. Thus, ξ' has the same set of orbits as ξ. We may easily verify that $\mu'(p) = \mu_{\xi'}(p) = 1$ for any $p \in M$. Hence, the group of diffeomorphisms of M generated by ξ' depends only equivalence classes of the real variable t mod 1. Consequently, it is a 1-dimensional compact Lie group. We call this group the *circle group* and denote it by S^1. The assertion (2) is evident. QED.

If a 2n-dimensional manifold admits a closed exterior 2-form with maximal rank Ω, i.e., a 2-form Ω such that $d\Omega = 0$, $(\Omega^n) \neq 0$, then we call the manifold a *symplectic manifold* with fundamental form Ω and the structure given by Ω the *symplectic structure*.

THEOREM 6.2. Let M be a (2n+1)-dimensional compact contact manifold with a regular contact form η. Then we have the following:

(1) M is a principal circle bundle over B;

(2) η defines a connection in this bundle;

(3) The base manifold B is a 2n-dimensional symplectic manifold whose fundamental form Ω is the curvature form of η, i.e., $d\Omega = \pi^*\Omega$ is the equation of structure of the connection;

(4) Ω determines an integral cocycle on B.

Proof. By Theorem 6.1 we can modify η so that ξ generates the circle group S^1.

(1) Since ξ is regular we can choose an open covering $\{U_i\}$ of M such that on U_i we have coordinates $(x^1,\ldots,x^{2n+1})$ with the integral curves of ξ being given by $x^1 = \text{const.},\ldots,x^{2n} = \text{const.}$ Then $\{\pi(U_i)\}$ is an open covering of B. We define local cross sections $s_i : \pi(U_i) \longrightarrow M$ with $\pi s_i = \text{id.}$ by $s_i(x^1,\ldots,x^{2n}) = (x^1,\ldots,x^{2n},c)$ for some constant c, $x^1,\ldots,x^{2n}$ being regarded as the coordinates on $\pi(U_i)$. We define the maps $f_i : \pi(U_i) \times S^1 \longrightarrow M$ by $f_i(p,t) = \phi_t s_i(p)$, $p \in M$, ϕ_t being elements of the 1-parameter group of diffeomorphisms generated by ξ. Then f_i are coordinate functions for the bundle.

(2) Since we have $L_\xi\eta = 0$ and $L_\xi d\eta = 0$, η and $d\eta$ are invariant under the action of S^1. Now, the Lie algebra $\mathfrak{S}^1$ of the group S^1 is 1-dimensional vector space. So, we identify $\mathfrak{S}^1$ with the line R, and

take $A = d/dt$ as a basis of $\mathfrak{S}^1$. Then $\tilde{\eta} = \eta A$ is an invariant form on M which has its values in $\mathfrak{S}^1$. For the sake of simplicity we write η instead of $\tilde{\eta}$ again. To prove that η defines a connection we must show that the following two conditions:

(a) $\eta(A^*) = A$,

where A^* is the fundamental vector field corresponds to A;

(b) $R_t^*\eta = (\mathrm{ad}\ t^{-1})\eta$,

where R_t is the right translation by $t \in S^1$, R_t^* is the dual map of R_t on the space of 1-forms on M.

First we have $\tilde{\eta}(A^*) = \eta(A^*)A = \eta(\xi)A = A$. This proves (a). Next, as $R_t(p) = \phi_t(p)$ and η is invariant under the group S^1, we have $R_t^*\eta = \eta$. Since S^1 is abelian, we see that $\mathrm{ad}(t^{-1}) = 1$. Thus we have (b).

(3) We denote by $\tilde{\Omega}$ the curvature form of the connection η. Then

$$d\eta(X,Y) = -\tfrac{1}{2}[\eta(X),\eta(Y)] + \tilde{\Omega}(X,Y).$$

However, as the group S^1 is abelian $[\eta(X),\eta(Y)] = 0$ and hence $d\eta = \tilde{\Omega}$. On the other hand, $d\eta$ is invariant and $d\eta(\xi,Y) = 0$. Therefore there exists a 2-form Ω on M such that $d\eta = \pi^*\Omega$. Now, $\pi^*d\Omega = d\pi^*\Omega = d^2\eta = 0$, so that $d\Omega = 0$ and $\pi^*(\Omega^n) = (\pi^*\Omega)^n = (d\eta)^n \neq 0$ giving $\Omega^n \neq 0$. Therefore M is symplectic.

(4) Finally, as the transition functions $f_{ij} : U_i \; U_j \longrightarrow S^1$ are real (mod 1) valued, one can check that $[\Omega] \in H^2(M,Z)$ (see Kobayashi [1] for details). QED.

Conversely we now prove the following

THEOREM 6.3. Let B be a 2n-dimensional symplectic manifold such that its fundamental form Ω determines an integral cohomology class. Then there is a principal circle bundle $\pi : M \longrightarrow B$ and a connection η on M such that

(1) η is a contact form on a (2n+1)-dimensional manifold M, and

(2) its associated vector field ξ generates the right translations of the structure group S^1 of the bundle.

Proof. We take $A = d/dt$ as the basis of the Lie algebra of S^1.

According to a theorem of Kobayashi [1], there exists a principal circle bundle M over B and a connection $\tilde{\eta}$ on M such that η defined by $\tilde{\eta} = \eta A$ satisfies $d\eta = \pi^*\eta$. Since π^* is an isomorphism, we have $(d\eta)^n = (\pi^*\Omega)^n = \pi^*\Omega^n \neq 0$. Therefore, if we denote the fundamental vector field corresponding to A by A* and 2n linearly independent horizontal vectors by $X_1,\ldots,X_{2n}$, then we have $(\eta \wedge (d\eta)^n)(A^*,X_1,\ldots,X_{2n}) \neq 0$. Thus, η is a contact form of M. Moreover, since $\tilde{\eta}(A^*) = A$, we have $\eta(A^*) = 1$. On the other hand, for arbitrary vector X of M we have $d\eta(A^*,X) = \pi^*\Omega(A^*,X) = \Omega(\pi A^*,\pi X)$. Hence, we get $d\ (A^*,X) = 0$. Therefore, A* is the vector field associated to η, i.e., $A^* = \xi$. Hence, ξ generates the group S^1 of right translations of the bundle M. QED.

Let $\pi : M \longrightarrow B$ be a Boothby-Wang fibration (Theorem 6.2). Since B carries a global symplectic form Ω, there exists a Riemannian metric h and a Tensor field J of type (1,1) such that (J,h) is an almost Kaehlerian structure on M with Ω as its fundamental 2-form. Let η be the contact form on M with $d\eta = \pi^*\Omega$ and ξ its associated vector field. We define a tensor field ϕ of type (1,1) on M by

$$\phi X = (J\pi_* X)^*, \quad X \in T_x(M),$$

where * denotes the horizontal lift with respect to η. Then

$$\phi^2 X = (J\pi_*(J\pi_* X)^*)^* = (J^2\pi_* X)^* = -(\pi_* X)^*$$

$$= -X + \eta(X)\xi,$$

that is,

$$\phi^2 = -I + \eta \otimes \xi.$$

Therefore, (ϕ,ξ,η) is an almost contact structure on M. We define a Riemannian metric g of M by

$$g(X,Y) = h(\pi_* X,\pi_* Y) + \eta(X)\eta(Y), \quad X,Y \in T_x(M),$$

that is,

$$g = \pi^*h + \eta \otimes \eta.$$

Clearly ξ is a unit Killing vector field with respect to g. Moreover we have

$$g(X,\phi Y) = h(\pi_*X,\pi_*(J\pi_*Y)^*) = h(\pi_*X,J\pi_*Y)$$

$$= \Omega(\pi_*X,\pi_*Y) = \pi^*\Omega(X,Y) = d\eta(X,Y)$$

and similarly

$$g(\phi X,\phi Y) = h(\pi_*X,\pi_*Y) = g(X,Y) - \eta(X)\eta(Y).$$

Thus we have (Hatakeyama [1])

THEOREM 6.4. Compact regular contact manifold carries a K-contact structure.

Let N_ϕ and N_J be the Nijenhuis torsions of ϕ and J respectively. Since we have $L_\xi\phi = 0$, we obtain

$$N_\phi(\xi,X) + 2d\eta(\xi,X)\xi = \phi^2[\xi,X] - \phi[\xi,\phi X] = 0.$$

On the other hand, for projectable horizontal vector fields X and Y, we have

$$\begin{aligned} &N_\phi(X,Y) + 2d\eta(X,Y)\xi \\ &= (J^2\pi_*[X,Y])^* + [(J\pi_*X)^*,(J\pi_*Y)^*] - (J\pi_*[(J\pi_*X)^*,Y])^* \\ &\quad - (J\pi_*[X,(J\pi_*Y)^*])^* + 2d\eta(X,Y)\xi \\ &= (J^2[\pi_*X,\pi_*Y])^* + [J\pi_*X,J\pi_*Y]^* + \eta([(J\pi_*X)^*,(J\pi_*Y)^*]) \\ &\quad - (J[J\pi_*X,\pi_*Y])^* - (J[\pi_*X,J\pi_*Y])^* + 2d\eta(X,Y)\xi \end{aligned}$$

$$= (N_J(\pi_*X,\pi_*Y))^* - 2(\Omega(J\pi_*X,J\pi_*Y))^* - (\Omega(\pi_*X,\pi_*Y))^*)$$

$$= (N_J(\pi_*X,\pi_*Y))^*.$$

Thus we see that K-contact structure (ϕ,ξ,η,g) is Sasakian if and only if M is Kaehlerian.

A *Hodge manifold* is by definition a compact Kaehlerian manifold such that the fundamental 2-form Ω determines an integral cocycle over the manifold. Thus, by the previous considerations, we have (Hatakeyama [1])

THEOREM 6.5. In order that a (2n+1)-dimensional manifold M with a regular contact structure admit a Sasakian structure is that the base manifold B of the Boothby-Wang fibration of M is a Hodge manifold.

7. BRIESKORN MANIFOLDS

Let C^{n+1} be the complex vector space of (n+1)-tuples of complex numbers $(z_0,z_1,\ldots,z_n)$. A *Brieskorn manifold* is by definition a (2n-1)-dimensional submanifold $B^{2n-1}(a_0,a_1,\ldots,a_n)$ in C^{n+1} defined by equations

$$z_0^{a_0} + z_1^{a_1} + \ldots + z_n^{a_n} = 0 \tag{7.1}$$

and

$$z_0\bar{z}_0 + z_1\bar{z}_1 + \ldots + z_n\bar{z}_n = 1, \tag{7.2}$$

where $a_0,a_1,\ldots,a_n$ are positive integers. We denote $B^{2n-1}(a_0,a_1,\ldots,a_n)$ by B^{2n-1} for simplicity. We also denote by $\bar{B}^{2n}$ the complex hypersurface in $C^{n+1} - \{0\}$ defined by (7.1). The Brieskorn manifold B^{2n-1} is the intersection of $\bar{B}^{2n}$ with the unit sphere S^{2n+1}.

Let us consider the C-action on C^{n+1} defined by

$$(7.3) \qquad z_j' = e^{mw/a_t} z_j,$$

where m is the least common multiple of integers $a_0, a_1, \ldots, a_n$ and w is a complex variable. We can easily see that the C-action fixes the origin O and transforms $\bar{B}^{2n}$ onto itself. Therefore, restricting w to its real part s and differentiating $z_j'(s)$ at s = 0, we see that

$$u_1 = (\frac{m}{a_j} z_j), \quad z \in \bar{B}^{2n}$$

is a tangent vector of $\bar{B}^{2n}$ at z. Similarly, restricting w to its purely imaginary part it ($t \in R$), we see that

$$u_2 = iu_1 = (\frac{m}{a_j} iz_j), \quad z \in \bar{B}^{2n}$$

is a tangent vector of $\bar{B}^{2n}$ at z orthogonal to u_1. When we restrict w to it, (7.3) gives a S^1-action on C^{n+1} and the S^1-action leaves $\bar{B}^{2n}$, S^{2n+1} and so B^{2n-1}. Therefore, if $z \in B^{2n-1}$, the orbit of z under this action lies on B^{2n-1} and so u_2 is a tangent vector of B^{2n-1}.

We denote by dz the differential at a point z on $\bar{B}^{2n}$. Then we have

$$\sum \frac{f}{z_j} dz_j = 0,$$

where $f(z_0, z_1, \ldots, z_n)$ is the polynomial on the left hand side of (7.1). This is equivalent to $\langle \overline{\partial f/\partial z}, dz \rangle = 0$, where the bracket means the inner product of two vectors $\overline{\partial f/\partial z}$ (the complex conjugate of $\partial f/\partial z$) and dz in C^{n+1}. Thus we have

$$\mathrm{Re}\langle \overline{\frac{\partial f}{\partial z}}, dz \rangle = 0 \quad \text{and} \quad \mathrm{Re}\langle i\overline{\frac{\partial f}{\partial z}}, dz \rangle = 0.$$

Therefore we see that

$$v_1 = (\overline{\frac{\partial f}{\partial z_j}}) = (a_j \bar{z}_j^{a_j - 1}), \quad v_2 = (i\overline{\frac{\partial f}{\partial z_j}}) = (ia_j \bar{z}_j^{a_j - 1}) = iv_1$$

are normal vectors of $\bar{B}^{2n}$ at the point z. We easily see that u_1, u_2, v_1 and v_2 are mutually orthogonal.

We now restricted the point z to the one on B^{2n-1}. Then the unit normal vector n of S^{2n+1} has z_j as its components. We see that v_1, v_2 and n are normals to B^{2n-1} in C^{n+1} and they are linearly independent. We put

$$v = n + \lambda v_1 + \mu v_2,$$

where

$$\lambda = -(\mathrm{Re}(\Sigma a_j z_j^{a_j})/\langle v_1, v_1\rangle), \qquad \mu = \mathrm{Im}(\Sigma a_j z_j^{a_j})/\langle v_2, v_2\rangle.$$

Then v_1, v_2 and v are normal vectors of B^{2n-1} in C^{n+1} orthogonal with each other, which shows that v is a normal vector of B^{2n-1} which tangent to $\bar{B}^{2n}$ at each point $z \in B^{2n-1}$.

$\bar{B}^{2n}$ inherits the complex structure from that of C^{n+1}. If we denote by $\langle\langle\ ,\ \rangle\rangle$ the Kaehlerian inner product, then

$$\langle\langle iv, dz\rangle\rangle = \mathrm{Re}\langle iv, dz\rangle.$$

Then we have

$$\langle\langle iv, dz\rangle\rangle = \tfrac{1}{2}i \sum_{j=0}^{n} (z_j d\bar{z}_j - \bar{z}_j dz_j).$$

We define a real 1-form η on B^{2n-1} by

$$\eta = \tfrac{1}{2}i \sum (z_j d\bar{z}_j - \bar{z}_j dz_j).$$

We notice that $\eta = \langle\langle iv, dz\rangle\rangle = \langle\langle in, dz\rangle\rangle$.

Sasaki-Hsu [1] proved that the form η is a contact form on B^{2n-1}.

THEOREM 7.1. Every Brieskorn manifold is a contact manifold.

Proof. We shall show that the 1-form η on B^{2n-1} is a contact form.

Since

$$d\eta = i\sum_{j=0}^{n} dz_j \wedge d\bar{z}_j,$$

we have

$$(7.4) \qquad \eta \wedge (d\eta)^{n-1} = \tfrac{1}{2}i^n\{\sum_{j=0}^{n}(z_j d\bar{z}_j - \bar{z}_j dz_j)\} \wedge (\sum_{k=0}^{n} dz_k \wedge d\bar{z}_k)^{n-1}$$

$$= \tfrac{1}{2}(n-1)!i^n[\{\sum_{j=0}^{n}(z_j d\bar{z}_j - \bar{z}_j dz_j)\}$$

$$\wedge \{\sum_{j<k}(dz_0 \wedge d\bar{z}_0) \wedge \ldots \wedge \widehat{(dz_j \wedge d\bar{z}_j)}$$

$$\wedge \ldots \wedge \widehat{(dz_k \wedge d\bar{z}_k)} \wedge \ldots \wedge (dz_n \wedge d\bar{z}_n)\}],$$

where roofs mean factors which should be omitted.

We may first restrict ourselves on the domain D_n on B^{2n-1} where $z_n \neq 0$. On D_n we have

$$(7.5) \qquad dz_n = -\sum_{p=0}^{n-1} f_p dz^p,$$

where we have put

$$f_p = t_p/t_n, \qquad t_j = a_j z_j^{a_j-1}.$$

We denote the equation complex conjugate to (7.5) by $(\overline{7.5})$. On the other hand, we have, by (7.2),

$$\sum_{j=0}^{n}(z_j d\bar{z}_j + \bar{z}_j dz_j) = 0.$$

From this and (7.5) we have

$$(7.6) \qquad \sum_{p=0}^{n-1}(m_p dz_p + \bar{m}_p d\bar{z}_p) = 0,$$

where we have put

$$m_p = \bar{z}_p - \bar{z}_n f_p, \qquad \bar{m}_p = z_p - z_n \bar{f}_p.$$

Since the functions $m_0, m_1, \ldots, m_{n-1}$ defined on D_n can not vanish simultaneously at any point of D_n, we may consider the subdomain $D_{n,n-1}$ in D_n such that $\bar{m}_{n-1} \neq 0$. Then

$$(7.7) \qquad d\bar{z}_{n-1} = -m\left(\sum_{p=0}^{n-1} m_p dz_p + \sum_{k=0}^{n-2} \bar{m}_k d\bar{z}_k\right)$$

on $D_{n,n-1}$, where we have put $m = 1/\bar{m}_{n-1}$.

Now, if we pay attention to the domain $D_{n,n-1}$ on B^{2n-1}, (7.4) can be written as

$$(7.8) \qquad \eta \wedge (d\eta)^{n-1} = \tfrac{1}{2}(n-1)!\,i^n(A + B + C),$$

where A, B and C are (2n-1)-forms defined as follows:

A: the sum of monomials each of which contains $z_k d\bar{z}_k - \bar{z}_k dz_k$ $(k = 0,1,\ldots,n-2)$ as its factor,

B; the sum of monomials each of which contains $z_{n-1} d\bar{z}_{n-1} - \bar{z}_{n-1} dz_{n-1}$ as its factor, and

C: the sum of monomials each of which contains $z_n d\bar{z}_n - \bar{z}_n dz_n$ as its factor.

We shall compute A, B and C on $D_{n,n-1}$. To simplify the notation, we put $w_j = dz_j \wedge d\bar{z}_j$.

(i) Calculation of A: If we fix the value of k, any non-zero monomial in (7.4) which contains $z_k d\bar{z}_k - \bar{z}_k dz_k$ does not contain w_k as its factor. So A can be written as

$$A = A_1 + A_2 + A_3,$$

where A_1, A_2 and A_3 are (2n-1)-forms with the following additional properties:

A_1: the sum of monomials each of which contains w_{n-1} as its factor, but does not contain w_n as its factor,

A_2: the sum of monomials each of which contains w_n as its factor, but does not contain w_{n-1} as its factor,

A_3: the sum of monomials each of which contains both w_{n-1} and w_n as its factor.

Then we see that

$$A_1 = \sum_{k=0}^{n-2} w_0 \wedge \cdots \wedge w_{k-1} \wedge (z_k d\bar{z}_k - \bar{z}_k dz_k) \wedge w_{k+1} \wedge \cdots \wedge w_{n-1}.$$

Substituting (7.7) into the equation above, we find

$$A_1 = -m \sum_{k=0}^{n-2} (z_k m_k + \bar{z}_k \bar{m}_k)\Omega,$$

where we have put

$$\Omega = w_0 \wedge w_1 \wedge \cdots \wedge w_{n-2} \wedge dz_{n-1}.$$

Next, we have

$$A_2 = \sum_{k=0}^{n-2} w_0 \wedge \cdots \wedge w_{k-1} \wedge (z_k d\bar{z}_k - \bar{z}_k dz_k) \wedge w_{k+1} \wedge \cdots \wedge w_{n-2} \wedge w_n.$$

Substituting (7.5) and $(\overline{7.5})$ into the equation above, we obtain

$$\begin{aligned} A_2 = \sum_{k=0}^{n-2} (&-z_k f_k \bar{f}_{n-1} w_0 \wedge \cdots \wedge w_{n-2} \wedge d\bar{z}_{n-1} \\ &+ z_k f_{n-1} \bar{f}_{n-1} w_0 \wedge \cdots \wedge w_{k-1} \wedge d\bar{z}_k \wedge w_{k+1} \wedge \cdots \wedge w_{n-1} \\ &+ \bar{z}_k f_{n-1} \bar{f}_{n-1} w_0 \wedge \cdots \wedge w_{n-2} \wedge dz_{n-1} \\ &- \bar{z}_k f_{n-1} \bar{f}_{n-1} w_0 \wedge \cdots \wedge w_{k-1} \wedge dz_k \wedge w_{k+1} \wedge \cdots \wedge w_{n-1}). \end{aligned}$$

Using (7.7), A_2 reduces to

$$A_2 = m\sum_{k=0}^{n-2}\{z_k\bar{f}_{n-1}(f_k m_{n-1} - f_{n-1}m_k) + \bar{z}_k f_{n-1}(\bar{f}_k\bar{m}_{n-1} - \bar{f}_{n-1}\bar{m}_k)\}\Omega.$$

We next compute A_3. We set

$$A_3 = A_3' + A_3'',$$

where we have put

$$A_3' = \sum_{k=0}^{n-2}\sum_{j=k+1}^{n-2} w_0\wedge\cdots\wedge w_{k-1}\wedge(z_k d\bar{z}_k - \bar{z}_k dz_k) \wedge w_{k+1}\wedge\cdots\wedge\hat{w}_j\wedge\cdots\wedge w_{n-2}\wedge w_{n-1}\wedge w_n,$$

$$A_3'' = \sum_{k=0}^{n-2}\sum_{h=0}^{k-1} w_0\wedge\cdots\wedge\hat{w}_h\wedge\cdots\wedge w_{k-1} \wedge(z_k d\bar{z}_k-\bar{z}_k dz_k)\wedge w_{k+1}\wedge\cdots\wedge w_{n-2}\wedge w_{n-1}\wedge w_n.$$

Substituting (7.5) and $(\overline{7.5})$ into A_3' and using (7.7), we find

$$A_3' = m\sum_{k=0}^{n-2}\sum_{j=k+1}^{n-2}\{z_k\bar{f}_j(f_k m_j - f_j m_k) + \bar{z}_k f_j(\bar{f}_k\bar{m}_j - \bar{m}_k\bar{f}_j)\}\Omega.$$

Similarly, we have

$$A_3'' = m\sum_{h=0}^{n-2}\sum_{k=h+1}^{n-2}\{z_k\bar{f}_h(f_k m_h - f_h m_k) + \bar{z}_k f_k(\bar{f}_k\bar{m}_h - \bar{f}_h\bar{m}_k)\}\Omega.$$

Changing indices h and k to k and j respectively, we have

$$A_3'' = m\sum_{k=0}^{n-2}\sum_{j=k+1}^{n-2}\{z_j\bar{f}_k(f_j m_k - f_k m_j) + \bar{z}_j f_k(\bar{f}_j\bar{m}_k - \bar{f}_k\bar{m}_j)\}\Omega.$$

Consequently, we have

$$A_3 = m\sum_{k=0}^{n-2}\sum_{j=k+1}^{n-2}\{(f_k m_j - f_j m_k)(z_k\bar{f}_j - z_j\bar{f}_k) + (\bar{f}_k\bar{m}_j - \bar{m}_k\bar{f}_j)(\bar{z}_k f_j - \bar{z}_j f_k)\}\Omega.$$

(ii) Calculation of B: Clearly B can be written as

$$B = B_1 + B_2,$$

where B_1 and B_2 are (2n-1)-forms with the following additional properties:

B_1: the monomial which contains $z_{n-1}d\bar{z}_{n-1} - \bar{z}_{n-1}dz_{n-1}$ as its factor, but does not contain w_n as its factor,

B_2: the sum of monomials each of which contains both of $z_{n-1}d\bar{z}_{n-1} - \bar{z}_{n-1}dz_{n-1}$ and w_n as its factors.

Substituting (7.7) into the equation

$$B_1 = w_0 \wedge w_1 \wedge \cdots \wedge w_{n-2} \wedge (z_{n-1}d\bar{z}_{n-1} - \bar{z}_{n-1}dz_{n-1}),$$

we obtain

$$B_1 = -m(z_{n-1}m_{n-1} + \bar{z}_{n-1}\bar{m}_{n-1})\Omega.$$

Moreover, we have

$$B_2 = \sum_{k=0}^{n-2} w_0 \wedge \cdots \wedge \widehat{w}_k \wedge \cdots \wedge w_{n-2} \wedge (z_{n-1}d\bar{z}_{n-1} - \bar{z}_{n-1}dz_{n-1}) \wedge w_n.$$

Thus (7.5), $(\overline{7.5})$ and (7.7) imply

$$B_2 = m\sum_{k=0}^{n-2} \{z_{n-1}\bar{f}_k(f_{n-1}m_k - f_k m_{n-1}) + \bar{z}_{n-1}f_k(\bar{f}_{n-1}\bar{m}_k - \bar{f}_k\bar{m}_{n-1})\}\Omega.$$

(iii) Calculation of C: C can be written as

$$C = C_1 + C_2,$$

where C_1 and C_2 are (2n-1)-forms with the following additional properties:

C_1: the monomial which contains $z_n d\bar{z}_n - \bar{z}dz_n$ as its factor, but does not contain w_{n-1},

C_2: the sum of monomials each of which contains both of $z_n d\bar{z}_n - \bar{z}_n dz_n$ and w_{n-1} as its factors.

First, we see that

$$C_1 = w_0 \wedge \cdots \wedge w_{n-2} \wedge (z_n d\bar{z}_n - \bar{z}_n dz_n).$$

Then, by (7.5), $(\overline{7.5})$ and (7.7), we find

$$C_1 = m(z_n \bar{f}_{n-1} m_{n-1} + \bar{z}_n f_{n-1} \bar{m}_{n-1})\Omega.$$

Similarly, from the equation

$$C_2 = \sum_{k=1}^{n-2} w_0 \wedge \cdots \wedge \hat{w}_k \wedge \cdots \wedge w_{n-2} \wedge w_{n-1} \wedge (z_n d\bar{z}_n - \bar{z}_n dz_n),$$

we obtain

$$C_2 = m \sum_{k=0}^{n-2} (z_n \bar{f}_k m_k + \bar{z}_n f_k \bar{m}_k) .$$

We now define a function F on $D_{n,n-1}$ by

$$\eta \quad (d\eta)^{n-1} = \tfrac{1}{2}(n-1)!\, i^n F\Omega.$$

Then we have

$$F\Omega = A + B + C = (A_1 + B_1) + (C_1 + C_2) + \{A_3 + (A_2 + B_2)\}.$$

To show $\eta \quad (d\eta)^{n-1} \neq 0$, it is sufficient to show that $F \neq 0$.

We have the following equations

$$A_1 + B_1 = -m \sum_{p=0}^{n-1} (z_p m_p + \bar{z}_p \bar{m}_p)\Omega,$$

$$C_1 + C_2 = m(z_n \sum_{p=0}^{n-1} \bar{f}_p m_p + \bar{z}_n \sum_{p=0}^{n-1} f_p \bar{m}_p)\Omega.$$

Moreover, we have

$$A_2 + B_2 = m\sum_{k=0}^{n-2}\{(f_k m_{n-1} - f_{n-1} m_k)(z_k \bar{f}_{n-1} - z_{n-1}\bar{f}_k) + (\bar{f}_k \bar{m}_{n-1} - \bar{f}_{n-1}\bar{m}_k)(\bar{z}_k f_{n-1} - \bar{z}_{n-1} f_k)\}\Omega,$$

from which

$$A_3 + (A_2 + B_2) = m\sum_{k=0}^{n-2}\sum_{j=k+1}^{n-1}\{(f_k m_j - f_j m_k)(z_k \bar{f}_j - z_j \bar{f}_k) + (\bar{f}_k \bar{m}_j - \bar{m}_k \bar{f}_j)(\bar{z}_k f_j - \bar{z}_j f_k)\}\Omega.$$

Consequently, we obtain

$$\tfrac{1}{2}F = -\sum_{p=0}^{n-1} z_p \bar{z}_p + \sum_{p=0}^{n-1} z_p f_p \bar{z}_n + \sum_{p=0}^{n-1} \bar{z}_p \bar{f}_p z_n - \sum_{p=0}^{n-1} f_p \bar{f}_p z_n \bar{z}_n - \sum_{k=0}^{n-2}\sum_{j=k+1}^{n-1}(z_k \bar{f}_j - z_j \bar{f}_k)(\bar{z}_k f_j - \bar{z}_j f_k).$$

Thus we have

$$\begin{aligned}\tfrac{1}{2}t_n \bar{t}_n F &= -\sum_{p=0}^{n-1}|t_n \bar{z}_p|^2 - \sum_{p=0}^{n-1}|\bar{t}_p z_n|^2 + 2\sum_{p=0}^{n-1}\mathrm{Re}((t_n \bar{z}_p)(\bar{t}_p z_n)) - \sum_{k=0}^{n-2}\sum_{j=k+1}^{n-1}|z_k \bar{t}_j - z_j \bar{t}_k|^2 \\ &= -\sum_{p=0}^{n-1}\{\mathrm{Re}(t_n \bar{z}_p) - \mathrm{Re}(\bar{t}_p z_n)\}^2 - \sum_{p=0}^{n-1}\{\mathrm{Im}(t_n \bar{z}_p) + \mathrm{Im}(\bar{t}_p z_n)\}^2 - \sum_{k=0}^{n-2}\sum_{j=k+1}^{n-1}|z_k \bar{t}_j - z_j \bar{t}_k|^2.\end{aligned}$$

Thus, we have $F \leq 0$ on $D_{n,n-1}$. If $F = 0$, then

$$\mathrm{Re}(t_n \bar{z}_p) = \mathrm{Re}(\bar{t}_p z_n), \qquad \mathrm{Im}(t_n \bar{z}_p) = -\mathrm{Im}(\bar{t}_p z_n)$$

for $p = 0,1,\dots,n-1$, and

$$z_k \bar{t}_j = z_j \bar{t}_k$$

for $k = 0,1,\dots,n-2$ and $j = k+1,\dots,n-1$. Then we have a contradiction. Therefore, $F < 0$ and hence $\eta \wedge (d\eta)^{n-1} \neq 0$ on $D_{n,n-1}$.

Quite similar argument can be performed for other domains $D_{n,k}$ $(k = 0,1,\dots,n-2)$ similarly defined as $D_{n,n-1}$. Thus $\eta \wedge (d\eta)^{n-1} \neq 0$ on D_n. In the same way, we can prove that $\eta \wedge (d\eta)^{n-1} \neq 0$ for domains $D_0, D_1, \dots, D_{n-1}$. Consequently, η is a contact form on B^{2n-1}. QED.

In the sequel, following Takao Takahashi [1], we shall show that the Brieskorn manifold has a Sasakian structure. First of all, we prove the following

THEOREM 7.2. Let M be a Sasakian manifold with structure tensors (ϕ,ξ,η,g) and μ be a vector field on M which satisfies the following three conditions:

$$\text{(a) } L_\mu g = 0, \quad \text{(b) } [\mu,\xi] = 0, \quad \text{(c) } 1 + \eta(\mu) > 0.$$

Define new structure tensors $(\bar\phi,\bar\xi,\bar\eta,\bar g)$ by

$$\bar\phi X = \phi(X - \bar\eta(X)\bar\xi), \quad \bar\eta = (1 + \eta(\mu))^{-1}, \quad \bar\xi = \xi + \mu,$$

$$\bar g(X,Y) = (1 + \eta(\mu))^{-1} g(X-\bar\eta(X)\bar\xi, Y-\bar\eta(Y)\bar\xi) + \bar\eta(X)\bar\eta(Y)$$

for any vector fields X and Y on M. Then $(\bar\phi,\bar\xi,\bar\eta,\bar g)$ is a Sasakian structure on M.

Proof. Let D be the distribution defined by $\eta = 0$. By the definition of $\bar\eta$ we see that D is also defined by $\bar\eta$. From the definition of structure tensors we have $\bar\eta(\bar\xi) = 1$ and $\bar g(\bar\xi,\bar\xi) = 1$. For any vector field X of D we have $\bar\eta(X) = 0$, $\bar\eta(\phi X) = 0$ and $\phi^2 X = -X$, hence

$$\bar\phi^2 X = \bar\phi(\phi(X-\bar\eta(X)\xi)) = \bar\phi(\phi X) = \phi(\phi X-\bar\eta(\phi X)\bar\xi) = \phi^2 X = -X.$$

Because $\bar\phi\bar\xi = 0$ and the equation above holds for any vector field of D, we conclude that

$$\bar\phi^2 = -I + \bar\eta \otimes \bar\xi.$$

Let X and Y be vector fields which belong to D. Then we see that $\phi X = \bar{\phi}X$, $\phi Y = \bar{\phi}Y$ and $\bar{\eta}(X) = \bar{\eta}(Y) = 0$. Thus we have

$$N_{\bar{\phi}}(X,Y) + 2d\bar{\eta}(X,Y)\bar{\xi}$$

$$= \bar{\phi}^2[x,y] + [\phi X,\phi Y] - \bar{\phi}[\phi X,Y] - \bar{\phi}[X,\phi Y] - \bar{\eta}([X,Y])\bar{\xi}$$

$$= N_{\phi}(X,Y) + \phi(\bar{\eta}([X,Y])\bar{\xi} + \bar{\eta}([X,Y])\bar{\xi})$$

$$= (1+\eta(\mu))^{-1}(\eta([\phi X,Y]) + \eta([X,\phi Y]))\phi\bar{\xi}$$

$$= (1+\eta(\mu))^{-1}N^{(2)}(Y,X)\phi\bar{\xi} = 0,$$

by Lemma 3.1. Moreover, we easily see that

$$N_{\bar{\phi}}(X,\bar{\xi}) + 2d\bar{\eta}(X,\bar{\xi})\bar{\xi} = 0.$$

Consequently, we obtain

$$N_{\bar{\phi}} + 2d\bar{\eta} \otimes \bar{\xi} = 0.$$

Next, we prove that $d\bar{\eta}(X,Y) = \bar{g}(X,\bar{\phi}Y)$. If X is a vector field in D, then

$$2d\bar{\eta}(X,\bar{\xi}) = -\bar{\eta}([\bar{\xi},X]) = 0.$$

If X and Y are vector fields in D, then

$$2d\bar{\eta}(X,Y) = -\bar{\eta}([X,Y]) = -(1+\eta(\mu))^{-1}\eta([X,Y])$$

$$= -2(1+\eta(\mu))^{-1}g(\phi X,Y) = 2\bar{g}(X,\bar{\phi}Y).$$

Therefore, we obtain $d\bar{\eta}(X,Y) = \bar{g}(X,\bar{\phi}Y)$. Consequently, $(\bar{\phi},\bar{\xi},\bar{\eta},\bar{g})$ is a Sasakian structure on M. QED.

Example 7.1. Let S^{2n+1} be the unit sphere with standard Sasakian structure (ϕ,ξ,η,g) in R^{2n+2} with coordinates $(x^1,y^1,\ldots,x^{n+1},y^{n+1})$.

At a point p on S^{2n+1} we put

$$\xi = \sum_{j=1}^{n+1} (x^j \partial y^j - y^j \partial x^j), \qquad \eta = \sum_{j=1}^{n+1} (x^j dy^j - y^j dx^j),$$

where we have put $\partial x^j = \partial/\partial x^j$, $\partial y^j = \partial/\partial y^j$. ϕ is defined to be the restriction of the almost complex structure J of R^{2n+2} which is the orthogonal complement of $\langle\xi\rangle$ in the tangent space of S^{2n+1} and 0 on $\langle\xi\rangle$. The Riemannian metric g on S^{2n+1} is induced from that of R^{2n+2}.

We put

$$\mu = \sum_{j=1}^{n+1} r_j (x^j \partial y^j - y^j \partial x^j),$$

where (r_j) is a (n+1)-tuple of real numbers such that

$$1 + \sum_{j=1}^{n+1} r_j ((x^j)^2 + (y^j)^2) > 0$$

on S^{2n+1}. Then μ satisfies the conditions of Theorem 7.2. The new trajectriy of $\bar{\xi}$ with the initial condition p is given by

$$x^j(t) = x^j \cos(1+r_j)t - y^j \sin(1+r_j)t,$$
$$y^j(t) = x^j \sin(1+r_j)t + y^j \cos(1+r_j)t$$

for j = 1,2,...,n+1.

THEOREM 7.3. Every Brieskorn manifold admits many Sasakian structures.

Proof. We define a mapping F of C^{n+1} onto C by

$$F(z_0, z_1, \dots, z_n) = (z_0)^2 + (z_1)^2 + \dots + (z_n)^2.$$

Then the Brieskorn manifold B^{2n-1} is given by $B^{2n-1} = S^{2n+1} \cap F^{-1}(0)$ and B^{2n-1} is a submanifold in S^{2n+1}. Let $x_0, y_0, \dots, x_n, y_n$ be the real coordinates of C^{n+1} such that $z_j = x_j + iy_j$ (j = 0,1,...,n). We define a real vector field $\bar{\xi}$ on C^{n+1} by

$$\bar{\xi} = \sum_{j=0}^{n} A_j(x_j\partial y_j - y_j\partial x_j),$$

where $A_j = (a_j)^{-1}A$ for a positive constant A, $(j = 0,1,\dots,n)$. $\bar{\xi}$ is tangent to S^{2n+1}. We put $\mu = \bar{\xi} - \xi$. Then μ satisfies the conditions of Theorem 7.2 because of $1 + \eta(\mu) = \eta(\bar{\xi}) > 0$ (see Example 7.1). Then we have a Sasakian structure $(\bar{\phi},\bar{\xi},\bar{\eta},\bar{g})$ on S^{2n+1} with respect to μ. The explicit formula of $\bar{\eta}$ is given by

$$\bar{\eta} = \sum_{j=0}^{n} (K)^{-1}(x_j dy_j - y_j dx_j),$$

where $K = A_j((x_j)^2+(y_j)^2)$.

Let P and Q be real valued functions on C^{n+1} defined by

$$P(x_0,y_0,\dots,x_n,y_n) = \text{the real part of } F(z_0,\dots,z_n),$$

$$Q(x_0,y_0,\dots,x_n,y_n) = \text{the imaginary part of } F(z_0,\dots,z_n).$$

Then it is easy to see that $\bar{\xi}P = -AQ$ and $\bar{\xi}Q = AP$. These show that the restriction of $\bar{\xi}$ to B^{2n-1} is tangent to B^{2n-1}. Since the restrictions of ϕ, $\bar{\phi}$ and J to D are the same mapping and $F^{-1}(0) - \{0\}$ is a complex submanifold of C^{n+1}, for any vector X tangent to B^{2n-1} which belongs to D on S^{2n+1}, we get

$$(\bar{\phi}X)(F) = (JX)(F) = iX(F) = 0$$

because of the Cauchy-Riemann equations for the complex analytic function F. Since we have the orthogonal decomposition, with respect to the induced metric on B^{2n-1},

$$T_p(B^{2n-1}) = (D_p + \langle\bar{\xi}_p\rangle) \cap T_p(B^{2n-1}) = (D_p \cap T_p(B^{2n-1})) + \langle\bar{\xi}_p\rangle,$$

where $\langle\bar{\xi}_p\rangle$ is the 1-dimensional space generated by ξ at p. Since $\bar{\phi}\bar{\xi} = 0$, we see that $\bar{\phi}$ maps $T_p(B^{2n-1})$ into itself. Therefore, B^{2n-1} is an invariant submanifold of S^{2n+1} with respect to $(\bar{\phi},\bar{\xi},\bar{\eta},\bar{g})$. Any invariant submanifold of a Sasakian manifold is also a Sasakian manifold (see §1 of Chapter VI). Consequently, B^{2n-1} admits a Sasakian structure.

QED.

EXERCISES

A. K-CONTACT RIEMANNIAN MANIFOLDS: Let M be a K-contact Riemannian manifold with structure tensors (ϕ,ξ,η,g). We denote by R and S the Riemannian curvature tensor and the Ricci tensor of M respectively. Then we have (Tanno [1])

THEOREM. Let M be a K-contact Riemannian manifold.
(1) If $R(X,\xi)S = 0$ for any X, then M is an Einstein manifold;
(2) If $R(X,\xi)R = 0$ for any X, then M is of constant curvature.

B. IMMERSED K-CONTACT RIEMANNIAN MANIFOLDS: Takahashi-Tanno [1] studied K-contact Riemannian manifolds immersed in a space form as hypersurfaces and proved the following theorems.

THEOREM 1. If an n-dimensional K-contact Riemannian manifold M is isometrically immersed in an (n+1)-dimensional space form $\bar{M}$, then M is a Sasakian manifold.

This theorem gives a sufficient condition for a K-contact Riemannian manifold to be Sasakian.

THEOREM 2. Let M be an n-dimensional K-contact Riemannian manifold isometrically immersed in an (n+1)-dimensional space form $\bar{M}$ of constant curvature 1. Then

(1) the type number $t \leq 2$, and

(2) M is of constant curvature 1 if and only if the scalar curvature $r = n(n-1)$.

THEOREM 3. Let M be an n-dimensional K-contact Riemannian manifold isometrically immersed in an (n+1)-dimensional space form $\bar{M}$ of constant curvature $c \neq 1$. Then $c < 1$ and M is of constant curvature 1.

C. PRODUCT OF ALMOST CONTACT MANIFOLDS: Let M and $\bar{M}$ be manifolds of dimension 2n+1 and 2m+1 and let $\Sigma = (\phi,\xi,\eta)$ and $\bar{\Sigma} = (\bar{\phi},\bar{\xi},\bar{\eta})$ be almost contact structures on M and $\bar{M}$, respectively. We define J by

$$J(X,\bar{X}) = (\phi X-\bar{\eta}(\bar{X})\xi,\bar{\phi}\bar{X}+\eta(X)\bar{\xi}).$$

Then, it is easily seen that $J^2 = -I$, which shows that J is an almost complex structure on $M \times \bar{M}$. We call J the induced almost complex structure on $M \times \bar{M}$ by Σ and $\bar{\Sigma}$.

If the induced almost complex structure J on $M \times M$ by Σ is integrable (i.e., complex structure), we call Σ is integrable. Morimoto [1] proved the following theorems.

THEOREM 1. The induced almost complex structure on $M \times \bar{M}$ is integrable if and only if Σ and $\bar{\Sigma}$ are both integrable.

THEOREM 2. An almost contact structure Σ is integrable if and only if Σ is normal.

Since an odd dimensional sphere is normal, we obtain (Calabi-Eckmann [1])

THEOREM 3. The direct product of two spheres of odd dimension has a complex structure.

D. TRIPLE OF KILLING VECTORS: Assume that a Riemannian manifold M admits three unit Killing vectors ξ, η and ζ which are mutually orthogonal and satisfy

$$\xi = \tfrac{1}{2}[\eta,\zeta], \qquad \eta = \tfrac{1}{2}[\zeta,\xi], \qquad \zeta = \tfrac{1}{2}[\xi,\eta].$$

We call such a set $\{\xi,\eta,\zeta\}$ a triple of Killing vectors. A triple $\{\xi,\eta,\zeta\}$ of Killing vectors is called a K-contact 3-structure, if ξ, η and ζ are K-contact structures. A K-contact 3-structure is called a *Sasakian 3-structure* if all of ξ, η and ζ are Sasakian structures. If, for a K-contact 3-structure $\{\xi,\eta,\zeta\}$, any two of ξ, η and ζ are Sasakian structures, then $\{\xi,\eta,\zeta\}$ is necessarily a Sasakian 3-structure (see Kuo [1], Kuo-Tachibana [1], Tachibana-Yu [1]). Kashiwada [1] proved

THEOREM. A Riemannian manifold with Sasakian 3-structure is an Einstein manifold.

E. SASAKIAN ϕ-SYMMETRIC MANIFOLDS: In view Proposition 5.2 a symmetric manifold condition is too strong for a Sasakian manifold. Toshio Takahashi [1] introduced the notion of Sasakian ϕ-symmetric manifolds.

Let M be a Sasakian manifold with structure tensors (ϕ,ξ,η,g). We denote by R the Riemannian curvature tensor of M. A Sasakian manifold M is said to be a *locally ϕ-symmetric manifold* if

$$\phi^2[(\nabla_V R)(X,Y)Z] = 0$$

for any horizontal vectors X, Y, Z and V, where a horizontal vector means that it is horizontal with respect to the connection form η of the local fibering; namely a horizontal vector is nothing but a vector which is orthogonal to ξ. Takahashi [1] proved the following

THEOREM 1. A Sasakian manifold is a locally ϕ-symmetric manifold if and only if each Kaehlerian manifold, which is a base space of local fibering, is a Hermitian locally symmetric manifold.

A complete and simply connected Sasakian locally ϕ-symmetric manifold is called a *globally ϕ-symmetric manifold*.

THEOREM 2. A Sasakian globally ϕ-symmetric manifold is a principal G^1-bundle over a Hermitian globally symmetric space, where G^1 is a 1-dimensional Lie group which is isomorphic to the 1-parameter group of global transformations generated by ξ.

F. CONTACT BOCHNER CURVATURE TENSOR: Let M be a (2n+1)-dimensional Sasakian manifold with structure tensors (ϕ,ξ,η,g). We denote by R, Q and r the Riemannian curvature tensor, the Ricci operator and the scalr curvature of M respectively. Then the *contact Bochner curvature tensor* B of M is defined by (Matsumoto-Chūman [1])

$$
\begin{aligned}
B(X,Y) = R(X,Y) &+ \frac{1}{m+4}(QY \wedge X - QX \wedge Y + Q\phi Y \wedge \phi X - Q\phi X \wedge \phi Y \\
&+ 2g(Q\phi X,Y)\phi + 2g(\phi X,Y)Q\phi + \eta(Y)\phi X \wedge \xi + \eta(X)\xi \wedge \phi Y) \\
&- \frac{k+m}{m+4}(\phi Y \wedge \phi X - 2g(\phi X,Y)\phi) - \frac{k-4}{m+4}Y \wedge X \\
&+ \frac{k}{m+4}(\eta(Y)\xi \wedge X + \eta(X)Y \wedge \xi),
\end{aligned}
$$

where we have put $m = 2n$, $k = (r+m)/(m+2)$ and $(X \wedge Y)Z = g(Y,Z)X - g(X,Z)Y$. Then we have (Matsumoto-Chūman [1])

THEOREM 1. Let M be a Sasakian manifold with vanishing contact Bochner curvature tensor. If M is an η-Einstein manifold, then M is of constant ϕ-sectional curvature.

We define the contact Ricci tensor L of M by setting

$$L(X,Y) = g(QX,Y) + 2g(X,Y) - 2(n+1)\eta(X)\eta(Y).$$

Then we have (Ikawa-Kon [1])

THEOREM 2. Let M be a $(2n+1)$-dimensional Sasakian manifold with constant scalar curvature and vanishing Bochner curvature tensor. If the contact Ricci tensor of M is positive semi-definite or negative semi-definite, then M is of constant ϕ-sectional curvature.

G. FIBRATION OF COMPACT NORMAL ALMOST CONTACT MANIFOLDS: Similar to the Boothby-Wang fibration of compact regular contact manifolds, we can construct a fibration of compact normal almost contact manifolds with ξ regular (see Hatakeyama [1], Morimoto [2]).

Conversely we have the following (Morimoto [2])

THEOREM. Let M be a $(2n+1)$-dimensional compact normal almost contact manifold with structure tensors (ϕ,ξ,η) and suppose that ξ is a regular vector field. Then M is the bundle space of a principal circle bundle $\pi : M \longrightarrow B$ over a real $2n$-dimensional almost complex manifold B. Moreover η is a connection form and 2-form Ψ on M such that $d\eta = \pi^*\Psi$ is of bidegree $(1,1)$.

H. BRIESKORN MANIFOLDS: Let B^{2n-1} be the Brieskorn manifold with contact form $\eta = \frac{1}{2}i\Sigma(z_j d\bar{z}_j - \bar{z}_j dz_j)$. We denote by g the metric tensor field of B^{2n-1} induced from the natural metric of C^{n+1}. Then η is given by $\eta(X) = g(Jv,X)$, $X \in T_x(B^{2n-1})$, where $v = n + \lambda v_1 + \mu v_2$, and J is an almost complex structure of C^{n+1}. We put $N = v/|v|$ and $\eta' = \eta/|v|$. Kashiwada [2] proved the following

THEOREM. For a Brieskorn manifold $B^{2n-1}(a_0,a_1,\ldots,a_n)$ the following conditions are equivalent:

(1) The induced structure (ϕ,JN,η',g) of B^{2n-1} is Sasakian;

(2) B^{2n-1} is η-umbilical in the Kaehlerian manifold $\bar{B}^{2n}$;

(3) $a_0 = a_1 = \ldots = a_n$;

(4) B^{2n-1} is umbilical in the Kaehlerian manifold $\bar{B}^{2n}$.

CHAPTER VI

SUBMANIFOLDS OF SASAKIAN MANIFOLDS

In this chapter, we study submanifolds of Sasakian manifolds, especially those of Sasakian space forms. In §1, we study invariant submanifolds of Sasakian manifolds. We compute the Laplacian for the second fundamental form and the Ricci tensor of an invariant submanifold of a Sasakian space form, and give some integral formulas. As applications of these integral formulas we prove the classification theorem of invariant submanifolds of codimension 2 of Sasakian space forms with η-parallel Ricci tensor. We also study invariant submanifolds of codimension 2 with constant scalar curvature. In §2, we discuss anti-invariant submanifolds, tangent to the structure vector field, of Sasakian manifolds. We give some examples of anti-invariant submanifolds tangent to the structure vector field of Sasakian space forms and prove some theorems which characterize these examples. In §3, we study submanifolds of Sasakian manifolds which are normal to the structure vector field. Then the submanifolds are anti-invariant submanifolds. §4 is devoted to the study of contact CR submanifolds of Sasakian manifolds. We construct some examples of contact CR submanifolds of a unit sphere, which have parallel mean curvature vector and flat normal connection. We also prove some theorems concerning these examples. In the last §5, we discuss the structure induced on submanifolds.

1. INVARIANT SUBMANIFOLDS OF SASAKIAN MANIFOLDS

Let $\bar{M}$ be a (2m+1)-dimensional Sasakian manifold with structure tensors (ϕ,ξ,η,g). A (2n+1)-dimensional submanifold M of $\bar{M}$ is said to be *invariant* if the structure vector field ξ is tangent to M everywhere on M and ϕX is tangent to M for any vector X tangent to M a every point of M, that is, $\phi T_x(M) \subset T_x(M)$ for all $x \in M$. We easily see that any invariant submanifold M with induced structure tensors, which will be denoted by the same letters (ϕ,ξ,η,g) as $\bar{M}$, is also a Sasakian manifold.

Let $\bar{\nabla}$ (resp. ∇) be the operator of covariant differentiation with respect to the Levi-Civita connection of $\bar{M}$ (resp. M). Since ξ is tangent to M, for any vector field X tangent to M, we have $-\phi X = \bar{\nabla}_X\xi = \nabla_X\xi + B(X,\xi)$. Thus we obtain $B(X,\xi) = 0$. For any vector fields X and Y tangent to M we obtain

$$\bar{\nabla}_X Y = \nabla_X\phi Y + B(X,\phi Y) = (\nabla_X\phi)Y + \phi\nabla_X Y + B(X,\phi Y)$$

$$= g(X,Y)\xi - \eta(Y)X + \phi\nabla_X Y + B(X,\phi Y),$$

$$\bar{\nabla}_X\phi Y = (\bar{\nabla}_X\phi)Y + \phi\bar{\nabla}_X Y$$

$$= g(X,Y)\xi - \eta(Y)X + \phi\nabla_X Y + \phi B(X,Y).$$

From these equations we have $\phi B(X,Y) = B(X,\phi Y)$. Since B is symmetric, we also have $\phi B(X,Y) = B(\phi X,Y)$. Consequently, we have the following lemma for the second fundamental form B of M, or equivalently, for the associated second fundamental form A of M.

LEMMA 1.1. Let M be an invariant submanifold of a Sasakian manifold $\bar{M}$. Then

(1.1) $$B(X,\xi) = 0, \qquad A_V\xi = 0,$$

(1.2) $$B(X,\phi Y) = B(\phi X,Y) = \phi B(X,Y),$$

(1.3) $$\phi A_V X = -A_V\phi X = A_{\phi V}X.$$

PROPOSITION 1.1. Any invariant submanifold M of a Sasakian manifold $\bar{M}$ is a minimal submanifold.

Proof. Let $e_1,\ldots,e_n,\phi e_1,\ldots,\phi e_n,\xi$ be a ϕ-basis of $T_x(M)$. By (1.1) and (1.2) we see that

$$\sum[B(e_i,e_i) + B(\phi e_i,\phi e_i)] + B(\xi,\xi) = 0$$

because of $B(\phi e_i,\phi e_i) = \phi^2 B(e_i,e_i) = -B(e_i,e_i)$. This means that M is a minimal submanifold of $\bar{M}$. QED.

PROPOSITION 1.2. If the second fundamental form of an invariant submanifold M of a Sasakian manifold $\bar{M}$ is parallel, then M is totally geodesic.

Proof. By the assumption on B we have

$$0 = (\nabla_X B)(Y,\xi) = B(Y,\phi X) = \phi B(X,Y),$$

from which

$$B(X,Y) = -\phi^2 B(X,Y) - \eta(B(X,Y))\xi = 0.$$

Therefore, M is totally geodesic. QED.

Let $\bar{R}$ (resp. R) be the Riemannian curvature tensor of $\bar{M}$ (resp. M). From equation of Gauss and (1.2) we have

$$g(R(X,\phi X)\phi X,X) = g(\bar{R}(X,\phi X)\phi X,X) - 2g(B(X,X),B(X,X)). \tag{1.4}$$

Thus we have the following

PROPOSITION 1.3. Let M be an invariant submanifold of a Sasakian space form $\bar{M}(c)$. Then M is totally geodesic if and only if M is of constant ϕ-sectional curvature c.

We now suppose that the ambient manifold $\bar{M}$ is a Sasakian space form with constant ϕ-sectional curvature c. Then we have

$$(1.5)\quad R(X,Y)Z = \tfrac{1}{4}(c+3)[g(Y,Z)X-g(X,Z)Y] + \tfrac{1}{4}(c-1)[\eta(X)\eta(Z)Y -\eta(Y)\eta(Z)X+g(X,Z)\eta(Y)\xi-g(Y,Z)\eta(X)\xi+g(\phi Y,Z)\phi X -g(\phi X,Z)\phi Y+2g(X,\phi Y)\phi Z] + A_{B(Y,Z)}X - A_{B(X,Z)}Y,$$

$$(1.6)\quad (\nabla_X B)(Y,Z) - (\nabla_Y B)(X,Z) = 0.$$

We denote by S and r the Ricci tensor and the scalar curvature of M respectively. Then, from (1.5) we have

$$(1.7)\quad S(X,Y) = \tfrac{1}{2}(n(c+3)+(c-1))g(X,Y) - \tfrac{1}{2}(n+1)(c-1)\eta(X)\eta(Y) - \sum_i g(B(X,e_i),B(Y,e_i)),$$

$$(1.8)\quad r = n^2(c+3)+n(c+1) - \sum_{i,j} g(B(e_i,e_j),B(e_i,e_j)),$$

where $\{e_i\}$ is an orthonormal basis of M.

In view of Proposition 1.2, the notion of parallel second fundamental form of an invariant submanifold of a Sasakian manifold is not essential. So, we need the following.

Definition. Let M be an invariant submanifold of a Sasakian manifold $\bar{M}$. The second fundamental form B of M is said to be η-*parallel* if $(\nabla_{\phi X}B)(\phi Y,\phi Z) = 0$ for all vector fields X, Y and Z tangent to M.

It will be proved that the notion of η-parallel second fundamental form corresponds to that of parallel second fundamental form for invariant submanifolds of Kaehlerian manifolds. We easily see that the second fundamental form B is η-parallel if and only if

$$(1.9)\quad (\nabla_X B)(Y,Z) = \eta(X)\phi B(Y,Z) + \eta(Y)\phi B(X,Z) + \eta(Z)\phi B(X,Y).$$

LEMMA 1.2. Let M be an invariant submanifold of a Sasakian manifold $\bar{M}$. Then the second fundamental form of M is η-parallel if and only if $|\nabla A|^2 = 3|A|^2$.

Proof. For the proof of this we take a ϕ-basis $e_1,\ldots,e_{2n+1}$ ($e_{n+t} = \phi e_t$, $e_{2n+1} = \xi$). We also take a basis $v_1,\ldots,v_{2p}$ of the normal space of M, where $2p = \text{codim } M$. Then we have

$$g((\nabla_{e_i} A)_a\xi,(\nabla_{e_i} A)_a\xi) = g(A_a e_i, A_a e_i),$$

$$\begin{aligned} g((\nabla_{e_i} A)_a\phi e_j,(\nabla_{e_i} A)_a\phi e_j) \\ = g(e_i,A_a e_j)^2 + g(\phi(\nabla_{e_i} A)_a e_j,\phi(\nabla_{e_i} A)_a e_j), \end{aligned}$$

where $A_a = A_{v_a}$ $(a = 1,\ldots,2p)$. These equations imply

$$\begin{aligned} |\nabla A|^2 &= \sum_{a=1}^{2p}\sum_{i,j=1}^{2n+1} g((\nabla_{e_i} A)_a e_j,(\nabla_{e_i} A)_a e_j) \\ &= \sum_{a=1}^{2p}\sum_{i,j=1}^{2n} g((\nabla_{e_i} A)_a \phi e_j,(\nabla_{e_i} A)_a \phi e_j) + 2|A|^2 \\ &= \sum_{a=1}^{2p}\sum_{i,j=1}^{2n} g(\phi(\nabla_{e_i} A)_a e_j,\phi(\nabla_{e_i} A)_a e_j) + 3|A|^2. \end{aligned}$$

Therefore we have

$$(1.10) \qquad |\nabla A|^2 - 3|A|^2 = \sum_{a=1}^{2p}\sum_{i,j=1}^{2n} g(\phi(\nabla_{e_i} A)_a e_j,\phi(\nabla_{e_i} A)_a e_j) \geq 0,$$

which proves our asertion. QED.

LEMMA 1.3. If the second fundamental form of an invariant submanifold M of a Sasakian manifold $\bar{M}$ is η-parallel, then the square of the length of the second fundamental form of M is constant, that is, $|A|^2$ is constant.

Proof. Let $\{e_i\}$ be an orthonormal basis of M. Then we have the following equation.

$$\nabla_X|A|^2 = \nabla_X[\sum_{i,j} g(B(e_i,e_j),B(e_i,e_j))] = 2\sum_{i,j} g((\nabla_X B)(e_i,e_j),B(e_i,e_j))$$

$$= 2\sum_{i,j}[\eta(X)g(B(e_i,e_j),B(e_i,e_j))+\eta(e_i)g(\phi B(X,e_j),B(e_i,e_j))$$
$$+\eta(e_j)g(\phi B(X,e_i),B(e_i,e_j))]$$

$$= 2\sum_i[g(\phi B(X,e_i),B(\xi,e_i)) + g(\phi B(X,e_i),B(\xi,e_i))] = 0.$$

Therefore, $|A|^2$ is a constant. QED.

In the sequel we give an integral formula, so-called the Simons' type formula, for an invariant submanifold of a Sasakian space form by computing the Laplacian of the square of the length of the second fundamental form.

Let M be a (2n+1)-dimensional invariant submanifold of a Sasakian space form $\bar{M}^{2n+1}(c)$. Hereafter we put p = m-n. We take an orthonormal basis $e_1,\ldots,e_{2n+1}$ of M such that $e_{n+t} = \phi e_t$ (t = 1,...,n), $e_{2n+1} = \xi$. We take also an orthonormal basis $v_1,\ldots,v_{2p}$ of the normal space of M such that $v_{p+s} = \phi v_s$ (s = 1,...,p).

In view of Proposition 3.1 of Chapter II, by a straightforward computation, we have

$$(1.11)\quad -g(\nabla^2 A,A) - 3|A|^2 = \sum_{a,b=1}^{2p}(\mathrm{Tr}A_aA_b)^2 + \sum_{a,b=1}^{2p}|[A_a,A_b]|^2$$
$$- \tfrac{1}{2}(n+2)(c+3)|A|^2.$$

LEMMA 1.4. Let M be a (2n+1)-dimensional invariant submanifold of a (2m+1)-dimensional Sasakian manifold $\bar{M}$. Then we have

$$(1.12)\qquad \frac{1}{2p}|A|^4 \leq \sum_{a,b=1}^{2p}(\mathrm{Tr}A_aA_b)^2 \leq \tfrac{1}{2}|A|^4,$$

$$(1.13)\qquad \frac{1}{n}|A|^4 \leq \sum_{a,b=1}^{2p}|[A_a,A_b]|^2 \leq |A|^4.$$

If $\bar{M}$ is a Sasakian space form, then M is η-Einstein if and only if $\Sigma|[A_a,A_b]|^2 = |A|^4/n$.

Proof. We put $A^* = \sum_{a=1}^{2p} A_a^2$. Then we have

$$\sum_{a,b=1}^{2p} |[A_a,A_b]|^2 = 2\mathrm{Tr}(A^*)^2.$$

Since A^* is symmetric, positive semi-definite and $\phi A^* = A^*\phi$, choosing a suitable basis, A^* is represented by a matrix form

$$A^* = \begin{pmatrix} \lambda_1 & & & & \\ & \cdot & & 0 & \\ & & \cdot & & \\ & & & \cdot & \\ & 0 & & & \cdot \\ & & & & \lambda_{2n} \end{pmatrix}, \quad \lambda_{n+t} = \lambda_t, \quad \lambda_t \geq 0.$$

Then we have

$$2\mathrm{Tr}(A^*)^2 = |A|^4 - 4\sum_{t\neq s}^{n} \lambda_t\lambda_s,$$

$$2\mathrm{Tr}(A^*)^2 = \frac{1}{n}|A|^4 + \frac{1}{n}\sum_{i>j}^{2n}(\lambda_i - \lambda_j)^2.$$

From these equations we obtain (1.13). If $\Sigma|[A_a,A_b]|^2 = |A|^4/n$, then we have $\lambda_i = \lambda_j$ for all i, j. From this and (1.7) we easily see that M is η-Einstein if and only if the equality above holds.

In the next place, choosing a suitable basis $\{v_a\}$ we have

$$\sum_{a,b=1}^{2p} (\mathrm{Tr}A_aA_b)^2 = \sum_{a=1}^{2p} (\mathrm{Tr}A_a^2)^2.$$

From this we obtain

$$\sum_{a,b=1}^{2p} (\mathrm{Tr}A_aA_b)^2 = \tfrac{1}{2}|A|^4 - \sum_{a\neq b}^{p} \mathrm{Tr}A_a^2\mathrm{Tr}A_b^2,$$

$$\sum_{a,b=1}^{2p} (\mathrm{Tr}A_aA_b)^2 = \sum_{a=1}^{2p} (\mathrm{Tr}A_a^2)^2 \geq \frac{1}{2p}(\sum_{a=1}^{2p} \mathrm{Tr}A_a^2) = \frac{1}{2p}|A|^4.$$

Therefore we have (1.12). QED.

From (1.11) and Lemma 1.4 we have

$$(1.14) \qquad -g(\nabla^2 A,A) - 3|A|^2 \leq \tfrac{1}{2}[3|A|^2 - (n+2)(c+3)]|A|^2.$$

Thus we have the following theorems (Kon [4], [8]).

THEOREM 1.1. Let M be a (2n+1)-dimensional compact invariant submanifold of a Sasakian space form $\bar{M}^{2m+1}(c)$. Then

$$0 \leq \int_M(|\nabla A|^2 - 3|A|^2)*1 \leq \tfrac{1}{2}\int_M[3|A|^2 - (n+2)(c+3)]|A|^2*1.$$

THEOREM 1.2. Let M be a (2n+1)-dimensional compact invariant submanifold of a Sasakian space form $\bar{M}^{2m+1}(c)$. Then either M is totally geodesic, or $|A|^2 = (n+2)(c+3)/3$, or at some point x of M, $|A|^2(x) > (n+2)(c+3)/3$.

THEOREM 1.3. Let M be a (2n+1)-dimensional η-Einstein invariant submanifold of a Sasakian space form $\bar{M}^{2m+1}(c)$. Then

$$0 \leq |\nabla A|^2 - 3|A|^2 \leq \tfrac{1}{2}(n+2)[\frac{1}{n}|A|^2 - (c+3)]|A|^2.$$

Proof. If M is an η-Einstein manifold, then the scalar curvature r of M is constant. Thus, by (1.8), $|A|^2$ is also constant and hence $-g(\nabla^2 A,A) = |\nabla A|^2$. Therefore, (1.11) and Lemma 1.4 imply our inequality. QED.

THEOREM 1.4. Let M be a (2n+1)-dimensional η-Einstein invariant submanifold of a Sasakian space form $\bar{M}^{2m+1}(c)$. Then either M is totally geodesic, or $|A|^2 \geq n(c+3)$.

If an invariant submanifold M is of constant ϕ-sectional curvature k, by using the Gauss equation, we have

$$(1.15) \qquad \sum_{a,b}(\mathrm{Tr}A_aA_b)^2 = (c-k)|A|^2 = \frac{1}{n(n+1)}|A|^4.$$

Therefore we have

THEOREM 1.5. Let M be a (2n+1)-dimensional invariant submanifold of a Sasakian space form $\bar{M}^{2m+1}(c)$. If M is of constant ϕ-sectional curvature k, then

$$0 \leq |\nabla A|^2 - 3|A|^2 = n(n+1)(n+2)(c-k)(\tfrac{1}{2}(c+3)-(k+3)).$$

THEOREM 1.6. Let M be a (2n+1)-dimensional invariant submanifold of a Sasakian space form $\bar{M}^{2m+1}(c)$. If M is of constant ϕ-sectional curvature k, and if $c > -3$, then either M is totally geodesic, that is, $c = k$, or $(c+3) \geq 2(k+3)$.

THEOREM 1.7. Let M be a (2n+1)-dimensional invariant submanifold of a Sasakian space form $\bar{M}^{2m+1}(c)$ with constant ϕ-sectional curvature k. If the second fundamental form of M is η-parallel, then M is totally geodesic, or $(c+3) = 2(k+3)$, the latter case arising only when $c > -3$.

From (1.12) and (1.15) we have

THEOREM 1.8. Let M be a (2n+1)-dimensional invariant submanifold of a Sasakian space form $\bar{M}^{2m+1}(c)$ with constant ϕ-sectional curvature k. If $p < n(n+1)/2$, then M is totally geodesic, where $p = m-n$.

In the next place, we consider a (2n+1)-dimensional invariant submanifold M of a Sasakian space form $\bar{M}^{2m+1}(c)$ with respect to the Ricci tensor S of M. First of all we compute the Laplacian for the Ricci operator Q of M (Kon [8]).

We put

$$T = Q - aI - b\eta \otimes \xi,$$

a, b being constant and $a+b = 2n$, $(2n+1)a+b = r$. We notice that $T = 0$ if and only if M is η-Einstein.

LEMMA 1.5. Let M be a (2n+1)-dimensional invariant submanifold of a Sasakian space form $\bar{M}^{2m+1}(c)$. If the scalar curvature r of M is constant, then

$$-g(\nabla^2 Q,Q) - 2|Q|^2 - 16n^3 - 8n^2 + 8nr$$

$$= -n(c+3)|T|^2 + \sum_{a=1}^{2p} |[Q,A_a]|^2.$$

Proof. From Lemma 5.5 of Chapter V we obtain

$$(\nabla^2 S)(X,Y) = \sum_{i=1}^{2n+1} (\nabla_{e_i}\nabla_{e_i} S)(X,Y)$$

$$= \sum[(R(e_i,X)S)(e_i,Y) + (R(e_i,\phi Y)S)(e_i,\phi X)]$$

$$- 4S(X,Y) + 8ng(X,Y) + (3r-12n^2-6n)\eta(X)\eta(Y).$$

On the other hand, by equation of Gauss, we have

$$\sum(R(e_i,X)S)(e_i,Y) = -\sum[S(\bar{R}(e_i,X)e_i,Y) + S(e_i,\bar{R}(e_i,X)Y)$$

$$+ S(A_{B(X,e_i)}e_i,Y) - S(A_{B(e_i,Y)}X,e_i) + S(A_{B(X,Y)}e_i,e_i)].$$

Moreover, we have

$$-\sum S(A_{B(X,Y)}e_i,e_i) = \sum_{a=1}^{p}\sum_{i=1}^{2n+1} [g(A_a A_{B(X,Y)}A_a e_i,e_i)$$

$$+ g(\phi A_a A_{B(X,Y)}\phi A_a e_i,e_i)] = 0,$$

$$\sum_i[S(A_{B(e_i,Y)}X,e_i) - S(A_{B(X,e_i)}e_i,Y)]$$

$$= \sum_a[g(A_a QA_a X,Y) - g(QA_a A_a X,Y)]$$

From these equations and Theorem5.4 of Chapter V we obtain

$$\sum_i(R(e_i,X)S)(e_i,Y) = \tfrac{1}{2}(n(c+3)+2)S(X,Y) - \tfrac{1}{4}(c+3)rg(X,Y)$$

$$+ \tfrac{1}{2}n(c-1)g(X,Y) - \tfrac{1}{2}n(2n+1)(c-1)\eta(X)\eta(Y) + \tfrac{1}{4}(c-1)r\eta(X)\eta(Y)$$

$$+ \sum_a[g(A_a QA_a X,Y) - g(QA_a A_a X,Y)].$$

Similarly we have

$$\sum_i (R(e_i,\phi Y)S)(e_i,\phi X) = \tfrac{1}{2}(n(c+3)+2)S(X,Y) - \tfrac{1}{4}(c+3)rg(X,Y)$$
$$+ \tfrac{1}{2}n(c-1)g(X,Y) - n(n(c+3)+2)\eta(X)\eta(Y) + \tfrac{1}{4}(c+3)r\eta(X)\eta(Y)$$
$$- \tfrac{1}{2}n(c-1)\eta(X)\eta(Y) + \sum_a [g(A_aQA_aX,Y) - g(QA_aA_aX,Y)].$$

From these equations we have our assertion. QED.

If the Ricci tensor S of a Sasakian manifold M satisfies $(\nabla_X S)(\phi Y,\phi Z) = 0$ for all vector fields X, Y and Z on M, then we say that the Ricci tensor S of M is η-*parallel*.

We now prepare the following lemmas.

LEMMA 1.6. Let M be a Sasakian manifold with η-parallel Ricci tensor. Then we have the following:

(1) The scalar curvature r of M is constant;

(2) The square of the length of the Ricci operator Q of M is constant, that is, $|Q|^2$ = constant.

Proof. The Ricci tensor S of M is η-parallel if and only if

$$(\nabla_X S)(Y,Z) = 2n[g(\phi X,Y)\eta(Z) + g(\phi X,Z)\eta(Y)]$$
$$+ \eta(Y)S(X,\phi Z) + \eta(Z)S(X,\phi Y).$$

Thus we have $\nabla_X r = \Sigma(\nabla_X S)(e_i,e_i) = 0$, which shows that r is constant. Moreover, since $g(QX,Y) = S(X,Y)$, we obtain $\nabla_X|Q|^2 = 2\Sigma g((\nabla_X Q)e_i,Qe_i)$ $= 0$ and hence $|Q|^2$ is constant. QED.

LEMMA 1.7. The Ricci tensor S of a (2n+1)-dimensional Sasakian manifold M is η-parallel if and only if the following equation is satisfied:

$$|\nabla Q|^2 - 2|Q|^2 - 16n^3 - 8n^2 + 8nr = 0.$$

Proof. By using a ϕ-basis $e_1,\ldots,e_{2n+1}$ ($e_{n+t} = \phi e_t$, $e_{2n+1} = \xi$), we have

$$
\begin{aligned}
|\nabla Q|^2 &= \sum_{i,j=1}^{2n+1} g((\nabla_{e_i} Q)e_j,(\nabla_{e_i} Q)e_j) \\
&= \sum_{i=1}^{2n+1}\sum_{j=1}^{2n} g((\nabla_{e_i} Q)e_j,(\nabla_{e_i} Q)e_j) + \sum_{i=1}^{2n+1} g((\nabla_{e_i} Q)\xi,(\nabla_{e_i} Q)\xi) \\
&= \sum_{i=1}^{2n+1}\sum_{j=1}^{2n} g((\nabla_{e_i} Q)\phi e_j,(\nabla_{e_i} Q)\phi e_j) + \sum_{i=1}^{2n+1} g((\nabla_{e_i} Q)\xi,(\nabla_{e_i} Q)\xi) \\
&= 2|Q|^2 + 16n^3 + 8n^2 - 8nr + \sum_{i=1}^{2n+1}\sum_{j=1}^{2n} g(\phi(\nabla_{e_i} Q)e_j,\phi(\nabla_{e_i} Q)e_j).
\end{aligned}
$$

On the other hand, we can see easily that the Ricci tensor S of M is η-parallel if and only if $\Sigma\Sigma g(\phi(\nabla_{e_i} Q)e_j,\phi(\nabla_{e_i} Q)e_j) = 0$. Thus we have our assertion. QED.

Here we notice that

$$|T|^2 = |Q|^2 - r^2/2n + 2r - 4n^2 - 2n.$$

From this we also have

$$\sum|[A_a,A_b]|^2 = 2\mathrm{Tr}(A^*)^2 = \frac{1}{n}|A|^4 + 2|T|^2.$$

PROPOSITION 1.4. Let M be a (2n+1)-dimensional invariant submanifold of a Sasakian space form $\bar{M}^{2m+1}(c)$ ($c < -3$). If the Ricci tensor S of M is η-parallel, then M is η-Einstein.

Proof. From Lemma 1.6, $|Q|^2$ is constant. Thus we have $-g(\nabla^2 Q,Q) = |\nabla Q|^2$. Therefore Lemmas 1.5 and 1.7 imply

$$-n(c+3)|T|^2 + \sum|[Q,A_a]|^2 = 0.$$

Since $c < -3$, we have $|T|^2 = 0$ and hence $T = 0$, which shows that M is η-Einstein. QED.

PROPOSITION 1.5. Let M be a $(2n+1)$-dimensional compact invariant submanifold of a Sasakian space form $\bar{M}^{2m+1}(c)$ $(c > -3)$ with constant scalar curvature. If $QA_a = A_aQ$ $(a = 1,\dots,p)$, then M is η-Einstein.

Proof. By the assumption and Lemma 1.5 we easily see that $|T|^2 = 0$. Therefore M is η-Einstein. QED.

LEMMA 1.8. Let M be an invariant submanifold of codimension 2 of a Sasakian space form $\bar{M}^{2n+3}(c)$. Then the Ricci tensor S of M is η-parallel if and only if the second fundamental form of M is η-parallel.

Proof. Suppose that the second fundamental form of M is η-parallel. Then it is easy to see that S is η-parallel.

Conversely, we assume that S is η-parallel. We shall prove that B is η-parallel. To prove this it sufficies to show that, at each point x of M,

$$\phi(\nabla_{\phi X}A)_v\phi Y = \phi(\nabla_{\phi X}A)_{\phi v}\phi Y = 0$$

for any $X, Y \in T_x(M)$, where $\{v,\phi v\}$ is an orthonormal basis of the normal space of M. There exists a 1-form s such that $D_Xv = s(X)\phi v$. From (1.7) we have

$$\phi(\nabla_XA_v)^2\phi Y = \phi(\nabla_XA)_vA_v\phi Y + s(X)\phi^2A_v^2\phi Y$$
$$+ \phi A_v(\nabla_XA)_v\phi Y + s(X)\phi A_v\phi A_v\phi Y = 0.$$

Since $\phi A_v = -A_v\phi$, the above equation reduces to

$$\phi(\nabla_XA)_vA_v\phi Y + \phi A_v(\nabla_XA)_v\phi Y = 0. \tag{1.16}$$

We take any two λ, μ of characteristic roots of A_v at a point x of M. We now define spaces by setting

$$T_\lambda = \{X \in T_x(M)\colon A_vX = \lambda X\}, \quad T_\mu = \{X \in T_x(M)\colon A_vX = \mu X\}.$$

Then $T_\lambda \cap T_\mu = \{0\}$ when $\lambda \neq \mu$. If $Y \in T_\mu$, then $\phi Y \in T_{-\mu}$. Let $Y \in T_\mu$. Then (1.16) implies

$$A_v\phi(\nabla_X A)_v\phi Y = -\mu\phi(\nabla_X A)_v\phi Y,$$

which shows that $\phi(\nabla_X A)_v\phi Y \in T_{-\mu}$. Let $X \in T_\lambda$, $Y \in T_\mu$ and $\lambda \neq \mu$. Then, from equation of Codazzi, we obtain

$$\phi(\nabla_{\phi X}A)_v\phi Y \in T_{-\lambda} \cap T_{-\mu}.$$

Thus we have $\phi(\nabla_{\phi X}A)_v\phi Y = 0$. On the other hand, if $X \in T_\lambda$, then $\phi^2 X \in T_\lambda$. Let $X \in T_\lambda$, $Y \in T_\mu$ and $\lambda = \mu \neq 0$. Then we find

$$\phi^2(\nabla_{\phi X}A)_v\phi Y \in T_\lambda.$$

From this and equation of Codazzi we have

$$\phi(\nabla_{\phi X}A)_v\phi^2(\nabla_{\phi X}A)_v\phi Y \in T_\lambda \cap T_{-\lambda},$$

which means that $\phi(\nabla_{\phi X}A)_v\phi^2(\nabla_{\phi X}A)_v\phi Y = 0$. From this we have $\phi(\nabla_{\phi X}A)_v\phi Y = 0$.

Let $\lambda = \mu = 0$. We take $X, Y \in T_0$ at $x \in M$, and extend these to local vector fields on M which are covariant constant with respect to ∇ at x. Here we notice that if $X \in T_0$, then $\phi X \in T_0$. From (1.16), we obtain, at $x \in M$,

$$g(\phi(\nabla_{\phi X}A)_v(\nabla_{\phi X}A)_v\phi Y, Y) = 0,$$

and hence $(\nabla_{\phi X}A)_v\phi Y = 0$. Consequently we have $\phi(\nabla_{\phi X}A)_v\phi Y = 0$ for any cases. On the other hand, if $\phi(\nabla_{\phi X}A)_v\phi Y = 0$, then

$$(\nabla_{\phi X}A)_{\phi v}\phi Y = -g(\phi X, A_v\phi Y)\xi,$$

from which $\phi(\nabla_{\phi X}A)_{\phi v}\phi Y = 0$. This completes the proof of our assertion.

QED.

LEMMA 1.9. Let M be an invariant submanifold of codimension 2 of a Sasakian space form $\bar{M}^{2n+3}(c)$. If the Ricci tensor S of M is η-parallel, then M is an η-Einstein manifold.

Proof. From Lemmas 1.5, 1.6 and 1.7 we have

$$n(c+3)|T|^2 = \sum_{a=1}^{2} |[Q,A_a]|^2.$$

Since M is of codimension 2, we can take a basis $\{v,\phi v\}$ for $T_x(M)^\perp$. Then (1.7) implies

$$S(A_vX,Y) = \tfrac{1}{2}(n(c+3)+(c-1))g(A_vX,Y) - 2g(A_v^3X,Y) = S(X,A_vY),$$

which shows that $QA_v = A_vQ$ and hence $n(c+3)|T|^2 = 0$. Thus if $c \neq -3$, then M is η-Einstein. Let $c = -3$. Then (1.11) and Lemma 1.8 show that M is totally geodesic and hence M is η-Einstein. QED.

From these lemmas we have (Kon [8])

THEOREM 1.9. Let M be an invariant submanifold of codimension 2 of a Sasakian space form $\bar{M}^{2n+3}(c)$. Then the following conditions are equivalent:

(1) The Ricci tensor of M is η-parallel;

(2) The second fundamental form of M is η-parallel;

(3) M is η-Einstein.

THEOREM 1.10. Let M be an invariant submanifold of codimension 2 with η-parallel Ricci tensor of a Sasakian space form $\bar{M}^{2n+3}(c)$. If $c \leq -3$, then M is totally geodesic. If $c > -3$, then either M is totally geodesic, or an η-Einstein manifold with $|A|^2 = n(c+3)$ and hence $r = n(n(c+3)-2)$.

Proof. From Theorem 1.9, M is η-Einstein. Then (1.11) and Lemma 1.4 imply

$$\tfrac{1}{2}(n+2)[\frac{1}{n}|A|^2 - (c+3)]|A|^2 = 0.$$

This proves our theorem. QED.

THEOREM 1.11. Let M be a compact invariant submanifold of codimension 2 of a Sasakian space form $\bar{M}^{2n+3}(c)$ $(c > -3)$. If the scalar curvature r of M is constant, then either M is totally geodesic, or an η-Einstein manifold with the scalar curvature $r = n(n(c+3)-2)$.

Proof. In this case we have already seen that $QA_v = A_vQ$. Then Proposition 1.5 states that M is an η-Einstein manifold. Thus our theorem follows from Theorem 1.10. QED.

Here we consider the case that the second fundamental form of M satisfies $|A|^2 = (n+2)(c+3)/3$. We prove

THEOREM 1.12. Let M be a (2n+1)-dimensional invariant submanifold of a Sasakian space form $\bar{M}^{2m+1}(c)$ $(c > -3)$. If $|A|^2 = (n+2)(c+3)/3$, then M is an η-Einstein manifold of dimension 3 and has the scalar curvature $r = (c+1)$.

Proof. Since $|A|^2$ is constant, we have $-g(\nabla^2 A,A) = |\nabla A|^2$. Then (1.11) implies

$$\begin{aligned}0 \le |\nabla A|^2 - 3|A|^2 &= \sum(\mathrm{Tr}A_aA_b)^2 + \sum|[A_a,A_b]|^2 - \tfrac{1}{2}(n+2)(c+3)|A|^2\\ &= \tfrac{1}{2}[3|A|^2-(n+2)(c+3)]|A|^2\\ &\quad - 4\sum_{t\neq s}^{n}\lambda_t\lambda_s - \sum_{a\neq b}^{p}\mathrm{Tr}A_a^2\mathrm{Tr}A_b^2.\end{aligned}$$

From the assumption we have

$$\sum_{t\neq s}^{n}\lambda_t\lambda_s = 0 \qquad \text{and} \qquad \sum_{a\neq b}^{p}\mathrm{Tr}A_a^2\mathrm{Tr}A_b^2 = 0.$$

Thus we may assume that $\lambda_t = 0$ for $t = 2,\ldots,n$ and $A_a = 0$ for $a = 2,\ldots,p$. Therefore we have

$$QX = \frac{1}{2}(n(c+3)+(c-1))X - \frac{1}{2}(n+1)(c-1)\eta(X)\xi - 2A_1^2X.$$

From this we see that $QA_a = A_aQ$ for all a. From this and Lemma 1.5, M is η-Einstein because of the second fundamental form of M is η-parallel and hence Q is η-parallel. Then Theorem 1.4 implies

$$\frac{1}{3}(n+2)(c+3) = |A|^2 \geq n(c+3),$$

from which n = 1 and hence M is of dimension 3. Thus we have r = (c+1) by (1.8). QED.

We now assume that M is a regular Sasakian manifold. Let M/ξ denote the set of orbits of ξ. Then M/ξ is a real 2n-dimensional Kaehlerian manifold. Then there is a fibering $\pi : M \longrightarrow M/\xi$. Henceforth X*, Y* and Z* on M will be horizontal lifts of X, Y and Z over M/ξ respectively with respect to the connection η. Then we have

$$(S'(Y,Z))^* = S(Y^*,Z^*) + 2g(Y^*,Z^*),$$

where S' denotes the Ricci tensor of M/ξ. From this we have

$$((\nabla'_X S')(Y,Z))^* = (\nabla_{X^*}S)(Y^*,Z^*),$$

where ∇' denotes the operator of covariant differentiation in M/ξ. therefore we see that the Ricci tensor S' of M/ξ is parallel if and only if $(\nabla_{X^*}S)(Y^*,Z^*) = 0$ which is equivalent to $(\nabla_{\phi U}S)(\phi V,\phi W) = 0$ for any U, V, W ε $T_x(M)$ because of the horizontal space is spanned by $\{\phi U: U \in T_x(M)\}$. On the other hand, $(\nabla_{\phi U}S)(\phi V,\phi W) = 0$ implies that $(\nabla_{\phi^2 U}S)(\phi V,\phi W) = -(\nabla_U S)(\phi V,\phi W) = 0$ and the converse is also true. Therefore the Ricci tensor S' is parallel if and only if $(\nabla_U S)(\phi V,\phi W) = 0$, which states the meaning of the definition of η-parallel Ricci tensor.

THEOREM 1.13. Let M be a regular Sasakian manifold. Then the Ricci tensor S of M is η-parallel if and only if the Ricci tensor S' of M/ξ is parallel.

Example 1.1. Let R^{2n+3} be a (2n+3)-dimensional Euclidean space with standard Sasakian structure of constant ϕ-sectional curvature c = -3. A (2n+3)-dimensional unit sphere S^{2n+3} has the standard Sasakian structure of constant ϕ-sectional curvature c > -3. By CD^{n+1}, R and (R,CD^{n+1}) we denote the open uni ball in a complex (n+1)-dimensional Euclidean space C^{n+1}, a real line and the product bundle $R \times CD^{n+1}$. Then (R,CD^{n+1}) also has a Sasakian structure with constant ϕ-sectional curvature c < -3. Let Q^n be an n-dimensional complex quadric in a complex projectivae space CP^{n+1}. We denote by (S,Q^n) a circle bundle over Q^n. Then (S,Q^n) define a Sasakian structure which is η-Einstein and (S,Q^n) is an invariant submanifold of S^{2n+3}. R^{2n+1}, (R,CD^n) and S^{2n+1} are totally geodesic invariant submanifolds of R^{2n+3}, (R,CD^{n+1}) and S^{2n+3} respectively.

We prove the following (Kon [8])

THEOREM 1.14. (1) S^{2n+1} and (S,Q^n) are the only connected complete invariant submanifolds of codimension 2 in S^{2n+3} which have η-parallel Ricci tensor;

(2) (R,CD^n) (resp. R^{2n+1}) is the only connected complete invariant submanifold of codimension 2 in (R,CD^{n+1}) (resp. R^{2n+3}) which has η-parallel Ricci tensor.

Proof. Let $\bar{M}$ be one of the R^{2n+3}, S^{2n+3} and (R,CD^{n+1}) and $\bar{B}$ be C^{n+1} (if $\bar{M} = R^{2n+3}$), CP^{n+1} (if $\bar{M} = S^{2n+3}$) and CD^{n+1} (if $\bar{M} = (R,CD^{n+1})$). Then $\bar{M}$ is a principal G^1-bundle over $\bar{B}$, where G^1 is a circle or a line. Then an invariant submanifold M of codimension 2 of $\bar{M}$ is also regular. We can consider the following commutative diagram:

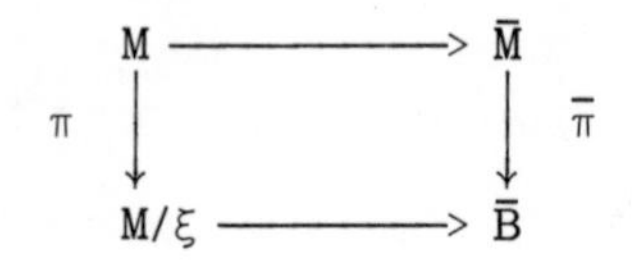

By Theorem 1.13, the Ricci tensor S' of M/ξ is parallel. Since M/ξ is an invariant submanifold of $\bar{B}$ of codimension 2, by Theorem 1.6 of Chapter IV, M/ξ is Einstein. From Theorem 1.10 of Chapter IV our theorem reduces to the following lemma.

LEMMA 1.10. Let M be a non-totally geodesic connected complete η-Einstein invariant submanifold of codimension 2 of S^{2n+3}. Then there is an automorphism $\bar{\theta}$ of S^{2n+3} such that $\bar{\theta}M = (S,Q^n)$.

Proof. By Theorem 1.10 of Chapter IV, $B = \pi M$ is holomorphically isometric to Q^n. From Theorem 1.8 of Chapter IV, there is a holomorphic isometry θ of CP^{n+1} such that $\theta B = Q^n$. Let $x \in B$, $\theta x = y$ and $\tau(t)$, $0 \leq t \leq 1$, be a curve joining x and y. Then we have a continuous family of J-basis $(\tau(t), e_i(t), Je_i(t))$, $i = 1,\dots,n+1$, on $\tau(t)$ such that $e_i(1) = \theta_*(e_i(0))$. Thus θ is contained in the connected component of the automorphism group of CP^{n+1}. Hence there are finite numbers of infinitesimal automorphisms $X_1, X_2, \dots, X_s$ of CP^{n+1} such that $\theta = \exp t_s X_s \dots \exp t_1 X_1$. By Lemma 5.1 of Tanno [4] there are infinitesimal automorphisms $Y_1,\dots,Y_s$ of S^{2n+3} such that $\pi Y_k = X_k$, $\pi(\exp t_k Y_k)(u) = \exp t_k X_k(\pi u)$, $u \in M$, $k = 1,\dots,s$. Putting $\bar{\theta} = \exp t_s Y_s \dots \exp t_1 Y_1$, we have $\bar{\theta}M = Q^n$. Since $\bar{\theta}M$ and (S,Q^n) have the same fibre, we have $\bar{\theta}M = (S,Q^n)$. QED.

By Theorem 1.11 and Theorem 1.14 we have (Kon [8])

THEOREM 1.15. S^{2n+1} and (S,Q^n) are the only compact invariant submanifolds of codimension 2 in S^{2n+3} which have constant scalar curvature.

2. ANTI-INVARIANT SUBMANIFOLDS TANGENT TO THE STRUCTURE VECTOR FIELD OF SASAKIAN MANIFOLDS

Let $\bar{M}$ be a (2m+1)-dimensional almost contact metric manifold with structure tensors (ϕ,ξ,η,g). An n-dimensional submanifold M immersed in $\bar{M}$ is said to be *anti-invariant* in $\bar{M}$ if $\phi T_x(M) \subset T_x(M)^\perp$ for each $x \in M$. Then we see that, ϕ being of rank 2m, $n \leq m+1$. When $n = m+1$ we have

PROPOSITION 2.1. Let $\bar{M}$ be an almost contact metric manifold of dimension 2n+1 and let M be an anti-invariant submanifold of $\bar{M}$ of dimension n+1. Then the structure vector field ξ is tangent to M.

Proof. By the assumption we have $\phi T_x(M) = T_x(M)^{\perp}$ at each point x of M. For any vector field X tangent to M we have $g(\xi,\phi X) = -g(\phi\xi,X) = 0$, which shows that the structure vector field ξ is tangent to M. QED.

PROPOSITION 2.2. Let M be a submanifold tangent to the structure vector field ξ of a Sasakian manifold $\bar{M}$. Then ξ is parallel with respect to the induced connection on M, i.e., $\nabla\xi = 0$, if and only if M is an anti-invariant submanifold of $\bar{M}$.

Proof. Since ξ is tangent to M, we have $\phi X = -\nabla_X\xi - B(X,\xi)$. Thus ξ is parallel if and only if $\phi X = -B(X,\xi)$ and hence ϕX is normal to M. QED.

In this section we sall study anti-invariant submanifolds tangent to the structure vector field of a Sasakian manifold. Therefore, in this section, we mean by an anti-invariant submanifold M of a Sasakian manifold $\bar{M}$ an anti-invariant submanifold M tangent to the structure vector field ξ of $\bar{M}$.

PROPOSITION 2.3. Let M be an (n+1)-dimensional anti-invariant submanifold of a (2n+1)-dimensional Sasakian manifold $\bar{M}$. If $n \geq 1$, then M is not totally umbilical.

Proof. If M is totally umbilical, then $B(X,Y) = g(X,Y)\mu$, μ being the mean curvature vector of M. Then $B(\xi,\xi) = 0$ implies $\mu = 0$ and hence M is totally geodesic. This contradicts to the fact that $\phi X = -B(X,\xi) \neq 0$. QED.

Here we choose a local field of orthonormal frames $e_0=\xi, e_1,\dots,e_n$; $e_{n+1},\dots,e_m; e_{1^*}= e_1,\dots,e_{n^*}= e_n; e_{(n+1)^*}= e_{n+1},\dots,e_{m^*}= e_m$ in $\bar{M}$ such that, restricted to M, $e_0,e_1,\dots,e_n$ are tangent to M. With respect to this frame field of $\bar{M}$, let $\omega^0=\eta,\omega^1,\dots,\omega^n;\omega^{n+1},\dots,\omega^m;\omega^{1^*},\dots,\omega^{n^*}$; $\omega^{(n+1)^*},\dots,\omega^{m^*}$ be the dual frames. Unless otherwise stated, we use the conventions that the ranges of indices are respectively:

$$A,B,C,D = 0,1,\dots,m,1^*,\dots,m^*,$$
$$i,j,k,l,s,t = 0,1,\dots,n,$$
$$x,y,z,v,w = 1,\dots,n,$$
$$a,b,c,d = n+1,\dots,m,1^*,\dots,m^*,$$
$$\lambda,\mu,\nu = n+1,\dots,m,(n+1)^*,\dots,m^*,$$
$$\alpha,\beta,\gamma = n+1,\dots,m.$$

Then we have the following equations:

$$\omega_y^x = \omega_{y^*}^{x^*}, \quad \omega_y^{x^*} = \omega_x^{y^*}, \quad \omega^x = -\omega_0^{x^*}, \quad \omega^{x^*} = \omega_0^x,$$

$$\omega_\beta^\alpha = \omega_{\beta^*}^{\alpha^*}, \quad \omega_\beta^{\alpha^*} = \omega_\alpha^{\beta^*}, \quad \omega^\alpha = -\omega_0^{\alpha^*}, \quad \omega^{\alpha^*} = \omega_0^\alpha,$$

$$\omega_\alpha^x = \omega_{\alpha^*}^{x^*}, \quad \omega_\alpha^{x^*} = \omega_x^{\alpha^*}.$$

We restrict these forms to M and then we have $\omega^a = 0$. Thus we have $\omega_0^x = \omega_0^{\alpha^*} = \omega_0^\alpha = 0$. Therefore (2.13) of Chapter II implies

$$h_{yz}^x = h_{xz}^y = h_{xy}^z, \quad h_{0i}^\lambda = 0, \quad h_{00}^x = 0, \quad h_{0i}^x = -\delta_{xi},$$

where we use h_{ij}^x in place of $h_{ij}^{x^*}$ to simplify the notation. For each a, the second fundamental form A_a is represented by a symmetric (n+1,n+1)-matrix $A_a = (h_{ij}^a)$. Then we have

$$A_x = \begin{array}{c} \\ \\ \\ x \\ \\ \end{array}\left(\begin{array}{c|ccc} 0 & 0 & \overset{x}{-1} & 0 \\ \hline 0 & & & \\ & & & \\ -1 & & h_{yz}^x & \\ & & & \\ 0 & & & \end{array}\right) \qquad \text{for all } x,$$

$$A_\lambda = \left(\begin{array}{c|c} 0 & 0 \\ \hline & \\ 0 & h_{yz}^\lambda \\ & \end{array}\right) \qquad \text{for all } \lambda.$$

Hereafter we put $H_a = (h^a_{yz})$, which is a symmetric (n,n)-matrix. We notice that

$$|A|^2 = |H|^2 + 2n,$$

where $|H|^2 = \sum_a \mathrm{Tr} H_a^2$. We also have

$$\mathrm{Tr} A_a = \mathrm{Tr} H_a$$

for all a, and hence M is minimal if and only if $\mathrm{Tr} H_a = 0$ for all a.

Since $\phi T_x(M) \subset T_x(M)^\perp$ at each point x of M, we have the following decomposition:

$$T_x(M)^\perp = \phi T_x(M) \oplus N_x(M),$$

where $N_x(M)$ is the orthogonal complement of $\phi T_x(M)$ in $T_x(M)^\perp$. If $V \in N_x(M)$, then $\phi V \in N_x(M)$.

For any vector field V normal to M we put

$$\phi V = tV + fV,$$

where tV is the tangential part of ϕV and fV the normal part of ϕV. Then t is a tangent bundle valued 1-form on the normal bundle $T(M)^\perp$ and f is an endomorphism on $T(M)^\perp$. Then we have

$$tfV = 0, \qquad f^2 V = -V - \phi tV,$$

$$t\phi X = -X + \eta(X)\xi, \qquad f\phi X = 0.$$

From these we also have

$$f^3 + f = 0.$$

Since f is of constant rank (see Stong [1]), if f does not vanish, it defines an f-structure in the normal bundle $T(M)^{\perp}$ (see Chapter VII). By the Gauss and Weingarten formulas we have

$$(\nabla_X f)V = -B(X,tV) - \phi A_V X.$$

If $(\nabla_X f)V = 0$ for all X and V, then the f-structure f in the normal bundle of M is said to be *parallel*.

LEMMA 2.1. Let M be an (n+1)-dimensional anti-invariant submanifold of a (2m+1)-dimensional Sasakian manifold $\bar{M}$. If the f-structure f in the normal bundle of M is parallel, then $A_V = 0$ for $V \in N_x(M)$, or equivalently, $A_\lambda = 0$.

Proof. If $V \in N_x(M)$, then $tV = 0$. Thus we have $\phi A_V X = 0$ and hence $\phi^2 A_V X = -A_V X + \eta(A_V X)\xi = 0$. On the other hand, we see that $\eta(A_V X) = g(B(X,\xi),V) = -g(\phi X,V) = 0$. Thus we have $A_V = 0$. QED.

We can easily see that if $H_a = 0$ for all a, then f is parallel.

We denote by $\bar{R}$ and R the Riemannian curvature tensors of $\bar{M}$ and M respectively. Then we have

$$\bar{R}^{x*}_{y*ij} = \bar{R}^{x}_{yij} - (\delta_{ix}\delta_{jy} - \delta_{iy}\delta_{jx}), \tag{2.1}$$

$$R^{0}_{ijk} = R^{i}_{0jk} = 0. \tag{2.2}$$

Suppose that the ambient manifold $\bar{M}$ is of constant ϕ-sectional curvature c. Then the Gauss equation is given by

$$\begin{aligned} R^{i}_{jkl} = {} & \tfrac{1}{4}(c+3)(\delta_{ik}\delta_{jl} - \delta_{il}\delta_{jk}) + \tfrac{1}{4}(c-1)(\eta_j\eta_k\delta_{il} - \eta_j\eta_l\delta_{ik} \\ & + \eta_i\eta_l\delta_{jk} - \eta_i\eta_k\delta_{jl}) + \sum_a (h^a_{ik}h^a_{jl} - h^a_{il}h^a_{jk}), \end{aligned} \tag{2.3}$$

from which we find the Ricci tensor R_{ij} of M is given by

$$\begin{aligned} R_{ij} = {} & \tfrac{1}{4}[n(c+3)-(c-1)]\delta_{ij} - \tfrac{1}{4}(n-1)(c-1)\eta_i\eta_j \\ & + \sum_{a,k} (h^a_{kk}h^a_{ij} - h^a_{ik}h^a_{jk}). \end{aligned} \tag{2.4}$$

On the other hand, we have

$$\sum_{a,k}(h^a_{kk}h^a_{ij} - h^a_{ik}h^a_{jk}) = \sum_{a,x}(h^a_{xx}h^a_{ij} - h^a_{ix}h^a_{jx}) - \delta_{ij} + \eta_i\eta_j.$$

Thus (2.4) reduces to

$$R_{ij} = \tfrac{1}{4}(n-1)(c+3)\delta_{ij} - \tfrac{1}{4}[(n-1)c+(n+3)]\eta_i\eta_j + \sum_{a,x}(h^a_{xx}h^a_{ij} - h^a_{ix}h^a_{jx}). \tag{2.5}$$

Since we have

$$\sum_{a,i,x}(h^a_{ix})^2 = \sum_{a,x,y}(h^a_{xy})^2 + n,$$

the scalar curvature r of M is given by

$$r = \tfrac{1}{4}n(n-1)(c+3) + \sum_{a,x,y}(h^a_{xx}h^a_{yy} - h^a_{xy}h^a_{xy}). \tag{2.6}$$

From (2.5) and (2.6) we have

PROPOSITION 2.4. Let M be an (n+1)-dimensional anti-invariant submanifold of a Sasakian space form $\bar{M}^{2m+1}(c)$. If M is minimal, then the Ricci tensor S and the scalar curvature r of M satisfiy

(1) $S - \frac{1}{4}(n-1)(c+3)g + \frac{1}{4}[(n-1)c-(n+3)]\eta\otimes\eta$ is negative semi-definite;

(2) $r \leq \frac{1}{4}n(n-1)(c+3)$.

PROPOSITION 2.5. Let M be an (n+1)-dimensional anti-invariant submanifold of a (2n+1)-dimensional Sasakian manifold $\bar{M}$. Then M is flat if and only if the normal connection of M is flat.

Proof. First of all we have

$$R^{x*}_{y*ij} = \bar{R}^x_{yij} + \sum_z(h^z_{ix}h^z_{jy} - h^z_{jx}h^z_{iy}) = R^x_{yij}$$

because of $h^z_{ix} = h^x_{iz}$ and (2.1). This combined with (2.2) proves our assertion. QED.

In the following we compute the Laplacian of the square of the length of the second fundamental form of M. Since $h^a_{ijk} - h^a_{ikj} = 0$, (3.1) of Chapter II reduces to

$$(\nabla^2 B)(X,Y) = \sum_i (R(e_i,X)B)(e_i,Y) + D_X D_Y(\mathrm{Tr}B),$$

from which

$$\sum_{a,i,j} h^a_{ij}\Delta h^a_{ij} = \sum_{a,i,j,k} (h^a_{ij}h^a_{kkij} + h^a_{ij}h^a_{kt}R^t_{ijk} + h^a_{ij}h^a_{ti}R^t_{kjk} - h^a_{ij}h^b_{ki}R^a_{bjk}).$$

Therefore, (2.16) and (2.18) of Chapter II imply

$$\sum_{a,i,j} h^a_{ij}\Delta h^a_{ij} = \sum_{a,i,j,k} (h^a_{ij}h^a_{kkij} + \bar{R}^t_{ijk}h^a_{ij}h^a_{kt} + \bar{R}^t_{kjk}h^a_{ij}h^a_{ti} - \bar{R}^a_{bjk}h^a_{ij}h^b_{ki}) - \sum_{a,b,i,j,k,t} [(h^a_{ik}h^b_{kj} - h^a_{jk}h^b_{ki})(h^a_{it}h^b_{tj} - h^a_{jt}h^b_{ti}) + h^a_{ij}h^b_{ij}h^a_{tk}h^b_{tk} - h^a_{ij}h^b_{jt}h^a_{ti}h^b_{kk}].$$

Consequently we obtain

$$\begin{aligned}(2.7)\quad \sum h^a_{ij}\Delta h^a_{ij} &= \sum h^a_{ij}h^a_{kkij} + \tfrac{1}{4}(n+1)(c+3)\sum \mathrm{Tr}A_a^2 - \tfrac{1}{4}(c-1)\sum \mathrm{Tr}A^2 \\ &- \tfrac{1}{2}(c+1)\sum(\mathrm{Tr}A_x)^2 - \tfrac{1}{4}(c+3)\sum(\mathrm{Tr}A\ )^2 - \tfrac{1}{2}n(n+1)(c-1) \\ &+ \sum[\mathrm{Tr}(A_aA_b - A_bA_a)^2 - (\mathrm{Tr}A_aA_b)^2 + \mathrm{Tr}A_b\mathrm{Tr}A_a^2A_b].\end{aligned}$$

PROPOSITION 2.6. Let M be an (n+1)-dimensional anti-invariant submanifold of a Sasakian space form $\bar{M}^{2m+1}(c)$. Then

$$\begin{aligned}\tfrac{1}{2}\Delta|A|^2 - \sum(h^a_{xyz})^2 &= \sum h^a_{xj}h^a_{kkxj} + \tfrac{1}{4}(c+3)\sum[n\mathrm{Tr}H_a^2 - (\mathrm{Tr}H_a)^2] \\ &+ \tfrac{1}{4}(c+3)\sum[\mathrm{Tr}H_x^2 - (\mathrm{Tr}H_x)^2] + \sum[\mathrm{Tr}(H_aH_b - H_bH_a)^2 \\ &- (\mathrm{Tr}H_aH_b)^2 + \mathrm{Tr}H_b\mathrm{Tr}H_a^2H_b].\end{aligned}$$

Proof. First of all, we have

$$\mathrm{Tr}A_x^2 = \mathrm{Tr}H_x^2 + 2, \quad \mathrm{Tr}A_\lambda^2 = \mathrm{Tr}H_\lambda^2, \quad \mathrm{Tr}A_a = \mathrm{Tr}H_a.$$

We also have the following equations:

$$\sum \mathrm{Tr}(A_aA_b - A_bA_a)^2 = \sum \mathrm{Tr}(H_aH_b - H_bH_a)^2 - 4\sum(\mathrm{Tr}H_x)^2$$
$$- 4\sum \mathrm{Tr}H_a^2 + 8\sum \mathrm{Tr}H_x^2 - 2n(n-1),$$

$$\sum(\mathrm{Tr}A_aA_b)^2 = \sum(\mathrm{Tr}A_a^2)^2 = \sum(\mathrm{Tr}H_a^2)^2 + 4\sum \mathrm{Tr}H_x^2 + 4n,$$

$$\sum \mathrm{Tr}A_b \mathrm{Tr}A_a^2A_b = \sum \mathrm{Tr}H_b \mathrm{Tr}H_a^2H_b + 2\sum(\mathrm{Tr}H_x^2)^2 + \sum(\mathrm{Tr}H_a)^2.$$

Moreover, we obtain

$$(2.8) \qquad \sum h_{ij}^a \Delta h_{ij}^a = \tfrac{1}{2}\Delta|A|^2 - \sum(h_{ijk}^a)^2$$
$$= \tfrac{1}{2}\Delta|A|^2 - \sum(h_{xyz}^a)^2 - 3\sum \mathrm{Tr}H_\lambda^2,$$

$$(2.9) \qquad \sum h_{ij}^a h_{kkij}^a = \sum h_{xj}^a h_{kkxj}^a - \sum(\mathrm{Tr}H_\lambda)^2.$$

Substituting these equations into (2.7), we have our equation. QED.

From (2.8) and (2.9) we also have

PROPOSITION 2.7. Let M be an (n+1)-dimensional anti-invariant submanifold of a (2m+1)-dimensional Sasakian manifold $\bar{M}$.

(1) If the second fundamental form of M is parallel, then $H_\lambda = 0$ for all λ;

(2) If the mean curvature vector of M is parallel, then $\mathrm{Tr}H_\lambda = 0$ for all λ.

We now put

$$T_{ab} = \sum_{x,y} h_{xy}^a h_{xy}^b, \quad T_a = T_{aa}, \quad T = \sum_a T_a = |H|^2.$$

THEOREM 2.1. Let M be an (n+1)-dimensional compact anti-invariant minimal submanifold of a Sasakian space form $\bar{M}^{2n+1}(c)$. Then

$$0 \le \int_M |\nabla A|^2 *1 \le \int_M [(2 - \frac{1}{n})T - \frac{1}{4}(n+1)(c+3)]T*1.$$

Proof. By the assumption on the dimensions, f vanishes identically. Thus Proposition 2.6 implies

$$\frac{1}{2}\Delta|A|^2 - |\nabla A|^2 = \frac{1}{4}(n+1)(c+3)T - \sum T_x^2 + \sum \mathrm{Tr}(H_xH_y - H_yH_x)^2$$

because of $\Sigma(h^a_{xyz})^2 = \Sigma(h^a_{ijk})^2$ by (2.8). Applying Lemma 5.1 of Chapter II, we obtain

$$\begin{aligned} -\sum \mathrm{Tr}(H_xH_y - H_yH_x)^2 &+ \sum T_x^2 - \frac{1}{4}(n+1)(c+3)T \\ &\le 2\sum_{x\neq y} T_xT_y + \sum T_x^2 - \frac{1}{4}(n+1)(c+3)T \\ &= [(2 - \frac{1}{n})T - \frac{1}{4}(n+1)(c+3)]T - \frac{1}{n}\sum_{x>y}(T_x - T_y)^2. \end{aligned}$$

Therefore, we have our assertion. QED.

THEOREM 2.2. Let M be an (n+1)-dimensional compact anti-invariant minimal submanifold of a Sasakian space form $\bar{M}^{2n+1}(c)$. Then, either $T = 0$, or $T = n(n+1)(c+3)/4(2n-1)$, or at some point x of M, $T(x) > n(n+1)(c+3)/4(2n-1)$.

THEOREM 2.3. Let M be an (n+1)-dimensional anti-invariant minimal submanifold of a Sasakian space form $\bar{M}^{2n+1}(1)$. If $|A|^2 = (5n^2-n)/(2n-1)$, then $n = 2$ and M is flat. With respect to an adapted dual orthonormal frame field $\omega^0, \omega^1, \omega^2, \omega^{1*}, \omega^{2*}$, the connection form (ω^A_B) of $\bar{M}^5(1)$, restricted to M, is given by

$$\begin{pmatrix} 0 & 0 & 0 & -\omega^1 & -\omega^2 \\ 0 & 0 & 0 & \omega^0-\lambda\omega^2 & -\lambda\omega^1 \\ 0 & 0 & 0 & -\lambda\omega^1 & -\omega^0-\lambda\omega^2 \\ \omega^1 & -\omega^0+\lambda\omega^2 & \lambda\omega^1 & 0 & 0 \\ \omega^2 & \omega^0+\lambda\omega^1 & \lambda\omega^2 & 0 & 0 \end{pmatrix}, \quad \lambda = 1/\sqrt{2}.$$

Proof. Since $|A|^2$ is constant, we have

$$0 \leq |\nabla A|^2 \leq [(2 - \frac{1}{n})T - (n+1)]T.$$

Thus, by the assumption and $T = |A|^2 - 2n$, we see that the second fundamental form of M is parallel. Moreover, we have, from Lemma 5.1 of Chapter II,

$$\sum_{x>y} (T_x - T_y)^2 = 0,$$

$$- \mathrm{Tr}(H_x H_y - H_y H_x)^2 = 2\mathrm{Tr}H_x^2 \mathrm{Tr}H_y^2,$$

and hence $T_x = T_y$ for all x, y and we may assume that $H_x = 0$ for $x = 3,\ldots,n$. Therefore, we must have $n = 2$ and we obtain

$$H_1 = \lambda \begin{pmatrix} 0 & 1 \\ 1 & 0 \end{pmatrix}, \qquad H_2 = \lambda \begin{pmatrix} 1 & 0 \\ 0 & -1 \end{pmatrix}$$

by putting $h^1_{12} = h^2_{11} = \lambda$. Then we have

$$A_1 = \begin{pmatrix} 0 & -1 & 0 \\ -1 & 0 & \lambda \\ 0 & \lambda & 0 \end{pmatrix}, \qquad A_2 = \begin{pmatrix} 0 & 0 & -1 \\ 0 & \lambda & 0 \\ -1 & 0 & -\lambda \end{pmatrix}.$$

On the other hand, we find

$$dh^x_{ab} = h^x_{ad}\omega^d_b + h^x_{db}\omega^d_a - h^y_{ab}\omega^x_y.$$

Putting $x = 1$, $a = 1$ and $b = 0$, we see that $d\lambda = \omega^2_0 = 0$, which shows that λ is a constant. Since $T = 2$, we get $4\lambda^2 = 2$. Thus we may assume that $\lambda = 1/\sqrt{2}$. Moreover, we have

$$\omega^0_1 = -\omega^{1*} = 0, \quad \omega^0_2 = -\omega^{2*} = 0, \quad \omega^0_{1*} = \omega^1, \quad \omega^0_{2*} = \omega^2,$$

$$\omega^{2*}_1 = \lambda\omega^1, \quad \omega^{1*}_1 = -\omega^0 + \lambda\omega^2, \quad \omega^{2*}_2 = -\omega^0 - \lambda\omega^2, \quad \omega^2_1 = 0.$$

On the other hand, from equation of Gauss, we easily see that M is flat and hence the normal connection of M also flat. These prove our theorem. QED.

Example 2.1. Let $J = (a_{ts})$ $(t,s = 1,\dots,6)$ be the almost complex structure of C^3 such that $a_{2i,2i-1} = 1$, $a_{2i-1,2i} = -1$ $(i = 1,2,3)$ and the other components being zero. Let $S^1(1/\sqrt{3}) = \{z \in C: |z|^2 = 1/3\}$, a plane circle of radius $1/\sqrt{3}$. We consider

$$M^3 = S^1(1/\sqrt{3}) \times S^1(1/\sqrt{3}) \times S^1(1/\sqrt{3})$$

in S^5 in C^3, which is obviously flat. The position vector X of M^3 in S^5 in C^3 has components given by

$$X = (1/\sqrt{3})(\cos u^1, \sin u^1, \cos u^2, \sin u^2, \cos u^3, \sin u^3),$$

u^1, u^2, u^3 being parameters on each $S^1(1/\sqrt{3})$. Putting $X_i = \partial_i X/\partial u^i$, we have

$$X_1 = (1/\sqrt{3})(-\sin u^1, \cos u^1, 0,0,0,0),$$

$$X_2 = (1/\sqrt{3})(0,0,-\sin u^2, \cos u^2, 0,0),$$

$$X_3 = (1/\sqrt{3})(0,0,0,0,-\sin u^3, \cos u^3).$$

The structure vector field ξ on S^5 is given by

$$\xi = -JX = (1/\sqrt{3})(\sin u^1, -\cos u^1, \sin u^2, -\cos u^2, \sin u^3, -\cos u^3).$$

Since $\xi = -(X_1 + X_2 + X_3)$, ξ is tangent to M^3. On the other hand, the structure tensors (ϕ,ξ,η) of S^5 satisfy

$$\phi X_i = JX_i - \eta(X_i)X, \quad i = 1,2,3,$$

which shows that ϕX_i is normal to M^3 for all i. Therefore, M^3 is an anti-invariant submanifold of S^5. Moreover, M^3 is a minimal submanifold of S^5 with $|A|^2 = 6$ and the normal connection of M^3 is flat.

Since the connection form (ω^A_B) of S^5, restricted to M^3, coincides with that in Theorem 3.2, we have

THEOREM 2.4. Let M be an (n+1)-dimensional compact anti-invariant minimal submanifold of S^{2n+1}. If $|A|^2 = (5n^2-n)/(2n-1)$, then M is

$$S^1(1/\sqrt{3}) \times S^1(1/\sqrt{3}) \times S^1(1/\sqrt{3}) \quad \text{in } S^5.$$

THEOREM 2.5. Let M be an (n+1)-dimensional compact anti-invariant minimal submanifold of S^{2m+1}. If the second fundamental form of M is parallel and if $|A|^2 = (5n^2-n)/(2n-1)$, then M is

$$S^1(1/\sqrt{3}) \times S^1(1/\sqrt{3}) \times S^1(1/\sqrt{3}) \quad \text{in an } S^5 \text{ in } S^{2m+1}.$$

Proof. Since the second fundamental form of M is parallel, by Proposition 2.7, $A_\lambda = 0$ for all λ. Thus we have the same inequality as in the proof of Theorem 2.1 and then we have $n = 2$. We also have $T_1 = T_2 \neq 0$. Thus the first normal space of M is spanned by e_{1*}, e_{2*}, that is, the first normal space of M is $\phi T_x(M)$. For any vector fields X and Y tangent to M and any vector field V in N(M) we have

$$g(D_X\phi Y,V) = g(\bar{\nabla}_X\phi Y,V) = g((\bar{\nabla}_X\phi)Y,V) + g(\phi\bar{\nabla}_X Y,V)$$

$$= g(\phi\nabla_X Y,V) + g(\phi B(X,Y),V) = -g(A_{\phi V}X,Y) = 0,$$

which shows that the first normal space of M is parallel. Thus M is in an S^5 in S^{2m+1} and M is anti-invariant in S^5. Therefore our theorem follows from Theorem 2.4. QED.

Example 2.2. Let $S^1(r_i) = \{z_i \in C\colon |z_i|^2 = r_i^2\}$, $i = 1,\ldots,n+1$. We consider

$$M^{n+1} = S^1(r_1) \times \ldots \times S^1(r_{n+1})$$

in C^{n+1} such that $r_1^2 + \dots + r_{n+1}^2 = 1$. Then M^{n+1} is a flat submanifold of S^{2n+1} with parallel mean curvature vector and with flat normal connection. The position vector X of M^{n+1} in C^{n+1} has components given by

$$X = (r_1\cos u^1, r_1\sin u^1, \dots, r_{n+1}\cos u^{n+1}, r_{n+1}\sin u^{n+1}).$$

Then X is an outward unit normal vector of S^{2n+1} in C^{n+1}. Putting $X_i = \partial_i X = \partial X/\partial u^i$, we have

$$X_1 = r_1(-\sin u^1, \cos u^1, 0, \dots, 0),$$
$$\dots\dots\dots\dots\dots\dots\dots\dots$$
$$X_{n+1} = r_{n+1}(0, \dots, 0, -\sin u^{n+1}, \cos u^{n+1}).$$

The structure vector field ξ on S^{2n+1} is given by its components

$$\xi = -JX = (r_1\sin u^1, -r_1\cos u^1, \dots, r_{n+1}\sin u^{n+1}, -r_{n+1}\cos u^{n+1}).$$

Therefore we see that $\xi = -(X_1 + \dots + X_{n+1})$, which means that the vector field ξ is tangent to M. So the structure tensors (ϕ, ξ, η, g) of S^{2n+1} satisfy

$$\phi X_i = JX_i - \eta(X_i)X, \quad i = 1, \dots, n+1.$$

Thus ϕX_i is normal to M for all i. Therefore M^{n+1} is anti-invariant in S^{2n+1}.

THEOREM 2.6. Let M be an $(n+1)$-dimensional compact anti-invariant submanifold of S^{2n+1} with parallel mean curvature vector. If the normal connection of M is flat, then M is

$$S^1(r_1) \times \dots \times S^1(r_{n+1}), \quad \Sigma r_i^2 = 1.$$

Proof. Since the normal connection of M is flat, Proposition 2.5 implies that M is flat. On the other hand, as the mean curvature vector of M is parallel, $|A|^2$ is constant. From these and Proposition 4.4 of Chapter II we see that the second fundamental form of M is parallel. Therefore our theorem follows from Theorem 4.4 of Chapter II. QED.

THEOREM 2.7. Let M be an (n+1)-dimensional compact anti-invariant submanifold with flat normal connection of S^{2m+1}. If the second fundamental form of M is parallel, then M is

$$S^1(r_1) \times \dots \times S^1(r_{n+1}) \quad \text{in an } S^{2n+1} \text{ in } S^{2m+1},$$

where $\Sigma r_i^2 = 1$.

Proof. From the assumption we have $A_\lambda = 0$ for all λ and we see that M is flat. We shall show that the first normal space of M is of dimension n and parallel with respect to the connection induced in the normal bundle. Since M is flat, by equation of Gauss, we have

$$0 = (\delta_{ik}\delta_{jl} - \delta_{il}\delta_{jk}) + \sum_a (h^a_{ik}h^a_{jl} - h^a_{il}h^a_{jk}).$$

If the dimension of the first normal space $N_1(M)$ of M is less than n, then for some x, $A_x = 0$. Thus we have

$$\begin{aligned} \sum_a (h^a_{ix}h^a_{jl} - h^a_{il}h^a_{jx}) &= \sum_y (h^y_{ix}h^y_{jl} - h^y_{il}h^y_{jx}) \\ &= \sum_y (h^x_{iy}h^y_{jl} - h^y_{il}h^x_{jy}) = 0. \end{aligned}$$

Therefore we have $0 = (\delta_{ix}\delta_{jl} - \delta_{il}\delta_{jx})$, which is a contradiction.

We also see that $N_1(M)$ is parallel by the similar method used in the proof of Theorem 2.5. QED.

THEOREM 2.8. Let M be an (n+1)-dimensional compact anti-invariant minimal submanifold of a Sasakian space form $\bar{M}^{2m+1}(c)$ $(c > -3)$. If $T \leq \frac{1}{4}nq(c+3)/(2q-1)$ $(q = 2m-n)$, then $T = 0$.

Proof. From Proposition 2.6 and Lemma 5.1 of Chapter II we have

$$\sum(h^a_{xyz})^2 - \tfrac{1}{2}\Delta|A|^2 = -\sum \mathrm{Tr}(H_aH_b - H_bH_a)^2 + \sum T_a^2$$
$$- \tfrac{1}{4}n(c+3)T - \tfrac{1}{4}(c+3)\sum \mathrm{Tr}H_x^2$$
$$\leq [(2 - \frac{1}{q})T - \tfrac{1}{4}n(c+3)]T - \frac{1}{q}\sum_{a>b}(T_a - T_b)^2 - \tfrac{1}{4}(c+3)\sum T_x.$$

Therefore, by the assumption, we obtain

$$\sum(T_a - T_b)^2 = 0, \qquad \sum T_x = 0.$$

Thus we see that $T_x = 0$ for all x and $T_a = T_b$ for all a and b. Hence we have $T_a = 0$ for all a. Consequently, we obtain $T = 0$. QED.

We give examples of anti-invariant submanifolds with $T = 0$.

Example 2.3. Let S^{2n+1} be a unit sphere of dimension 2n+1 with standard Sasakian structure and let CP^n be a complex projective space of real dimension 2n with constant holomorphic sectional curvature 4. A real projective space RP^n of dimension n with constant curvature 1 is imbedded in CP^n as an anti-invariant and totally geodesic submanifold (see Abe [1]). We consider the following commutative diagram:

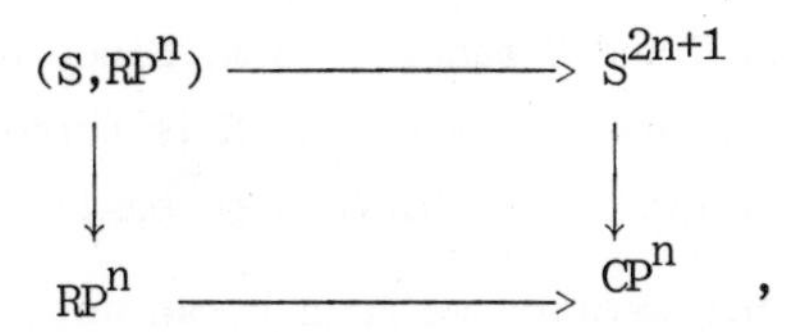

where (S,RP^n) denotes a circle bundle over RP^n. Then (S,RP^n) is an anti-invariant submanifold with $T = 0$ of S^{2n+1}.

Example 2.4. Let R^{2n+1} be an Euclidean space with cartesian coordinates $(x^1,\dots,x^n,y^1,\dots,y^n,z)$. As in Example 5.5 of Chapter V we derive the standard Sasakian structure in R^{2n+1} with constant ϕ-sectional curvature -3. We consider the following natural imbedding of R^{n+1} into R^{2n+1}:

$$(x^1,\ldots,x^n,z) \longrightarrow (x^1,\ldots,x^n,0,\ldots,0,z).$$

Then we easily see that R^{n+1} is an anti-invariant submanifold of R^{2n+1} which has $T = 0$.

3. ANTI-INVARIANT SUBMANIFOLDS NORMAL TO THE STRUCTURE VECTOR FIELD OF SASAKIAN MANIFOLDS

From the consideration of §4 of Chapter V it is interesting to study the following submanifolds of contact manifolds, especially those of Sasakian manifolds.

First of all, we prove

THEOREM 3.1. Let M be an n-dimensional submanifold of a (2m+1)-dimensional K-contact manifold $\bar{M}$. If the structure vector field ξ is normal to M, then M is an anti-invariant submanifold of $\bar{M}$, and $n \leq m$.

Proof. From the Weingarten formula we have

$$g(\phi X,Y) = -g(\bar{\nabla}_X\xi,Y) = g(A_\xi X,Y)$$

for any vector fields X and Y tangent to M. Since A_ξ is symmetric and ϕ is skew-symmetric, we have $A_\xi = 0$ and X is normal to M. Thus M is an anti-invariant submanifold of $\bar{M}$. We also easily see that $n \leq m$. QED.

Throughout in this section we mean by an anti-invariant submanifold M of an almost contact metric manifold $\bar{M}$, a submanifold M of $\bar{M}$ normal to the structure vector field ξ of $\bar{M}$. Especially we consider a submanifold M of a Sasakian manifold $\bar{M}$ normal to ξ.

We state some fundamental properties of the second fundamental form of an anti-invariant submanifold M of a Sasakian manifold $\bar{M}$. We already have the following

$$A_\xi = 0. \tag{3.1}$$

Moreover, we easily see

$$A_{\phi X}Y = A_{\phi Y}X \tag{3.2}$$

for any vector fields X and Y tangent to M.

LEMMA 3.1. Let M be an n-dimensional anti-invariant submanifold of a (2m+1)-dimensional Sasakian manifold $\bar{M}$. If the second fundamental form of M is parallel, then

$$A_{\phi X} = 0$$

for any vector field X tangent to M. If moreover n = m, then M is totally geodesic in $\bar{M}$.

Proof. From the assumption on the second fundamental form of M we obtain

$$0 = g((\nabla_X B)(Y,Z),\xi) = g(B(Y,Z),\phi X) = g(A_{\phi X}Y,Z),$$

from which $A_{\phi X} = 0$. If n = m, the normal space $T_x(M)^\perp$ is equal to $\phi T_x(M) \oplus \{\xi\}$. Therefore M is totally geodesic in $\bar{M}$. QED.

Since $\phi T_x(M) \subset T_x(M)^\perp$ at each point x of M, we have the decomposition of $T_x(M)^\perp$ into the direct sum

$$T_x(M)^\perp = \phi T_x(M) \oplus N_x(M),$$

$N_x(M)$ being the orthogonal complement of $\phi T_x(M)$ in $T_x(M)^\perp$. We have $N_x(M) = \phi N_x(M) \oplus \{\xi\}$. For any vector field V normal to M we put

$$\phi V = tV + fV,$$

where tV is the tangential part of ϕV and fV the normal part of ϕV. Then t is a tangent bundle valued 1-form on the normal bundle of M and f is an endomorphism of the normal bundle of M. We find

$$tfV = 0, \qquad f^2V = -V - \phi tV + \eta(V)\xi,$$

$$t\phi X = -X, \qquad f\phi X = 0.$$

Moreover, we have

$$f^3 + f = 0,$$

which means that f defines an f-structure in the normal bundle of M.

LEMMA 3.2. If the f-structure f in the normal bundle of M is parallel, then

$$A_V = 0 \qquad \text{for } V \in N_x(M).$$

Proof. First of all, we have $(\nabla_X f)V = -B(X,tV) - \phi A_V X$. Since $\nabla f = 0$ and $tV = 0$, we obtain $\phi A_V X = 0$. Thus we have

$$\phi^2 A_V X = -A_V X + \eta(A_V X)\xi = -A_V X = 0,$$

which proves our equation. QED.

Let μ be the mean curvature vector of M. Then, from (3.1), we have

$$g(D_X\mu,\xi) = g(\mu,\phi X).$$

From this we have the following

LEMMA 3.3. Let M be an n-dimensional anti-invariant submanifold of a (2m+1)-dimensional Sasakian manifold $\bar{M}$. If the mean curvature vector μ of M is parallel, then $\mu \in N_x(M)$ at each point x of M. Moreover, if n = m, then M is minimal.

From Lemmas 3.1, 3.2 and 3.3 we have

PROPOSITION 3.1. Let M be an n-dimensional anti-invariant submanifold of a (2m+1)-dimensional Sasakian manifold $\bar{M}$ with parallel f-structure f in the normal bundle of M.

(1) If the second fundamental form of M is parallel, then M is totally geodesic;

(2) If the mean curvature vector of M is parallel, then M is minimal.

We choose a local field of orthonormal frames $e_1,\dots,e_n;e_{n+1},\dots,e_m;e_{0*}=\xi,e_{1*}=\phi e_1,\dots,e_{n*}=\phi e_n;e_{(n+1)*}=\phi e_{n+1},\dots,e_{m*}=\phi e_m$ in $\bar{M}$ in such a way that, restricted to M, $e_1,\dots,e_n$ are tangent to M. Unless otherwise stated, we use the conventions that the ranges of indices are respectively:

$$i,j,k,l,t,s = 1,\dots,n,$$
$$a,b,c,d = n+1,\dots,m,0*,1*,\dots,m*,$$
$$p,q,r = n+1,\dots,m,1*,\dots,m*.$$

In the following we put $h^0_{ij} = h^{0*}_{ij} = g(A_\xi e_i,e_j)$ and $h^t_{ij} = h^{t*}_{ij} = g(A_{e_{t*}}e_i,e_j)$ to simplify the notation.

We now assume that the ambient manifold $\bar{M}$ is a Sasakian space form $\bar{M}^{2m+1}(c)$. Then the Gauss equation of M is given by

$$(3.3)\qquad R^i_{jkl} = \tfrac{1}{4}(c+3)(\delta_{ik}\delta_{jl} - \delta_{il}\delta_{jk}) + \sum_a(h^a_{ik}h^a_{jl} - h^a_{il}h^a_{jk}),$$

from which

$$(3.4)\qquad R_{ij} = \tfrac{1}{4}(n-1)(c+3)\delta_{ij} + \sum_{a,k}(h^a_{ij}h^a_{kk} - h^a_{ik}h^a_{jk}),$$

$$(3.5)\qquad r = \tfrac{1}{4}n(n-1)(c+3) + \sum_{a,i,j}(h^a_{ii}h^a_{jj} - h^a_{ij}h^a_{ij}).$$

If M is minimal, we have $\Sigma_i h^a_{ii} = 0$ for all a, and hence we have the following

PROPOSITION 3.2. Let M be an n-dimensional anti-invariant submanifold of a Sasakian space form $\bar{M}^{2m+1}(c)$. Then M is totally geodesic if and only if M satisfies one of the following conditions:

(1) M is of constant curvature $\frac{1}{4}(c+3)$;

(2) $S = \frac{1}{4}(n-1)(c+3)g$;

(3) $r = \frac{1}{4}n(n-1)(c+3)$,

where S and r be the Ricci tensor and scalar curvature of M respectively.

Let M be an n-dimensional anti-invariant submanifold of a Sasakian space form $\bar{M}^{2m+1}(c)$. Then we have (see §2)

$$\sum_{a,i,j} h^a_{ij}\Delta h^a_{ij} = \sum_{a,i,j,k} h^a_{ij}h^a_{kkij} + \tfrac{1}{4}(c+3)\sum_a[n\mathrm{Tr}A_a^2 - (\mathrm{Tr}A_a)^2]$$
$$+ \tfrac{1}{4}(c-1)\sum_t[\mathrm{Tr}A_t^2 - (\mathrm{Tr}A_t)^2] + \sum_{a,b}[\mathrm{Tr}(A_aA_b - A_bA_a)^2$$
$$- (\mathrm{Tr}A_aA_b)^2 + \mathrm{Tr}A_b\mathrm{Tr}A_a^2A_b].$$

On the other hand, we have

$$\sum_{a,i,j} h^a_{ij}\Delta h^a_{ij} = \tfrac{1}{2}\Delta|A|^2 - \sum_{a,i,j,k}(h^a_{ijk})^2$$
$$= \tfrac{1}{2}\Delta|A|^2 - \sum_{p,i,j,k}(h^p_{ijk})^2 - \sum_t \mathrm{Tr}A_t^2.$$

From these equations we obtain

PROPOSITION 3.3. Let M be an n-dimensional anti-invariant submanifold of a Sasakian space form $\bar{M}^{2m+1}(c)$. Then

$$\tfrac{1}{2}\Delta|A|^2 - \sum(h^p_{ijk})^2 = \sum h^a_{ij}h^a_{kkij} + \tfrac{1}{4}n(c+3)|A|^2 - \sum(\mathrm{Tr}A_a^2)^2$$
$$- \tfrac{1}{4}(c+3)\sum(\mathrm{Tr}A_a)^2 + \tfrac{1}{4}(c+3)\sum\mathrm{Tr}A_t^2 - \tfrac{1}{4}(c-1)\sum(\mathrm{Tr}A_t)^2$$
$$+ \sum[\mathrm{Tr}(A_aA_b - A_bA_a)^2 + \mathrm{Tr}A_b\mathrm{Tr}A_a^2A_b].$$

THEOREM 3.2. Let M be an n-dimensional compact anti-invariant submanifold with parallel mean curvature vector of a Sasakian space form $\bar{M}^{2n+1}(c)$. Then

$$0 \leq \int_M \sum_{p,i,j,k} (h^p_{ijk})^2 *1 \leq \int_M [(2 - \frac{1}{n})|A|^2 - \tfrac{1}{4}(n+1)(c+3)]|A|^2 *1.$$

Proof. From Lemma 3.3 we see that M is minimal. Thus Proposition 3.3 implies

$$\tfrac{1}{2}\Delta|A|^2 - \sum(h^p_{ijk})^2 = \tfrac{1}{4}(n+1)(c+3)|A|^2 - \sum(\mathrm{Tr}A_t^2)^2 + \sum\mathrm{Tr}(A_tA_s - A_sA_t)^2.$$

On the other hand, we obtain

$$-\sum\mathrm{Tr}(A_tA_s - A_sA_t)^2 + \sum(\mathrm{Tr}A_t^2)^2 - \tfrac{1}{4}(n+1)(c+3)|A|^2$$

$$\leq 2\sum_{t\neq s}\mathrm{Tr}A_t^2A_s^2 + \sum(\mathrm{Tr}A_t^2)^2 - \tfrac{1}{4}(n+1)(c+3)|A|^2$$

$$= [(2 - \frac{1}{n})|A|^2 - \tfrac{1}{4}(n+1)(c+3)]|A|^2 - \frac{1}{n}\sum_{t>s}(\mathrm{Tr}A_t^2 - \mathrm{Tr}A_s^2)^2.$$

Thus we have our inequality. QED.

THEOREM 3.3. Let M be an n-dimensional compact anti-invariant submanifold with parallel mean curvature vector of a Sasakian space form $\bar{M}^{2n+1}(c)$. Then either M is totally geodesic, or $|A|^2 = n(n+1)(c+3)/(4(2n-1)$, or at some point x of M, $|A|^2(x) > n(n+1)(c+3)/4(2n-1)$.

We now take a unit sphere S^{2n+1} as an ambient manifold. Then we have

THEOREM 3.4. Let M be an n-dimensional anti-invariant submanifold with parallel mean curvature vector of S^{2n+1} $(n > 1)$. If $|A|^2 = n(n+1)/(2n-1)$, then $n = 2$ and M is a flat surface of S^5.

Proof. From the assumption we have

$$\sum(\mathrm{Tr}A_t^2 - \mathrm{Tr}A_s^2) = 0,$$

$$-\mathrm{Tr}(A_tA_s - A_sA_t) = 2\mathrm{Tr}A_t^2\mathrm{Tr}A_s^2.$$

Therefore, $\mathrm{Tr}A_t^2 = \mathrm{Tr}A_s^2$ for all t, s. From Lemma 5.1 of Chapter II we may assume that $A_t = 0$ for $t = 3,\ldots,n$. Then we must have $n = 2$ and we can put

$$A_0 = 0, \qquad A_1 = \lambda\begin{pmatrix} 0 & 1 \\ 1 & 0 \end{pmatrix}, \qquad A_2 = \lambda\begin{pmatrix} 1 & 0 \\ 0 & -1 \end{pmatrix}.$$

Since $|A|^2 = 2$, we obtain $2\lambda^2 = 1$. Thus we may assume that $\lambda = 1/\sqrt{2}$. Moreover, the Gauss equation (3.3) shows that M is flat.

Example 3.1. Let C^{n+1} be a complex (n+1)-dimensional number space with almost complex structure J and let S^{2n+1} be a (2n+1)-dimensional unit sphere in C^{n+1} with standard Sasakian structure (ϕ,ξ,η,g). Let S^1 be a circle of radius 1. Let us consider

$$T^n = S^1 \times \ldots \times S^1.$$

Then we can constract an isometric minimal immersion of T^n into S^{2n+1} which is anti-invariant in the following way.

Let $X: T^n \longrightarrow S^{2n+1}$ be a minimal immersion represented by

$$X = \frac{1}{n+1}(\cos u^1, \sin u^1, \ldots, \cos u^n, \sin u^n, \cos u^{n+1}, \sin u^{n+1}),$$

where we have put $u^{n+1} = -(u^1 + \ldots + u^n)$. We may regard X as a position vector of S^{2n+1} in C^{n+1}. The structure vector field ξ of S^{2n+1}, restricted to T^n, is then given by

$$\xi = -JX = \frac{1}{n+1}(\sin u^1, -\cos u^1, \ldots, \sin u^{n+1}, -\cos u^{n+1}).$$

Putting $X_i = \partial X/\partial u^i$, we have

$$X_i = \frac{1}{n+1}(0,\ldots,0,-\sin u^i, \cos u^i, 0,\ldots,0, \sin u^{n+1}, -\cos u^{n+1}),$$

$i = 1,\dots,n$. Thus $X_1,\dots,X_n$ are linearly independent and $\eta(X_i) = 0$ for $i = 1,\dots,n$. Therefore the immersion X is anti-invariant. Moreover, the immersion X is a minimal immersion with $|A|^2 = n(n-1)$.

From these considerations we have (Yano-Kon [4])

THEOREM 3.5. Let M be an n-dimensional compact anti-invariant submanifold with parallel maen curvature vector of S^{2n+1} $(n > 1)$. If $|A|^2 = n(n+1)/(2n-1)$, then M is $S^1 \times S^1$.

4. CONTACT CR SUBMANIFOLDS

Let $\bar{M}$ be a (2m+1)-dimensional Sasakian manifold with structure tensors (ϕ,ξ,η,g). We consider a Riemannian manifold M isometrically immersed in $\bar{M}$ with induced metric tensor field g.

Throughout in this section, we assume that the submanifold M is tangent to the structure vector field ξ of a Sasakian manifold $\bar{M}$.

For any vector field X tangent to M, we put

$$\phi X = PX + FX, \tag{4.1}$$

where PX is the tangential part of ϕX and FX the normal part of ϕX. Similarly, for any vector field V normal to M, we put

$$\phi V = tV + fV, \tag{4.2}$$

where tV is the tangential part of ϕV and fV the normal part of ϕV. We easily see that P and f are skew-symmetric. From (4.1) and (4.2) we have

$$g(FX,V) + g(X,tV) = 0.$$

Moreover, we obtain

$$P^2 = -I - tF + \eta \otimes \xi, \qquad FP + fF = 0,$$

$$Pt + tf = 0, \qquad f^2 = -I - Ft.$$

Since we have $\phi\xi = P\xi + F\xi = 0$, we find $P\xi = 0$ and $F\xi = 0$. For any vector field X tangent to M, we have $\bar{\nabla}_X\xi = -\phi X = \nabla_X\xi + B(X,\xi)$. Thus, we have the following

$$(4.3) \qquad \nabla_X\xi = -PX, \qquad B(X,\xi) = -FX, \qquad A_V\xi = tV.$$

Especially, we have $B(\xi,\xi) = 0$. Let X and Y be vector fields tangent to M. Then we obtain

$$(4.4) \qquad (\nabla_X P)Y = A_{FY}X + tB(X,Y) + g(X,Y)\xi - \eta(Y)\xi,$$

$$(4.5) \qquad (\nabla_X F)Y = -B(X,PY) + fB(X,Y).$$

For any vector field X tangent to M and any vector field V normal to M, we also have

$$(4.6) \qquad (\nabla_X t)V = A_{fV}X - PA_VX,$$

$$(4.7) \qquad (\nabla_X f)V = -FA_VX - B(X,tV).$$

Let M be an (n+1)-dimensional submanifold of a Sasakian space form $\bar{M}^{2m+1}(c)$. Then we have the following equations of Gauss and Codazzi respectively:

$$(4.8) \qquad R(X,Y)Z = \tfrac{1}{4}(c+3)[g(Y,Z)X-g(X,Z)Y] + \tfrac{1}{4}(c-1)[\eta(X)\eta(Z)Y$$
$$-\eta(Y)\eta(Z)X+g(X,Z)\eta(Y)\xi-g(Y,Z)\eta(X)\xi+g(PY,Z)PX-g(PX,Z)PY$$
$$+2g(X,PY)PZ] + A_{B(Y,Z)}X - A_{B(X,Z)}Y,$$

$$(4.9) \qquad (\nabla_X B)(Y,Z)-(\nabla_Y B)(X,Z) = \tfrac{1}{4}(c-1)[g(PY,Z)FX-g(PX,Z)FY+2g(X,PY)FZ].$$

Moreover, we have equation of Ricci

$$(4.10)\quad g(R^{\perp}(X,Y)U,V) + g([A_V,A_U]X,Y)$$
$$= \tfrac{1}{4}(c-1)[g(FY,U)g(FX,V)-g(FX,U)g(FY,V)+2g(X,PY)g(fU,V)].$$

Definition. Let M be a submanifold tangent to the structure vector field ξ isometrically immersed in a Sasakian manifold $\bar{M}$. Then M is called a *contact* CR *submanifold* of $\bar{M}$ if there exists a differentiable distribution $D : x \longrightarrow D_x \subset T_x(M)$ on M satisfying the following conditions:

(1) D is invariant with respect to ϕ, i.e., $\phi D_x \subset D_x$ for each $x \in M$, and

(2) the complementary orthogonal distribution $D^{\perp}: x \longrightarrow D^{\perp}_x \subset T_x(M)$ is anti-invariant with respect to ϕ, i.e., $\phi D^{\perp}_x \subset T_x(M)^{\perp}$ for each $x \in M$.

In the sequel, we put

$$\dim \bar{M} = 2m+1, \quad \dim M = 2n+1, \quad \dim D = h,$$
$$\dim D^{\perp} = p, \quad \operatorname{codim} M = 2m-n = q.$$

If p = 0, then a contact CR submanifold is an invariant submanifold of $\bar{M}$, and if h = 0, then M is an anti-invariant submanifold of $\bar{M}$ tangent to the structure vector field ξ. If q = p, then a contact CR submanifold M is called a *generic submanifold* of $\bar{M}$. In this case we have $\phi T_x(M)^{\perp} \subset T_x(M)$ for every point x of M. If h > 0 and p > 0, then a contact CR submanifold M is said to be *non-trivial* (*proper*).

Let M be a contact CR submanifold of a Sasakian manifold $\bar{M}$. We denote by l and $l^{\perp}$ the projection operators on D and $D^{\perp}$ respectively. Then we have

$$l + l^{\perp} = I, \quad l^2 = l, \quad l^{\perp 2} = l^{\perp}, \quad l l^{\perp} = l^{\perp} l = 0.$$

Since we have $\phi lX = PlX + FlX$, we obtain

$$l^{\perp}Pl = 0, \qquad Fl = 0.$$

From $\phi l^{\perp}X = Pl^{\perp}X + Fl^{\perp}X$ we have $Pl^{\perp} = 0$, and hence $Pl = P$. Moreover we have

$$FP = 0, \qquad fF = 0, \tag{4.11}$$

$$tf = 0, \qquad Pt = 0. \tag{4.12}$$

Thus we find

$$P^3 + P = 0, \tag{4.13}$$

$$f^3 + f = 0. \tag{4.14}$$

These equations show that P is an f-structure in M and f is an f-structure in the normal bundle.

Conversely, for a submanifold M, tangent to the structure vector field ξ, of a Sasakian manifold $\bar{M}$, we assume that we have $FP = 0$. Then we have $fF = 0$, (4.12), (4.13) and (4.14). We put

$$l = -P^2 + \eta \otimes \xi, \qquad l^{\perp} = I - l.$$

Then we can easily verify that

$$l + l^{\perp} = I, \quad l^2 = l, \quad l^{\perp 2} = l^{\perp}, \quad ll^{\perp} = l^{\perp}l = 0,$$

which mean that l and $l^{\perp}$ are complementary projection operators and consequently define orthogonal distributions D and $D^{\perp}$ respectively. From $l = -P^2 + \eta \otimes \xi$, we have $Pl = P$ because of $P^3 = -P$ and $P\xi = 0$. This equation can also be written as $Pl^{\perp} = 0$. But $g(PX,Y)$ is skew-symmetric and $g(l^{\perp}X,Y)$ is symmetric and consequently $l^{\perp}P = 0$. Thus we have $l^{\perp}Pl = 0$. On the other hand, we obtain $Fl = 0$ because of $FP = 0$

and $F\xi = 0$. Consequently, the distribution $\mathcal{D}$ is invariant and $\mathcal{D}^{\perp}$ is anti-invariant with respect to ϕ. Moreover, we have $l\xi = \xi$, $l^{\perp}\xi = 0$ and consequently the distribution $\mathcal{D}$ contains ξ.

On the other hand, putting

$$l = -P^2, \qquad l^{\perp} = I + P^2,$$

we still see that l and $l^{\perp}$ define complementary orthogonal distributions $\mathcal{D}$ and $\mathcal{D}^{\perp}$ respectively. We also have $Pl = P$, $l^{\perp}P = 0$, $Fl = 0$ and $Pl^{\perp} = 0$. Thus we see that $\mathcal{D}$ is invariant and $\mathcal{D}^{\perp}$ is anti-invariant with respect to ϕ and also that $l\xi = 0$, $l^{\perp}\xi = \xi$, which mean that $\mathcal{D}^{\perp}$ contains ξ.

From these considerations we have (Yano-Kon [16])

THEOREM 4.1. In order for a submanifold M, tangent to the structure vector field ξ, of a Sasakian manifold $\bar{M}$ to be a contact CR submanifold, it is necessary and sufficient that $FP = 0$.

THEOREM 4.2. Let M be a contact CR submanifold of a Sasakian manifold $\bar{M}$. Then P is an f-structure in M and f is an f-structure in the normal bundle.

LEMMA 4.1. Let M be a contact CR submanifold of $\bar{M}$. Then

$$A_{FX}Y - A_{FY}X = \eta(Y)X - \eta(X)Y \quad \text{for } X, Y \in \mathcal{D}^{\perp}.$$

Proof. Let X, Y be in $\mathcal{D}^{\perp}$. Then $PX = PY = 0$, and hence

$$g((\nabla_Z P)X,Y) = g(\nabla_Z PX,Y) - g(P\nabla_Z X,Y) = 0$$

for any vector field Z tangent to M. From this and (4.4) we find

$$g(A_{FX}Z,Y) + g(tB(Z,X),Y) = \eta(Y)g(Z,X) - \eta(X)g(Z,Y).$$

Thus we have our equation. QED.

THEOREM 4.3. Let M be an (n+1)-dimensional contact CR submanifold of a (2m+1)-dimensional Sasakian manifold $\bar{M}$. The distribution $D^\perp$ is completely integrable and its maximal integral submanifold is a q-dimensional anti-invariant submanifold of $\bar{M}$ normal to ξ or a (q+1)-dimensional anti-invariant submanifold of $\bar{M}$ tangent to ξ.

Proof. For any vector fields X and Y in $D^\perp$ we have

$$\phi[X,Y] = P[X,Y] + F[X,Y] = -(\nabla_X P)Y + (\nabla_Y P)X + F[X,Y]$$

$$= A_{FX}Y - A_{FY}X - \eta(Y)X + \eta(X)Y + F[X,Y] = F[X,Y].$$

Thus we have $\phi[X,Y] \in T(M)^\perp$, and consequently $[X,Y] \in D^\perp$. QED.

THEOREM 4.4. Let M be an (n+1)-dimensional contact CR submanifold of a (2m+1)-dimensional Sasakian manifold $\bar{M}$. Then the distribution D is completely integrable if and only if

$$B(X,PY) = B(Y,PX)$$

for any vector fields X, Y $\in D$, and then $\xi \in D$. Moreover, the maximal integral submanifold of D is an (n+1-q)-dimensional invariant submanifold of $\bar{M}$.

Proof. Let X, Y $\in D$. Then (4.5) implies

$$\phi[X,Y] = P[X,Y] + F[X,Y] = P[X,Y] + (\nabla_Y F)X - (\nabla_X F)Y$$

$$= P[X.Y] + B(X,PY) - B(Y,PX).$$

Thus we see that $[X,Y] \in D$ if and only if $B(X,PY) = B(Y,PX)$ for every X, Y $\in D$. If D is normal to ξ, then $g([X,Y],\xi) = 2g(PX,Y)$ for X,Y $\in D$. Thus, if D is completely integrable, we have $g(PX,Y) = 0$, which shows that dim D = 0. This proves our assertion. QED.

We now give examples of contact CR submanifolds and generic submanifolds of a Sasakian manifold.

Example 4.1. Let S^{2m+1} be a (2m+1)-dimensional unit sphere with standard Sasakian structure (ϕ,ξ,η,g). We denote by $S^m(r)$ an m-dimensional sphere with radius r. We consider the following immersion:

$$S^{m_1}(r_1) \times \ldots \times S^{m_k}(r_k) \longrightarrow S^{n+k}, \qquad n+1 = \sum_{i=1}^{k} m_i,$$

where $m_1,\ldots,m_k$ are odd numbers and $r_1^2 + \ldots + r_k^2 = 1$. Here n+k is also odd. Let v_i be a point of $S^{m_i}(r_i)$ in $R^{m_i+1} = C^{(m_i+1)/2}$. $S^{m_i}(r_i)$ is a real hypersurface of $C^{(m_i+1)/2}$ with unit normal $r_i^{-1}v_i$. Thus v $= (v_1,\ldots,v_k)$ is a unit vector in $R^{n+k+1} = C^{(n+k+1)/2}$. We restrict the almost complex structure of $C^{(n+k+1)/2}$ to $C^{(m_i+1)/2}$. Then each Jv_i is tangent to $S^{m_i}(r_i)$. Thus Jv is tangent to $M_{m_1,\ldots,m_k} = S^{m_1}(r_1)$ $\times \ldots \times S^{m_k}(r_k)$. We then consider the normal space of $M_{m_1,\ldots,m_k}$ in S^{n+k} which is the orthogonal complement of the 1-dimensional space $\langle v\rangle$ spanned by v in the space $\langle v_1,\ldots,v_k\rangle$ spanned by $v_1,\ldots,v_k$. That is,

$$\langle v\rangle \oplus T_x(M_{m_1,\ldots,m_k})^{\perp} = \langle v_1,\ldots,v_k\rangle \quad \text{in } C^{(n+k+1)/2}.$$

Let $w_1,\ldots,w_{k-1}$ be an orthonormal frame for $T_x(M_{m_1,\ldots,m_k})^{\perp}$. Then w_i is given by a linear combination od $v_1,\ldots,v_k$. Thus Jw_i is tangent to $M_{m_1,\ldots,m_k}$, and hence we see that

$$\phi w_i = Jw_i - \eta(w_i)v = Jw_i.$$

Therefore ϕw_i is tangent to $M_{m_1,\ldots,m_k}$ for all i = 1,...,k-1. Thus

$$\phi T_x(M_{m_1,\ldots,m_k})^{\perp} \subset T_x(M_{m_1,\ldots,m_k}).$$

Consequently, $M_{m_1,\ldots,m_k}$ is a generic submanifold of S^{n+k}. $M_{m_1,\ldots,m_k}$ has parallel second fundamental form and flat normal connection.

Furthermore, $M_{m_1,\ldots,m_k}$ is a contact CR submanifold of S^{2m+1}

($2m+1 > n+k$) with parallel second fundamental form and flat normal connection.

Example 4.2. In Example 4.1, if $r_i = (m_i/(n+1))^{1/2}$ ($i = 1,\dots,k$), then $M_{m_1,\dots,m_k}$ is a generic minimal submanifold of S^{n+k} and hence minimal contact CR submanifold of S^{2m+1} ($2m+1 > n+k$). Then the square of the length of the second fundamental form of M is given by $|A|^2 = (n+1)(k-1)$.

We notice that in Examples 4.1 and 4.2, if we put $m_i = 1$ for all $i = 1,\dots,k$ ($k = n+1$), then $M_{m_1,\dots,m_k}$ is an anti-invariant submanifold of S^{2n+1}.

In the following we consider a contact CR submanifold with flat normal connection. Let M be an $(n+1)$-dimensional contact CR submanifold of a $(2m+1)$-dimensional Sasakian manifold $\bar{M}$. Then we have the following decomposition of the tangent space $T_x(M)$ at each point x of M:

$$T_x(M) = H_x(M) \oplus \{\xi\} \oplus N_x(M),$$

where $H_x(M) = \phi H_x(M)$ and $N_x(M)$ is the orthogonal complement of $H_x(M) \oplus \{\xi\}$ in $T_x(M)$. Then $\phi N_x(M) = FN_x(M) \subset T_x(M)^\perp$. Similarly, we have

$$T_x(M)^\perp = FN_x(M) \oplus N_x(M)^\perp,$$

where $N_x(M)^\perp$ is the orthogonal complement of $FN_x(M)$ in $T_x(M)^\perp$. Then $\phi N_x(M)^\perp = fN_x(M)^\perp = N_x(M)^\perp$.

We now take an orthonormal basis $e_1,\dots,e_{2m+1}$ of $\bar{M}$ such that, restricted to M, $e_1,\dots,e_{n+1}$ are tangent to M. Then $e_1,\dots,e_{n+1}$ form an orthonormal basis of M. We can take $e_1,\dots,e_{n+1}$ such that $e_1,\dots,e_p$ form an orthonormal basis of $N_x(M)$ and $e_{p+1},\dots,e_n$ form an orthonormal basis of $H_x(M)$ and $e_{n+1} = \xi$, where $p = \dim N_x(M)$. Moreover, we can take $e_{n+2},\dots,e_{2m+1}$ of an orthonormal basis of $T_x(M)^\perp$ such that $e_{n+2},\dots,e_{n+1+p}$ form an orthonormal basis of $FN_x(M)$ and $e_{n+2+p},\dots,e_{2m+1}$ form an orthonormal basis of $N_x(M)^\perp$. In case of need, we can take $e_{n+2},\dots,e_{n+1+p}$ such that $e_{n+2}=Fe_1,\dots,e_{n+1+p}=Fe_p$.

Unless otherwise stated, we use the conventions that the ranges of indices are respectively:

$$i,j,k = 1,\dots,n+1; \qquad x,y,z = 1,\dots,p;$$
$$a,b,c = p+1,\dots,n; \qquad \alpha,\beta,\gamma = n+2,\dots,n+1+p.$$

Here, we take S^{2m+1} as an ambient manifold $\bar{M}$. Then we have

LEMMA 4.2. If the normal connection of M is flat, then

$$A_{fV} = 0$$

for any vector field V normal to M.

Proof. Since $R^{\perp} = 0$, (4.10) implies that $A_V A_U = A_U A_V$. Thus, from (4.3), we obtain $A_V tU = A_U tV$. From this and $tf = 0$, we see that $A_{fV}tU = 0$ and $A_{fV}\xi = 0$. Moreover, from (4.7), we obtain $(\nabla_X f)fV = 0$. Thus, from (4.5) and (4.14), we have

$$g((\nabla_X f)fV, FY) = -g(f^2V, (\nabla_X F)Y) = -g(A_{fV}X, Y) + g(A_{f^2V}X, PY) = 0.$$

From this and the fact that $A_{fV}A_{f^2V} = A_{f^2V}A_{fV}$, we have

$$\begin{aligned}\mathrm{Tr}A_{fV}^2 &= \mathrm{Tr}A_{fV}PA_{fV} = -\mathrm{Tr}A_{fV}PA_{f^2V} = -\mathrm{Tr}A_{f^2V}A_{fV}P \\ &= -\mathrm{Tr}A_{fV}A_{f^2V}P = -\mathrm{Tr}A_{f^2V}PA_{fV} = -\mathrm{Tr}A_{fV}^2.\end{aligned}$$

Consequently, we have $\mathrm{Tr}A_{fV}^2 = 0$ and hence $A_{fV} = 0$. QED.

LEMMA 4.3. Let M be an (n+1)-dimensional contact CR submanifold of S^{2m+1} with flat normal connection. If $PA_V = A_VP$ for any vector field V normal to M, then

$$g(A_UX, A_VY) = g(X,Y)g(tU,tV) - \sum_i g(A_UtV, e_i)g(A_{Fe_i}X, Y).$$

Proof. From the assumption we have $g(A_UPX, tV) = 0$, which implies

$$g((\nabla_Y A)_U PX,tV) + g(A_U(\nabla_Y P)X,tV) + g(A_U PX,(\nabla_Y t)V) = 0.$$

Thus, from (4.4) and (4.6), we have

$$g((\nabla_Y A)_U PX,tV) - g(X,Y)g(A_U\xi,tV) + \eta(X)g(A_U Y,tV)$$

$$+ g(A_U A_{FX}Y,tV) + g(A_U tB(Y,X),tV) + g(A_U PX,A_{fV}Y)$$

$$- g(A_U PX,PA_V Y) = 0,$$

from which and Lemma 4.2, we find

$$g((\nabla_{PY}A)_U PX,tV) + g(X,PY)g(tU,tV)$$

$$+ g(A_U tV,tB(PY,X)) - g(A_U PX,PA_V PY) = 0.$$

On the other hand, we have

$$g(A_U tV,tB(PY,X) = -\sum_i g(A_U tV,e_i)g(A_{Fe_i}X,PY),$$

$$-g(A_U PX,PA_V PY) = g(A_U PX,A_V Y).$$

From these equations we obtain

$$g((\nabla_{PY}A)_U PX,tV) + g(X,PY)g(tU,tV)$$

$$- \sum_i g(A_U tV,e_i)g(A_{Fe_i}X,PY) + g(A_U PX,A_V Y) = 0.$$

Therefore, the Codazzi equation implies

$$g(X,PY)g(tU,tV) - \sum_i g(A_U tV,e_i)g(A_{Fe_i}X,PY) + g(A_U PX,A_V Y) = 0,$$

from which

$$(4.15) \quad g(PX,PY)g(tU,tV) - \sum_i g(A_U tV,e_i)g(A_{Fe_i}PX,PY) + g(A_U P^2X,A_V Y) = 0.$$

On the other hand, we have

$$g(PX,PY)g(tU,tV) = g(X,Y)g(tU,tV) - \eta(X)\eta(Y)g(tU,tV) - g(FX,FY)g(tU,tV),$$

$$-\sum_i g(A_U tV,e_i)g(A_{Fe_i}PX,PY) = -\sum_i g(A_U tV,e_i)g(A_{Fe_i}X,Y) + \eta(Y)g(A_U tV,X) + \eta(X)\eta(Y)g(tU,tV) - \sum_i g(A_U tV,e_i)g(A_{Fe_i}X,tFY),$$

$$g(A_U P^2X,A_V Y) = -g(A_U X,A_V Y) - \eta(Y)g(A_U tV,X) - g(A_U X,A_V tFY).$$

Substituting these equations into (4.15), we find

$$g(X,Y)g(tU,tV) - \sum_i g(A_U tV,e_i)g(A_{Fe_i}X,Y) - g(A_U X,A_V Y) - g(FX,FY)g(tU,tV) - \sum_i g(A_U tV,e_i)g(A_{Fe_i}X,tFY) - g(A_U X,A_V tFY) = 0.$$

Moreover, we obtain

$$-\sum_i g(A_U tV,e_i)g(A_{Fe_i}X,tFY) = g(A_U tV,A_{FY}X) + g(FX,FY)g(tU,tV) - g(A_U X,A_V tFY) = - g(A_U tV,A_{FY}X).$$

From these, we obtain our equation. QED.

LEMMA 4.4. Let M be an (n+1)-dimensional contact CR submanifold of S^{2m+1} with flat normal connection. If the mean curvature vector of M is parallel, and if $PA_V = A_V P$ for any vector field V normal to M, then the square of the length of the second fundamental form of M is constant.

Proof. Since $A_{fV} = 0$, we have $|A|^2 = \Sigma_\alpha \mathrm{Tr}A_\alpha^2$. On the other hand,

Lemma 4.3 gives

$$|A|^2 = (n+1)p + \sum_{\alpha,\beta} g(A_\alpha te_\alpha, te_\beta)\mathrm{Tr}A_\beta.$$

Since the normal connection of M is flat, we can take $\{e_\alpha\}$ such that $De_\alpha = 0$ for each α, because, for any $V \in FN(M)$ we have $D_X V \in FN(M)$ by (4.7) and $A_V tU = A_U tV$. Then we have

$$\nabla_X |A|^2 = \sum_{\alpha,\beta} g((\nabla_X A)_\alpha te_\alpha, te_\beta)\mathrm{Tr}A_\beta = \sum_{\alpha,\beta} g((\nabla_{te_\alpha} A)_\beta te_\alpha, X)\mathrm{Tr}A_\beta$$

by using $\nabla_X(te_\alpha) = (\nabla_X T)e_\alpha = A_{fe_\alpha}X - PA_\alpha X$ and $Pt = 0$. On the other hand, using $PA_V = A_V P$, we have

$$\sum_i g((\nabla_{Pe_i} A)_\alpha Pe_i, X) = \sum_i [g((\nabla_{Pe_i} P)A_\alpha e_i, X) + g(P(\nabla_{Pe_i} A)_\alpha e_i, X) - g(A_\alpha(\nabla_{Pe_i} P)e_i, X)].$$

Since A_α is symmetric and P is skew-symmetric, using (4.4), we see that

$$\sum_i g((\nabla_{Pe_i} P)A_\alpha e_i, X) = 0, \quad \sum_i g(A_\alpha(\nabla_{Pe_i} P)e_i, X) = 0.$$

Therefore, we have

$$\begin{aligned}\sum_i g((\nabla_{Pe_i} A)_\alpha Pe_i, X) &= \sum_i g(P(\nabla_{Pe_i} A)_\alpha e_i, X) \\ &= -\sum_i g((\nabla_{Pe_i} A)_\alpha e_i, PX) = -\sum_i g((\nabla_{PX} A)_\alpha Pe_i, e_i) = 0,\end{aligned}$$

where we have used the Codazzi equation and the fact that $(\nabla_{PX} A)_\alpha$ is symmetric and P is skew-symmetric. Since we have $\Sigma(\nabla e_a A)_\alpha e_a = \Sigma(\nabla_{Pe_i} A)_\alpha Pe_i$, the equation above implies

$$\sum_a (\nabla_{e_a} A)_\alpha e_a = 0.$$

Moreover, we see that

$$(\nabla_\xi A)_\alpha \xi = 0.$$

Since the mean curvature vector of M is parallel, we have

$$0 = \sum_i (\nabla_{e_i} A)_\alpha e_i = \sum_a (\nabla_{e_a} A)_\alpha e_a + (\nabla_\xi A)\xi + \sum_x (\nabla_{e_x} A)_\alpha e_x$$
$$= \sum_x (\nabla_{e_x} A)_\alpha e_x = \sum_\beta (\nabla_{te_\beta} A)_\alpha te_\beta .$$

Therefore, the square of the length of the second fundamental form of M is constant. QED.

From Lemmas 4.2, 4.4 and Proposition 3.1 of Chapter II we have

LEMMA 4.5. Let M be an (n+1)-dimensional contact CR submanifold of S^{2m+1} with flat normal connection. If the mean curvature vector of M is parallel, and if $PA_V = A_V P$ for any vector field V normal to M, then

$$|\nabla A|^2 = -(n+1)\sum_\alpha \mathrm{Tr} A_\alpha^2 + \sum_\alpha (\mathrm{Tr} A_\alpha)^2$$
$$+ \sum_{\alpha,\beta} (\mathrm{Tr} A_\alpha A_\beta)^2 - \sum_{\alpha,\beta} \mathrm{Tr} A_\beta \mathrm{Tr} A_\alpha^2 A_\beta .$$

LEMMA 4.6. Under the same assumptions as those of Lemma 4.5, the second fundamental form of M is parallel.

Proof. From Lemma 4.3 we obtain

$$\mathrm{Tr} A_\alpha^2 A_\beta = \mathrm{Tr} A_\alpha g(e_\alpha, e_\beta) + \sum_\gamma \mathrm{Tr}(A_\gamma A_\alpha) g(A_\gamma te_\alpha, te_\beta),$$

$$\mathrm{Tr} A_\alpha A_\beta = (n+1) g(e_\alpha, e_\beta) + \sum_\gamma \mathrm{Tr} A_\gamma g(A_\gamma te_\alpha, te_\beta).$$

From these equations we have

$$\sum_{\alpha,\beta} (\mathrm{Tr} A_\alpha A_\beta)^2 = (n+1)\sum_\alpha \mathrm{Tr} A_\alpha^2 + \sum_{\alpha,\beta,\gamma} \mathrm{Tr} A_\alpha A_\beta \mathrm{Tr} A_\gamma g(A_\gamma te_\alpha, te_\beta),$$

$$-\sum_{\alpha,\beta} \mathrm{Tr}A_\beta \mathrm{Tr}A_\alpha^2 A_\beta = -\sum_\alpha (\mathrm{Tr}A_\alpha)^2 - \sum_{\alpha,\beta,\gamma} \mathrm{Tr}A_\alpha A_\beta \mathrm{Tr}A_\gamma g(A_\gamma te_\alpha, te_\beta).$$

Substituting these equations into the equation of Lemma 4.5, we find $|\nabla A|^2 = 0$, which shows that the second fundamental form of M is parallel. QED.

We prove the following theorems (Yano-Kon [16]).

THEOREM 4.5. Let M be an (n+1)-dimensional complete contact CR submanifold of S^{2m+1} with flat normal connection. If the mean curvature vector of M is parallel, and if $PA_V = A_V P$ for any vector field V normal to M, then M is S^{n+1} or

$$S^{m_1}(r_1) \times \dots \times S^{m_k}(r_k), \quad n+1 = \sum_i m_i, \quad \sum_i r_i^2 = 1, \quad 2 \leq k \leq n+1,$$

in some S^{n+1+p}, where $m_1, \dots, m_k$ are odd numbers.

Proof. We first assume that F = 0, thst is, M is an invariant submanifold of S^{2m+1}. Then we have $PA_V + A_V P = 0$. Thus we have $PA_V = 0$ and hence $A_V = 0$. Consequently, M is totally geodesic in S^{2m+1} and M is S^{n+1}.

We next assume that $F \neq 0$. Since the second fundamental form of M is parallel and the normal connection of M is falt, by Theorem 5.5 of Chapter II, the sectional curvature of M is non-negative. On the other hand, from Lemma 4.3, we see that $A_V \neq 0$ for any $V \in FN_x(M)$. Thus Lemma 4.2 shows that the first normal space of M is of dimension p. Therefore, by Theorem 4.3 of Chapter II and Example 4.1, we have our assertion. QED.

THEOREM 4.6. Let M be an (n+1)-dimensional complete generic submanifold of S^{2m+1} with flat normal connection. If the mean curvature vector of M is parallel, and if $PA_V = A_V P$ for any vector field V normal to M, then M is

$$S^{m_1}(r_1) \times \dots \times S^{m_k}(r_k), \quad n+1 = \sum_i m_i, \quad \sum_i r_i^2 = 1, \quad 2 \leq k \leq n+1,$$

where $m_1, \ldots, m_k$ are odd numbers.

Let M be an (n+1)-dimensional contact CR submanifold of S^{2m+1} with flat normal connection. Let V be a parallel vector field normal to M. Then we have $\nabla_X tV = -PA_V X$. Hence we have div $tV = -\mathrm{Tr}PA_V = 0$. Therefore, from Theorem 4.3 of Chapter I, we find

$$\mathrm{div}(\nabla_{tV} tV) = S(tV,tV) + \tfrac{1}{2}|L(tv)g|^2 - |\nabla tV|^2.$$

In the following we suppose that M is minimal. Then the Ricci tensor S of M is given by

$$S(X,Y) = ng(X,Y) - \sum_\alpha g(A_\alpha^2 X,Y).$$

On the other hand, we have

$$|\nabla tV|^2 = \mathrm{Tr}A_V^2 - g(tV,tV) - \sum_\alpha g(A_\alpha^2 tV,tV).$$

From these equations we obtain

$$\mathrm{div}(\nabla_{tV} tV) = (n+1)g(tV,tV) - \mathrm{Tr}A_V^2 + \tfrac{1}{2}|L(tV)g|^2.$$

Since the normal connection of M is flat, we can take $\{e_\alpha\}$ of FN(M) such that $De_\alpha = 0$ for each α. Thus we have

$$\mathrm{div}(\sum_\alpha \nabla_{te_\alpha} te_\alpha) = (n+1)p - |A|^2 + \tfrac{1}{2}\sum_\alpha |L(te_\alpha)g|^2.$$

THEOREM 4.7. Let M be an (n+1)-dimensional compact minimal contact CR submanifold of S^{2m+1} with flat normal connection. Then

$$0 \leq \tfrac{1}{2}\int_M \sum_\alpha |L(te_\alpha)g|^2 *1 = \int_M [|A|^2 - (n+1)p]*1.$$

As an application of Theorem 4.7, we have

THEOREM 4.8. Let M be an (n+1)-dimensional compact minimal contact CR submanifold of S^{2m+1} with flat normal connection. If $|A|^2 = (n+1)p$, then M is

$$S^{m_1}(r_1) \times \dots \times S^{m_k}(r_k), \quad r_i = (m_i/(n+1))^{1/2} \quad (i = 1,\dots,k),$$

$n+1 = \Sigma m_i$, $\Sigma r_i^2 = 1$, $2 \leq i \leq n+1$, in some S^{n+1+p}, where $m_1,\dots,m_k$ are odd numbers.

Proof. From the assumption we have $|L(te_\alpha)g| = 0$ for each α. Thus we have

$$\begin{aligned} 0 = (L(te_\alpha)g)(X,Y) &= g(\nabla_X te_\alpha, Y) + g(\nabla_Y te_\alpha, X) \\ &= g((A_\alpha P - PA_\alpha)X,Y) \end{aligned}$$

for each α. Consequently, $PA_V = A_V P$ for any vector field V normal to M. Therefore, our assertion follows from Theorem 4.5. QED.

5. INDUCED STRUCTURES ON SUBMANIFOLDS

Let M be an n-dimensional manifold. We assume that there exist on M a tensor field f of type (1,1), vector fields U and V, 1-forms u and v, and a function λ satisfying the conditions:

$$f^2X = -X + u(X)U + v(X)V, \tag{5.1}$$

$$u(fX) = \lambda v(X), \quad v(fX) = -\lambda u(X), \quad fU = -\lambda V, \quad fV = \lambda U, \tag{5.2}$$

$$u(U) = v(V) = 1 - \lambda^2, \quad u(V) = v(U) = 0, \tag{5.3}$$

for any vector field X on M. In this case, we say that the manifold M has an (f,U,V,u,v,λ)-structure. Then we have (Yano-Okumura [1])

THEOREM 5.1. A manifold M with (f,U,V,u,v,λ)-structure is of even dimensional.

Proof. Let x be a point of x at which $\lambda^2 \neq 1$. Then we see that $U \neq 0$ and $V \neq 0$ at x. Two vectors U and V are linearly independent. For, if there are two numbers a and b such that $aU + bV = 0$, then $u(aU + bV) = au(U) = a(1 - \lambda^2) = 0$, $v(aU + bV) = bv(V) = b(1 - \lambda^2) = 0$. Thus we have $a = b = 0$. Thus U and V being linearly independent at x, we can choose n linearly independent vectors $X_1 = U$, $X_2 = V$, $X_3, \ldots, X_n$ which span the tangent space $T_x(M)$ and such that $u(X_a) = 0$, $v(X_a) = 0$, for $a = 3,\ldots,n$. Consequently, we have $f^2X_a = -X_a$ for all a, which shows that f is an almost complex structure in the subspace E_x of $T_x(M)$ at x spanned by $X_3,\ldots,X_n$ and that E_x is even dimensional. Thus $T_x(M)$ is also even dimensional.

Next, let x be a point of M at which $\lambda^2 = 1$. In this case, we see that $u(U) = u(V) = v(U) = v(V) = 0$. We also see that if $u \neq 0$, then $v \neq 0$, and if $u = 0$, then $v = 0$. We first consider the case in which $u \neq 0$, $v \neq 0$. Then u and v are linearly independent. Because, if there are two numbers a and b such that $au + bv = 0$, then $(au + bv)(fX) = \lambda(bu - av)(X) = 0$ and hence $bu - av = 0$, λ being different from zero. Thus we have $(a^2 + b^2)u = 0$, from which $a = 0$, $b = 0$. Thus, u and v being linearly independent at x, we can choose n linearly independent covectors $w_1 = u$, $w_2 = v$, $w_3,\ldots,w_n$ which span the cotangent space $T_x(M)^*$ of M at x. We denote the dual basis by $X_1, X_2,\ldots,X_n$. If U and V are linearly independent at x, we can assume that $X_{n-1} = U$, $X_n = V$. Then we have

$$f^2X_a = -X_a + u(X_a)U + v(X_a)V = -X_a, \quad a = 3,\ldots,n$$

which shows that f is an almost complex structure in the subspace E_x of $T_x(M)$ at x spanned by $X_3,\ldots,X_n$ and that E_x is even dimensional and consiquently $T_x(M)$ is also even dimensional.

If U and V are linearly dependent, there exist two numbers a and b such that $aU + bV = 0$ and $a^2 + b^2 \neq 0$. Applying f to this equation, we find $\lambda(-aV + bU) = 0$, and hence $bU - aV = 0$. Thus, we must have

$U = V = 0$. Consequently, we have $f^2X = -X$ for any vector X in $T_x(M)$ and $T_x(M)$ is even dimensional.

If $u = 0$, $v = 0$, we also have $f^2X = -X$ for any vector X in $T_x(M)$ and consequently $T_x(M)$ is even dimensional. Thus we have our assertion. QED.

The structure (f,U,V,u,v,λ) is said to be *normal* if

$$S(X,Y) = N(X,Y) + du(X,Y)U + dv(X,Y)V = 0$$

for any vector fields X and Y on M, N being the Nijenhuis torsion of f. We consider a product manifold $M \times R^2$, where R^2 is a 2-dimensional Euclidean space. Then, (f,U,V,u,v,λ)-structure gives rise to an almost complex structure J on $M \times R^2$:

$$J = \begin{pmatrix} f & U & V \\ -u & 0 & -\lambda \\ -v & \lambda & 0 \end{pmatrix}$$

as we can easily check using (5.1), (5.2) and (5.3).

Computing the Nijenhuis torsion of J, we can easily prove

PROPOSITION 5.1. If J is integrable, then (f,U,V,u,v,λ)-structure is normal.

We assume that, in M with (f,U,V,u,v,λ)-structure, there exists a positive definite Riemannian metric g such that

$$u(X) = g(U,X), \qquad v(X) = g(V,X), \tag{5.4}$$

$$g(fX,fY) = g(X,Y) - u(X)u(Y) - v(X)v(Y). \tag{5.5}$$

We call such a structure a metric (f,U,V,u,v,λ)-structure and denote it by (f,g,u,v,λ).

Example 5.1. Let $\bar{M}$ be a $(2n+1)$-dimensional almost contact metric manifold with structure tensors (ϕ,ξ,η,g). Then the structure tensors satisfy

$$\phi^2 = -I + \eta \otimes \xi, \quad \phi\xi = 0, \quad \eta(\phi X) = 0, \quad \eta(\xi) = 1,$$

$$\eta(X) = g(\xi,X), \quad g(\phi X,\phi Y) = g(X,Y) - \eta(X)\eta(Y)$$

for any vector field X on $\bar{M}$.

Let M be a 2n-dimensional hypersurface of $\bar{M}$. We denote by the same g the induced metric tensor field on M. The unit normal of M in $\bar{M}$ will be denoted by C.

For any vector field X tangent to M we put

$$\phi X = fX + u(X)C, \quad \xi = V + \lambda C, \quad \phi C = -U,$$

$$v(X) = \eta(X), \quad \lambda = \eta(C) = g(\xi,C),$$

where f is a tensor field of type (1,1), u, v 1-forms, U, V vector fields and λ a scalar function on M. Then they satisfy

$$f^2X = -X + u(X)U + v(X)V, \quad u(fX) = \lambda v(X), \quad v(fX) = -\lambda u(X),$$

$$fU = -\lambda V, \quad fV = \lambda U, \quad u(V) = 0, \quad v(U) = 0,$$

$$u(U) = 1 - \lambda^2, \quad v(V) = 1 - \lambda^2.$$

Moreover, we have

$$g(U,X) = u(X), \quad g(V,X) = v(X), \quad g(fX,Y) = -g(X,fY),$$

$$g(fX,fY) = g(X,Y) - u(X)u(Y) - v(X)v(Y).$$

Therefore, the hypersurface M admits (f,g,u,v,λ)-structure.

Example 5.2. Let $\bar{M}$ be a real (2n+2)-dimensional almost Hermitian manifold with structure tensors (J,g). Then $J^2 = -I$ and g(JX,JY) =

$g(X,Y)$ for any vector fields X and Y on $\bar{M}$. Let M be a submanifold of codimension 2 of $\bar{M}$ with orthonormal frame C, D for $T_x(M)^{\perp}$. We put

$$JX = fX + u(X)C + v(X)D,$$

$$JC = -U + \lambda D, \qquad JD = -V - \lambda C.$$

Then the induced structure (f,g,u,v,λ) on M satisfies (5.1) ~ (5.5). Therefore, the submanifold M admits an (f,g,u,v,λ)-structure.

Example 5.3. Let $\bar{M}$ be a (2n+1)-dimensional Sasakian manifold and let M be a hypersurface of M. We have the Gauss and Weingarten formulas

$$\bar{\nabla}_X Y = \nabla_X Y + g(AX,Y)C, \qquad \bar{\nabla}_X C = -AX.$$

Then we have

$$(\nabla_X f)Y = -v(Y)X + u(Y)AX + g(X,Y)V - g(AX,Y)U,$$

$$\nabla_X U = \lambda X + fAX, \qquad \nabla_X V = -fX + \lambda AX.$$

On the other hand, we obtain

$$\begin{aligned} S(X,Y) &= (\nabla_{fX}F)Y - (\nabla_{fY}f)X + f(\nabla_Y f)X - f(\nabla_X f)Y \\ &\quad + g(\nabla_X U,Y)U - g(\nabla_Y U,X)U + g(\nabla_X V,Y)V - g(\nabla_Y V,X)V \\ &= u(Y)(AfX - fAX) - u(X)(AfY - fAY). \end{aligned}$$

Thus, if $Af = fA$, then the hypersurface M is normal. We shall prove the converse. Let $S(X,Y) = 0$ for all X and Y and put $PX = (Af-fA)X$. Then

$$u(U)PX = u(X)PU.$$

Also, it can be shown that $g(PX,Y) = g(X,PY)$ so that

$$u(X)g(PU,Y) = u(Y)g(PU,X),$$

that is to say,

$$g(PU,Y) = \alpha u(Y)$$

for some α. Thus we have

$$u(U)g(PX,Y) = u(X)g(PU,Y) = \alpha u(X)u(Y),$$

but since the trace of P is zero, we have $\alpha = 0$, i.e., $P = 0$, which means that $Af = fA$. Consequently, M is normal if and only if $Af = fA$.

EXERCISES

A. K-CONTACT SUBMANIFOLDS: Let $\bar{M}$ be a (2n+1)-dimensional K-contact Riemannian manifold with structure tensors (ϕ,ξ,η,g). A submanifold M of $\bar{M}$ is said to be invariant in $\bar{M}$ if ϕX is tangent to M for any tangent vector field X to M, and ξ is always tangent to M. Any invariant submanifold M of a K-contact Riemannian manifold $\bar{M}$ is also a K-contact Riemannian manifold with respect to the induced structure on M. Then we have (Endo [1])

THEOREM 1. Any invariant submanifold M of a K-contact Riemannian manifold $\bar{M}$ is a minimal submanifold.

On the other hand, we obtain (cf. Kon [1])

THEOREM 2. Let M be an invariant submanifold of a K-contact Riemannian manifold $\bar{M}$. Then M is totally geodesic if and only if the second fundamental form of M is parallel.

B. COSYMPLECTIC MANIFOLDS: Let $\bar{M}$ be a normal almost contact metric manifold such that the fundamental 2-form Φ is closed and $d\eta = 0$. Then $\bar{M}$ is called a *cosymplectic manifold*. The cosymplectic structure is characterized by $\bar{\nabla}_X\phi = 0$ and $\bar{\nabla}_X\eta = 0$. Let M be an invariant submanifold of a cosymplectic manifold $\bar{M}$. Then M is also a cosymplectic manifold with respect to the induced structure on M. Ludden [1] proved the following

THEOREM. If $\bar{M}$ is a cosymplectic manifold of constant ϕ-sectional curvature and M is an invariant submanifold of codimension 2 of $\bar{M}$ which is η-Einstein, then M is locally symmetric.

C. FLAT NORMAL CONNECTION OF INVARIANT SUBMANIFOLDS: Let $\bar{M}$ be a Sasakian space form of constant ϕ-sectional curvature k and M be an invariant submanifold of $\bar{M}$. Then we have

THEOREM. The following conditions are equivalent:

(a) The normal connection of M is flat, i.e., $R^{\perp} = 0$;

(b) $k = 1$ and M is totally geodesic in $\bar{M}$.

In the case of codimension 2, Kenmotsu [2] proved the above theorem. The theorem above was proved by Kon [3] when the codimension of M is greater than 2.

D. AXIOM OF ϕ-HOLOMORPHIC PLANES: Let M be a (2n+1)-dimensional Sasakian manifold with structure tensors (ϕ,ξ,η,g). We say that M admits the axiom of ϕ-holomorphic (2r+1)-planes if, for each point x of M and any (2r+1)-dimensional ϕ-holomorphic subspace S of $T_x(M)$, $1 < r < n$, there exists a (2r+1)-dimensional totally geodesic submanifold N passing through x and satisfying $T_x(N) = s$, where we mean a ϕ-holomorphic subspace S by a subspace of $T_x(M)$ satisfying $\phi S \subset S$. I. Ishihara [2] proved the following

THEOREM. A Sasakian manifold is of constant ϕ-sectional curvature if and only if the manifold satisfies the axiom of ϕ-holomorphic (2r+1)-planes.

E. REDUCTION THEOREMS OF ANTI-INVARIANT SUBMANIFOLDS: I. Ishihara [2] studied the reduction theorems of codimension of anti-invariant submanifolds of Sasakian space forms. Let M be an (n+1)-dimensional anti-invariant submanifold, tangent to the structure vector field ξ, of a (2m+1)-dimensional Sasakian manifold $\bar{M}$. If $\Sigma_i h^a_{iix} = 0$ for all indices a and x, then the mean curvature vector of M is said to be *pseudo-parallel*. (For the ranges of indices, see §2.) The normal connection of M is said to be *pseudo-flat* if $R^a_{bxy} = 0$ for all indices. Then

THEOREM 1. Let M be an (n+1)-dimensional ($n \geq 3$) anti-invariant submanifold, tangent to the structure vector field ξ, of a Sasakian space form $\bar{M}^{2m+1}(c)$ $(c \neq -3)$ with pseudo-parallel mean curvature vector. If the normal connection of M is pseudo-flat, then there is in $\bar{M}^{2m+1}(c)$ a totally geodesic and invariant submanifold $\bar{M}^{2n+1}(c)$ of dimension 2n+1 in such a way that M is immersed in $\bar{M}^{2n+1}(c)$ as a flat anti-invariant submanifold.

Let M be an n-dimensional anti-invariant submanifold, normal to the structure vector field ξ, of a (2m+1)-dimensional Sasakian manifold $\bar{M}$. If $\Sigma_k h^p_{kki} = 0$ for all indices i and p, then the mean curvature vector of M is said to be *η-parallel*. (For the ranges of indices, see §3.) For the normal curvature tensor $R^\perp$ we consider the condition

$$(*) \qquad g(R^\perp(X,Y)U,V) = g(\phi Y,V)g(\phi X,U) - g(\phi X,V)g(\phi Y,U)$$

for any vector fields X, Y tangent to M and any vector fields normal to M. Then we have (I. Ishihara [3])

THEOREM 2. Let M be an n-dimensional ($n \geq 3$) anti-invariant submanifold, normal to the structure vector field ξ, of a Sasakian space form $\bar{M}^{2m+1}(c)$ ($c \neq -3$) with η-parallel mean curvature vector. If $R^\perp$ satisfies (*), then there is in $\bar{M}^{2m+1}(c)$ a totally geodesic and invariant submanifold $\bar{M}^{2n+1}(c)$ of dimension 2n+1 in such a way that M is immersed in $\bar{M}^{2n+1}(c)$ as a flat anti-invariant minimal submanifold.

F. CONFORMALLY FLAT ANTI-INVARIANT SUBMANIFOLDS: Let M be an (n+1)-dimensional anti-invariant submanifold, tangent to the structure vector field ξ, of a (2m+1)-dimensional Sasakian manifold $\bar{M}$. If the second fundamental form B of M satisfies

$$B(X,Y) = [g(X,Y)-\eta(X)\eta(Y)]\alpha + \eta(X)B(X,\xi) + \eta(Y)B(X,\xi)$$

for any vector fields X, Y tangent to M, where α denotes a normal vector field to M, then M is said to be a *contact totally umbilical*. We have (Kon [10])

THEOREM 1. Let M be an (n+1)-dimensional ($n \geq 3$) contact totally umbilical, anti-invariant submanifold, tangent to the structure vector field ξ, of a (2m+1)-dimensional Sasakian manifold $\bar{M}$ with vanishing contact Bochner curvature tensor. Then M is locally a product of a conformally flat Riemannian manifold M^n and a 1-dimensional space M^1.

Yano [8] proved the following

THEOREM 2. Let M be an n-dimensional ($n \geq 3$) totally umbilical anti-invariant submanifold, normal to the structure vector field ξ, of a $(2m+1)$-dimensional Sasakian manifold $\bar{M}$ with vanishing contact Bochner curvature tensor. Then M is conformally flat.

G. GENERIC SUBMANIFOLDS: Yano-Kon [10] proved the following

THEOREM. Let M be an $(n+1)$-dimensional complete generic minimal submanifold of S^{2m+1} with parallel second fundamental form. If M is Einstein, then M is

$$S^q(r) \times \ldots \times S^q(r) \text{ (N-times)}, \quad r = (q/(n+1))^{1/2},$$

where q is an odd number and $2m-n = N-1$, $Nq = n+1$.

H. PSEUDO-UMBILICAL HYPERSURFACE: Let M be a hypersurface of a Sasakian manifold $\bar{M}$, tangent to the structure vector field. If the second fundamental form A of M is of the form

$$AX = a[X-\eta(X)\xi] + bu(X)U + \eta(X)U + u(X)\xi$$

for any vector field X tangent to M, a and b being functions, then M is called a *pseudo-umbilical hypersurface* of $\bar{M}$, where a vector field U and a 1-form u are defined to be $U = -\phi C$, $u(X) = g(U,X)$ respectively for a unit normal C of M. The notion of pseudo-umbilical hypersurfaces of Sasakian manifolds corresponds to that of η-umbilical real hypersurfaces of Kaehlerian manifolds.

If M is a pseudo-umbilical hypersurface of S^{2n+1} ($n \geq 2$), then M has two constant principal curvatures with multiplicities $2n-1$ and 1 respectively. Then we have (Yano-Kon [10])

THEOREM 1. Let M be a compact pseudo-umbilical hypersurface of S^{2n+1} ($n \geq 2$). Then M is

$$S^{2n-1}(r_1) \times S^1(r_2), \quad r_1^2 + r_2^2 = 1.$$

A Sasakian manifold M of dimension 2n+1 is said to satisfy the P-axiom if for each $x \in M$ and each 2n-dimensional subspace S of $T_x(M)$, $\xi \in S$, there exists a pseudo-umbilical hypersurface N such that $T_x(N) = S$, $x \in N$ and $g(AU,U) = a+b$ = constant. Yano-Kon [10] proved the following

THEOREM 2. If a (2n+1)-dimensional Sasakian manifold M ($n \geq 2$) satisfies the P-axiom, then M is a Sasakian space form.

To prove the theorem above, we need the following theorem of Tanno [7].

THEOREM 3. A (2n+1)-dimensional ($n \geq 2$) Sasakian manifold M is a Sasakian space form if and only if $R(X,\phi X)X$ is proportional to ϕX for any vector field X of M such that $\eta(X) = 0$.

I. PSEUDO-EINSTEIN HYPERSURFACES: Let M be a 2n-dimensional hypersurface of S^{2n+1} tangent to the structure vector field. If the Ricci tensor S of M is of the form

$$S(X,Y) = a[g(X,Y)-\eta(X)\eta(Y)] + bu(X)u(Y)$$
$$+ \eta(X)S(\xi,Y) + \eta(Y)S(\xi,X) - \eta(X)\eta(Y)S(\xi,\xi),$$

a and b being constant, then M is called a *pseudo-Einstein hypersurface* of S^{2n+1}. Yano-Kon [10] proved the following theorems (see also Kon [12]).

THEOREM 1. Let M be a pseudo-Einstein hypersurface of S^{2n+1} ($n \geq 3$). Then M has two constant principal curvatures or four constant principal curvatures.

We give examples of pseudo-Einstein hypersurfaces of S^{2n+1}. Let C^{n+1} be the space of (n+1)-tuples of complex numbers $(z_1,\dots,z_{n+1})$. Put $S^{2n+1} = \{(z_1,\dots,z_{n+1}) \in C^{n+1} : \Sigma|z_j|^2 = 1\}$. For a positive number r we denote by $M_0(2n,r)$ a hypersurface of S^{2n+1} defined by

$$\sum_{j=1}^{n} |z_j|^2 = r|z_{n+1}|^2, \qquad \sum_{j=1}^{n+1} |z_j|^2 = 1.$$

For an integer m ($2 \le m \le n-1$) and a positive number s a hypersurface $M(2n,m,s)$ of S^{2n+1} is defined by

$$\sum_{j=1}^{m} |z_j|^2 = s \sum_{j=m+1}^{n+1} |z_j|^2, \qquad \sum_{j=1}^{n+1} |z_j|^2 = 1.$$

For a number t ($0 < t < 1$) we denote by $M(2n,t)$ a hypersurface of S^{2n+1} defined by

$$\left|\sum_{j=1}^{n+1} z_j^2\right|^2 = t, \qquad \sum_{j=1}^{n+1} |z_j|^2 = 1.$$

$M_0(2n,r)$ and $M(2n,m,s)$ have two constant principal curvatures and $M(2n,t)$ has four constant principal curvatures (Nomizu [ ], R. Takagi [ ]). $M_0(2n,r)$ is always pseudo-Einstein. $M(2n,m,s)$ is pseudo-Einstein if $s = (m-1)/(n-m)$ and $M(2n,t)$ is pseudo-Einstein if $t = 1/(n-1)$.

THEOREM 2. If M is a complete pseudo-Einstein hypersurface in S^{2n+1} ($n \ge 3$), then M is congruent to some $M_0(2n,r)$ or to some $M(2n,m,(m-1)/(n-m))$ or to $M(2n,1/(n-1))$.

J. HYPERSURFACES WITH (f,g,u,v,λ)-STRUCTURE: Let M be a 2n-dimensional hypersurface of S^{2n+1}. Then M admits an (f,g,u,v,λ)-structure. Nakagawa-Yokote [1] proved the following

THEOREM 1. Let M be a compact hypersurface of S^{2n+1} satisfying $Af = fA = 2\alpha f$, α being function. If $n \ge 3$, then one of the following two assertions (a) and (b) is true:

(a) M is isometric to one of the following spaces:
 (1) the great sphere S^{2n};
 (2) the small sphere $S^{2n}(c)$, where $c = 1+\alpha^2$;
 (3) the product manifold $S^{2n-1}(c_1) \times S^1(c_2)$, where $c_1 = 1+\alpha^2$, $c_2 = 1+1/\alpha^2$;
 (4) the product manifold $S^n(c_1) \times S^n(c_2)$, where $c_1 = 2(1+\alpha^2+\alpha(1+\alpha^2)^{1/2})$, $c_2 = 2(1+\alpha^2-\alpha(1+\alpha^2)^{1/2})$;

(b) M has exactly four distinct constant principal curvatures $\alpha\pm(1+\alpha^2)^{1/2}$, $(-1\pm(1+\alpha^2)^{1/2})/\alpha$ with maltiplicities n-1, n-1, 1 and 1, respectively.

In the case that (b) holds, the hypersurface M is congruent to some M(2n,t) in Exercise (I) (see R. Takagi [3]).

When the second fundamental tensor A commutes to f, Kon [12] proved

THEOREM 2. Let M be a complete hypersurface of S^{2n+1} $(n \geq 2)$. If $Af = fA$, then M is congruent to $S^{2n}(\alpha^2+1)$, $\alpha = v(AV)/(1-\lambda^2)$, or to $S^{2p+1}(r_1) \times S^{2q+1}(r_2)$, $p+q = n-1$.

CHAPTER VII

f-STRUCTURES

In this chapter, we study a manifold which admits an f-structure. In §1, we define an f-structure on a manifold and give a necessary and sufficient condition for a manifold to admit an f-structure (Yano [4]). We also give integrability conditions of an f-structure (Ishihara-Yano [1]. §2 is devoted to the study of the normality of an f-structure (S. Ishihara [1]). In §3, we consider a manifold with a globally framed f-structure (Goldberg-Yano [2]). In the last §4, we discuss hypersurfaces of framed manifolds and give some theorems (Goldberg [2], Goldberg-Yano [1]).

1. f-STRUCTURE ON MANIFOLDS

A structure on an n-dimensional manifold M given by a non-null tensor field f satisfying

$$f^3 + f = 0$$

is called an f-*structure* (Yano [4]). Then the rank of f is a constant, say r (Stong [1]). If n = r, then an f-structure gives an almost complex structure of the manifold M and n = r is necessary even. If M is orientable and n-1 = r, then an f-structure gives an almost contact structure of the manifold M and n is necessary odd.

We put

$$l = -f^2, \qquad m = f^2 + I,$$

I denoting the identity operator, then we have

$$l + m = I, \quad l^2 = l, \quad m^2 = m,$$

$$fl = lf = f, \quad mf = mf = 0.$$

These equations show that the operators l and m applied to the tangent space at each point of the manifold M are complementary projection operators. Then there exist in M two distributions L and N corresponding to the projection operators l and m respectively. When the rank of f is r, L is r-dimensional and N (n-r)-dimensional.

We now introduce in M a local coordinate system and denote by f_i^h, l_i^h, m_i^h the local components of the tensors f, l, m respectively. We also introduce a positive definite Riemannian metric in M and take r mutually orthogonal unit vectors u_a^h (a,b,c,... = 1,2,...,r) in L and n-r mutually orthogonal unit vectors u_A^h (A,B,C,... = r+1,...,n) in M. We then have

$$(1.1) \qquad l_i^h u_b^i = u_b^h, \quad l_i^h u_B^i = 0, \qquad m_i^h u_b^i = 0, \quad m_i^h u_B^i = u_B^h.$$

From $fm = 0$, that is, $f_i^h m_j^i = 0$, we find, contracting with u_B^j and taking account of the last equation of (1.1),

$$f_i^h u_B^i = 0.$$

If we denote by (v_i^a, v_i^A) the matrix inverse to (u_b^h, u_B^h), then v_i^a and v_i^A are both components of linearly independent covariant vectors and satisfy

$$(1.2) \qquad v_i^a u_b^i = \delta_b^a, \quad v_i^a u_B^i = 0, \qquad v_i^A u_b^i = 0, \quad v_i^A u_B^i = \delta_B^A$$

and

$$(1.3) \qquad v_i^a u_a^h + v_i^A u_A^h = \delta_i^h.$$

Now from (1.1) and (1.3), we find

$$(l_i^h v_h^a) u_b^i = \delta_b^a, \qquad (l_i^h v_h^a) u_B^i = 0,$$

$$(m_i^h v_h^A) u_b^i = 0, \qquad (m_i^h v_h^A) u_B^i = \delta_B^A,$$

which show that

$$l_i^h v_h^a = v_i^a, \qquad m_i^h v_h^A = v_i^A,$$

from which

$$(1.4) \qquad \begin{aligned} l_i^h v_h^a &= v_i^a, \qquad & l_i^h v_h^A &= 0, \\ m_i^h v_h^a &= 0, \qquad & m_i^h v_h^A &= v_i^A. \end{aligned}$$

From $mf = 0$, that is, $f_i^h m_h^j = 0$, we find, contracting with v_j^A and taking account of the last equation of (1.4),

$$f_i^h v_h^A = 0.$$

On the other hand, from $l_j^h u_a^j = u_a^h$, we find

$$l_j^h v_i^a u_a^j = v_i^a u_a^h, \qquad l_j^h(\delta_i^j - v_i^A u_A^j) = v_i^a u_a^h,$$

that is,

$$(1.5) \qquad l_i^h = v_i^a u_a^h$$

by (1.1) and (1.3). Similarly, we get

(1.6) $$m_i^h = v_i^A u_A^h.$$

If we change u_b^h and u_B^h into $\bar{u}_b^h$ and $\bar{u}_B^h$ respectively by orthogonal transformations

$$\bar{u}_b^h = c_b^a u_a^h, \qquad \bar{u}_B^h = c_B^A u_A^h,$$

where

$$c_c^a c_b^a = \delta_{cb}, \qquad c_C^A c_B^A = \delta_{CB},$$

then v_i^a and v_i^A are change into $\bar{v}_i^a$ and $\bar{v}_i^A$ respectively by the rules

$$\bar{v}_i^a = c_a^b v_i^b, \qquad \bar{v}_i^A = c_A^B v_i^B,$$

and we have

$$\bar{v}_j^a \bar{v}_i^a = v_j^a v_i^a, \qquad \bar{v}_j^A \bar{v}_i^A = v_j^A v_i^A.$$

Consequently, if we put

(1.7) $$a_{ji} = v_j^a v_i^a + v_j^A v_i^A,$$

then a_{ji} is a globally defined positive definite Riemannian metric with respect to which (u_b^h, u_B^h) form an orthogonal frame such that

(1.8) $$v_j^a = a_{ji} u_a^i, \qquad v_j^A = a_{ji} u_A^i,$$

as we can easily verify it. If we put

$$l_{ji} = l_j^t a_{ti}, \qquad m_{ji} = m_j^t a_{ti},$$

we find, from (1.5) and (1.6),

$$l_{ji} = v_j^a v_i^a, \qquad m_{ji} = v_j^A v_i^A$$

because (1.8). Consequently

$$l_{ji} + m_{ji} = a_{ji},$$

that is, l_{ji} and m_{ji} are both symmetric and their sum is equal to a_{ji}.

We can easily verify the following relations:

$$l_j^t l_i^s a_{ts} = l_{ji}, \qquad l_j^t m_i^s a_{ts} = 0, \qquad m_j^t m_i^s a_{ts} = m_{ji}.$$

If we put

$$g_{ji} = \tfrac{1}{2}(a_{ji} + f_j^t f_i^s a_{ts} + m_{ji}),$$

then g_{ji} is again a globally defined positive definite Riemannian metric satisfying

$$v_j^A = g_{ji} u_A^i, \qquad m_{ji} = m_j^t g_{ti}.$$

Thus the distributions L and N which were orthogonal with respect to a_{ji} are still orthogonal with respect to g_{ji} and u_A^h which were mutually orthogonal unit vectors with respect to a_{ji} are still mutually orthogonal unit vectors with respect to g_{ji}. We can easily verify that g_{ji} satisfies

$$f_j^t f_t^i - m_j^i = -\delta_j^i.$$

If we put

$$f_i^s g_{st} = f_{it},$$

we have

$$f_j^t f_{it} + m_{ji} = g_{ji} \quad \text{and} \quad f_j^t f_{ti} - m_{ji} = -g_{ji}.$$

Hence we have

$$f_j^t (f_{it} + f_{ti}) = 0.$$

The rank of f being r and n-r linearly independent solutions of $f_j^t v_t = 0$ being given by v_t^A, these equation give

$$f_{it} + f_{ti} = v_t^A w_i^A$$

for certain w_i^A, from which $w_i^A = 0$, or

$$f_{it} + f_{ti} = 0$$

by $f_{it} u_B^t = f_i^t v_i^B = 0$ and $f_{ti} u_B^t = (f_t^j u_B^t) g_{ji} = 0$. Thus f_{ji} is skew-symmetric tensor of rank r, hence r must be even. Gathering the results, we have (Yano [4])

THEOREM 1.1. Let M be an n-dimensional manifold with f-structure f of rank r. Then there exist complementary distributions L of dimension r and N of dimension n-r and a positive definite Riemannian metric g with respect to which L and N are orthogonal and such that

$$f_j^t f_i^s g_{ts} + m_j^t g_{ti} = g_{ji},$$

$$f_{ji} = -f_{ij}, \qquad f_{ji} = f_j^t g_{ti}.$$

Thus the rank r of f must be even.

Take a vector u^h in the distribution L, then the vector $f_i^h u^i$ is also in L and orthogonal to u^h, and moreover has the same length as u^h with respect to g_{ji}. Consequently, we can choose in L r = 2m mutually orthogonal unit vectors u_b^h such that

$$u^h_{m+1} = f^h_i u^i_1,\ u^h_{m+2} = f^h_i u^i_2, \ldots\ldots,\ u^h_{2m} = f^h_i u^i_m.$$

Then with respect to the orthogonal frame (u^h_b, u^h_B), the tensors g_{ji} and f_{ji} have components:

$$(1.9) \qquad g = \begin{pmatrix} I_m & 0 & 0 \\ 0 & I_m & 0 \\ 0 & 0 & I_{n-2m} \end{pmatrix}, \quad f = \begin{pmatrix} 0 & I_m & 0 \\ -I_m & 0 & 0 \\ 0 & 0 & 0 \end{pmatrix},$$

I_m denoting (m,m)-identity matrix. We call such a frame an *adapted frame* of the structure f.

Now take another adapted frame $(\bar{u}^h_b, \bar{u}^h_B)$ with respect to which g_{ji} and f_{ji} have the same components as (1.9) and put

$$\bar{u}^h_b = r^a_b u^h_a, \qquad \bar{u}^h_B = r^A_B u^h_A,$$

then we can easily see that the orthogonal matrix

$$r = \begin{pmatrix} A_m & B_m & 0 \\ C_m & D_m & 0 \\ 0 & 0 & O_{n-2m} \end{pmatrix}$$

must have the form

$$r = \begin{pmatrix} A_m & B_m & 0 \\ -B_m & A_m & 0 \\ 0 & 0 & O_{n-2m} \end{pmatrix}.$$

Thus the group of the tangent bundle of M can be reduced to $U(m) \times O(n-2m)$.

Conversely, if the group of the tangent bundle of M can be reduced to $U(m) \times O(n-2m)$, then we can define a positive definite Riemannian metric g and a tensor f of type (1,1) and of rank 2m as tensors having (1.9) as components with respect to the adapted frames. Then we have $f^3 + f = 0$. Thus we have (Yano [4])

THEOREM 1.2. A necessary and sufficient condition for an n-dimensional manifold M admit an f-structure f of rank r is that r is even (r = 2m) and the group of tangent bundle of M be reduced to the group $U(m) \times O(n-2m)$.

In the next place we consider the integrability conditions of the distributions L and N.

The Nijenhuis tensor N_f of f is given by

$$N_f(X,Y) = [fX,fY] - f[fX,Y] - f[X,fY] - l[X,Y].$$

Then we have the following identities:

$$N_f(mX,mY) = lN_f(mX,mY) = -l[mX,mY],$$

$$mN_f(X,Y) = m[fX,fY], \qquad mN_f(lX,lY) = m[fX,fY],$$

$$mN_f(fX,fY) = m[lX,lY].$$

Since $l = -f^2$, $lf = f$, if we have $N_f(lX,lY) = 0$ for all vector fields X and Y, then $N_f(fX,fY) = 0$ and conversely.

We also see that the following three conditions are equivalent:

$$\text{(i) } mN_f(X,Y) = 0, \text{ (ii) } mN_f(lX,lY) = 0, \text{ (iii) } mN_f(fX,fY) = 0$$

for any vector fields X and Y. The Lie derivative $L_Y f$ is, by definition, given by

$$(L_Y f)X = f[X,Y] - [fX,Y].$$

We also have

$$N_f(lX,mY) = f(L_{mY}f)lX = f[l(L_{mY}f)lX],$$

from which

$$fN_f(lX,mY) = -l(L_{mY}f)lX.$$

The distribution N is integrable if and only if $l[mX,mY] = 0$ for any vector fields X and Y. Thus we have

PROPOSITION 1.1. A necessary and sufficient condition for N to be integrable is that

$$N_f(mX,mY) = 0 \quad \text{or equivalently} \quad lN_f(mX,mY) = 0$$

for any vector fields X and Y.

The distribution L is integrable if and only if $m[lX,lY] = 0$ for any vector fields X and Y. Thus we have

PROPOSITION 1.2. A necessary and sufficient condition for the distribution L to be integrable is that one of the conditions (i), (ii) or (iii) be satisfied.

Because of $l + m = I$, $N_f(X,Y)$ can be written in the form

$$N_f(X,Y) = lN_f(lX,lY) + mN_f(lX,lY) + N_f(lX,mY)$$
$$+ N_f(mX,lY) + N_f(mX,mY).$$

Thus we have

PROPOSITION 1.3. A necessary and sufficient condition for both of two distributions L and N to be integrable is that

$$N_f(X,Y) = lN_f(lX,lY) + N_f(lX,mY) + N_f(mX,lY)$$

for any vector fields X and Y.

Suppose that the distribution L is integrable and take an arbitrary vector field X' in an integral manifold of L. We define an operator f' by $f'X' = fX'$. Then f' leaves invariant tangent spaces of every integral manifolds of L, and f' is an almost complex structure on each integral manifold of L. For any vector fields X' and Y' tangent to integral manifold of L, we denote by $N'(X',Y')$ the vector-valued two form corresponding to the Nijenhuis tensor of the almost complex structure induced on each integral manifold of L from the structure f. Then we have

$$N_f(lX,lY) = N'(lX,lY)$$

for any vector fields X and Y on M.

If the distribution L is integrable and moreover if the almost complex structure f' induced from f on each integral manifold of L is integrable, then we say that the f-structure is *partially integrable*. We have the following theorem.

THEOREM 1.3. A necessary and sufficient condition for an f-structure to be partially integrable is that one of the following equivalent conditions be satisfied:

$$N_f(lX,lY) = 0, \quad \text{or} \quad N_f(fX,fY) = 0$$

for any vector fields X and Y on M.

Next, from Propositions 1.1, 1.3 and Theorem 1.3, we have

PROPOSITION 1.4. In order that the distribution N is integrable and the f-structure be partially integrable, it is necessary and sufficient that

$$N_f(X,Y) = N_f(lX,mY) + N_f(mX,lY)$$

for any vector fields X and Y.

PROPOSITION 1.5. The tensor field $l(L_{mY}f)l$ vanishes identically for any vector field Y if and only if

$$N_f(lX,mY) = 0$$

for any vector fields X and Y.

When two distributions L and N are both integrable, we can choose a local coordinate system in such a way that L's are represented by putting n-r local coordinates constnat and N's by putting the other r coordinates constant. Then the projection operators l and m can be supposed to have the components

$$l = \begin{pmatrix} I_r & 0 \\ 0 & 0 \end{pmatrix}, \qquad m = \begin{pmatrix} 0 & 0 \\ 0 & I_{n-r} \end{pmatrix}$$

respectively. Since f satisfies $fl = lf = l$ and $fm = mf = 0$, f has the components

$$f = \begin{pmatrix} f_r & 0 \\ 0 & 0 \end{pmatrix},$$

where f_r is an (r,r)-matrix. Thus, for a vector field mY in M, the Lie derivative $L_{mY}f$ has components

$$L_{mY}f = \begin{pmatrix} * & 0 \\ & \\ 0 & 0 \end{pmatrix}.$$

Hence, if we assume that the tensor field $l(L_{mY}f)l$ vanishes identically for any vector field Y, then we have $L_{mY}f = 0$, which means that the components of f are independent of the coordinates which are constant along the integral manifold of L in an adapted coordinate system. Conversely, if the components of f are independent of these coordinates, then we easily see that $l(L_{mY}f)l$ vanishes. Thus we have

PROPOSITION 1.6. Suppose that the two distributions L and N are both integrable and that an adapted coordinate system has been chosen. A necessary and sufficient condition for the local components of the f-structure to be functions independent of the coordinates which are constant along the integral manifold of L is that

$$N_f(lX,mY) = 0$$

for any vector fields X and Y.

From Propositions 1.3 and 1.6 we have

PROPOSITION 1.7. Suppose that L and N are both integrable and that an adapted coordinate system has been choosen. The components of f are independent of the coordinates which are constant along the integral manifold of L if and only if

$$N_f(X,Y) = lN_f(lX,lY)$$

for any vector fields X and Y.

We now assume the following:

(a) f is partially integrable, that is,

$$N_f(X,Y) = N_f(lX,mY) + N_f(mX,lY);$$

(b) The distribution N is integrable, that is,

$$N_f(mX, mY) = 0;$$

(c) The components of f are independent of the coordinates which are constant along the integral manifolds of L in an adapted coordinate system.

In this case we say that the structure f is *integrable*. Combining Propositions 1.1, 1.7 and Theorem 1.3 we have (Ishihara-Yano [1])

THEOREM 1.4. A necessary and sufficient condition for the structure f to be integrable is that

$$N_f(X,Y) = 0$$

for any vector fields X and Y on M.

When the structure f is integrable, the components of f have the form $\begin{pmatrix} f_r & 0 \\ 0 & 0 \end{pmatrix}$ in an adapted coordinate system and f_r is an (r,r)-matrix whose elements are functions independent of the coordinates which are constant along the integral manifolds of L. Since f_r defines a complex structure on an integral manifold of L, we can effect a change of adapted coordinate system in such a way that f_r becomes

$$f_r = \begin{pmatrix} 0 & -I_m \\ I_m & 0 \end{pmatrix},$$

where r = 2m. The converse being evident, we have (Ishihara-Yano [1])

THEOREM 1.5. A necessary and sufficient condition for an f-structure to be integrable is that there exist a coordinate system in which f has the constant components

$$f = \begin{pmatrix} 0 & -I_m & 0 \\ I_m & 0 & 0 \\ 0 & 0 & 0 \end{pmatrix},$$

r = 2m being rank of f.

2. NORMAL f-STRUCTURE

Let M be an n-dimensional manifold with an f-structure f. We denote by L and N the distributions corresponding to the projection operators l and m respectively. We put r = rank f. Then L is r-dimensional and N (n-r)-dimensional.

Let U be an arbitrary coordinate neighborhood of M. If we take in M arbitrary an adapted set $\{f_x\}$ of n-r vector fields f_x spanning the distribution N at each point, then there exists uniquely in U an ordered set $\{f^y\}$ of n-r 1-forms f^y such that

$$\sum_{x=n+1}^{2n-r} f^x \otimes f_x = m, \qquad f^x(f_y) = \delta_y^x,$$

where the indices x,y,z,... run over the range {n+1,n+2,...,2n-r}. We then have

$$f^y(fX) = 0, \quad ff_x = 0$$

for any vector X at each point of M. We call such an ordered set $\{f_x\}$ an (n-r)-frame and the ordered set $\{f^y\}$ an (n-r)-coframe being dual to $\{f_x\}$.

If a 1-form ϕ, global or local, satisfies $\phi X = 0$ for any $X \in L$, ϕ is said to be *transversal* to L. It is expressed uniquely by $\phi = \phi_y f^y$. Similarly, any vector field v in N is expressed uniquely by $v = v^x f_x$.

Denoting by f_b^a, f_y^a, f_b^x respectively the components of f, f_y, f^x with respect to local coordinates (η^a) in U, where the indices a,b,c,... run over the range {1,2,...,n}. Then we have

$$f_b^c f_c^a = f_b^y f_y^a - \delta_b^a, \qquad f_y^c f_c^a = 0,$$

$$f_a^c f_c^x = 0, \qquad f_y^c f_c^x = \delta_y^x.$$

The set of all tangent vectors belonging to the distribution N forms a vector bundle p : N(M) ——> M over M, which is a subbundle of the tangent bundle T(M) of M. Let N(M)* be the vector bundle dual to N(M). If we take an element $\tilde{\phi}$ belonging to the fibre N^*_x of N(M)* at $x \in M$, then there exists at x uniquely a 1-form ϕ of M, which is transversal to L, such that $\phi v = \tilde{\phi} v$ for any element v belonging to the fibre N_x of N(M) at x. Conversely, for any ϕ transversal to L at x, there exists uniquely $\tilde{\phi}$ of N^*_x such that $\phi v = \tilde{\phi} v$ for any $v \in N_x$. Thus N(M)* can be identified naturally with the set of all 1-forms transversal to L, and hence the bundle N*(M) can be regarded as a subbundle of the cotangent bundle T(M)* of M. In a coordinate neighborhood U, $\{f_x\}$ is a basis of N_x and $\{f^y\}$ is a basis of N^*_x. Let v be a vector field of N and ϕ be a transversal 1-form to L. Then we have $v = v^x f_x$, $\phi = \phi_y f^y$ in U with functions v^x and ϕ_y defined in U. We call (v^x) and (ϕ_y) the components of v and ϕ respectively with respect to f_x .

Let U and U' be two coordinate neighborhoods of M such that $U \cap U' \neq \emptyset$. If $\{f_x\}$ and $\{f_{x'}\}$ be (n-r)-frames in U and U' respectively, then we have

$$f_{y'} = A^y_{y'} f_y$$

in $U \cap U'$, where the matrix $(A^y_{y'})$ is a function in $U \cap U'$. Taking $v \in N$ and a 1-form ϕ transversal to L, we have

$$v = v^x f_x, \qquad \phi = \phi_y f^y, \qquad v = v^{x'} f_{x'}, \qquad \phi = \phi_{y'} f^{y'}$$

respectively in U and in U', and

$$v^x = A^x_{x'} v^{x'}, \qquad \phi_y = A^{y'}_y \phi_{y'}$$

in $U \cap U'$, where $(A^{x'}_x) = (A^x_{x'})^{-1}$.

Let there be given a connection ω^* in N(M). Then ω^* has $n(n-r)^2$ components $\Gamma_c{}^x{}_y$ with respect to local coordinates (η^a) of M and an (n-r)-frame $\{f_x\}$ in U. Denotong by $\Gamma_{a\ y}^{\ x}$ and $\Gamma_{a'\ y'}^{\ x'}$ the components of the given connection ω^* respectively with respect to $\{\eta^a, f_x\}$ and to $\{\eta^{a'}, f_{x'}\}$, we find in $U \cap U'$

$$\Gamma_{c'\ y'}^{\ x'} = \frac{\partial \eta^c}{\partial \eta^{c'}} A_x^{x'} (\Gamma_{c\ y}^{\ x} A_{y'}^{y} + \partial_c A_{y'}^{x}),$$

where $\partial_c = \partial/\partial\eta^c$. Taking a vector X and a 1-form ρ at a point x of M, we consider an element $T_x(X,\rho)$ of the fibre F_x of the vector bundle $N(M)^* \otimes N(M)$ at x and suppose for $T_x(X,\rho)$ to be bilinear with respect to its arguments X and ρ. The correspondence $T_x : (X,\rho) \longrightarrow T_x(X,\rho)$ is called an F_x-valued tensor of type (1,1) at x. If there is given a correspondence $T : x \longrightarrow T_x$, it is called an $N(M)^* \otimes N(M)$-valued tensor field of type (1,1) and its differentiability is naturally defined. Let v and ϕ be respectively a vector field and a 1-form and T an $N(M)^* \otimes N(M)$-valued tensor field of type (1,1). Denote by $T(V,\phi)$ a cross-section of $N(M)^* \otimes N(M)$ such that its value at x is given by $T_x(v_x,\phi_x)$, where v_x and ϕ_x are respectively the values of v and ϕ at $x \in M$. Then $T(\sigma v,\tau\phi) = \sigma\tau T(v,\phi)$ for any functions σ and τ. $T(X,\rho)$ is locally expressed by

$$T(X,\rho) = \sum_{b,c=1}^{n} X^c \rho_b T_c^b, \qquad T_c^b = \sum_{x,y=n+1}^{2n-r} T_{c\ y}^{b\ x} f^y \otimes f_x$$

with functions $T_{c\ y}^{b\ x}$ defined in U and X^a, ρ_b are respectively components of X and ρ with respect to (η^a).

For any vector field $v = v^x f_x \in N$ any transversal 1-form $\phi = \phi_y f^y$, if we put

$$\nabla_c v^x = \partial_c v^x + \Gamma_{c\ y}^{\ x} v^y, \qquad \nabla_c \phi_y = \partial_c \phi_y - \Gamma_{c\ y}^{\ x} \phi_x,$$

then it is easily verified that $(\nabla_c v^x) f_x$ and $(\nabla_c \phi_y) f^y$ are globally defined covariant vector fields in M which take their values respectively in N(M) and in N(M)*. In this sense, we call

$$\nabla_c v = \nabla_c v^x f_x \qquad \text{and} \qquad \nabla_c \phi = \nabla_c \phi_y f^y,$$

or simply $\nabla_c v^x$ and $\nabla_c \phi_y$, respectively the covariant derivatives of v and ϕ with respect to the connection ω^*.

Let there be given a linear connection ω in M and denote by $\Gamma_c{}^a{}_b$ its components with respect to local coordinates (η^a) in U. If we consider now an $N(M) \otimes N(M)^*$-valued vector field T^a, then we can put

$$T^a = T^a{}_y{}^x f_x \otimes f^y$$

in U with components $T^a{}_y{}^x$ with respect to $\{\eta^a, f_x\}$. On putting in U

$$\nabla_c T^a{}_y{}^x = \partial_c \Gamma^a{}_y{}^x + \Gamma_c{}^a{}_b T^b{}_y{}^x + \Gamma_c{}^x{}_z T^a{}_y{}^z - \Gamma_c{}^z{}_y T^a{}_z{}^x,$$

then the tensor field

$$\nabla_c T^a = (\nabla_c T^a{}_y{}^x) f_x \otimes f^y$$

defined in each neighborhood U determines globally in M a tensor field of type (1,1) which takes values in $N(M) \otimes N(M)^*$. In this sense, we call $\nabla_c T^a$, or $\nabla_c T^a{}_y{}^x$, the covariant derivative of T^a with respect to connections ω and ω^*.

In the same way, we can define the covariant derivatives of tensor field of any mixed type. Summing up, if there are given connections ω and ω^* respectively in M and N(M), we can introduce the covariant differentiation ∇_c operating on tensor fields of any mixed type. In general case, two connections ω and ω^* may be given independently. However, if there is given a linear connection ω in M, then there exists in N(M) a connection ω^* defined by components

$$\Gamma_c{}^x{}_y = (\partial_c f^e_y + \Gamma_c{}^e{}_d f^d_y) f^x_e,$$

where $\Gamma_c{}^a{}_b$ are the components of the given linear connection ω.

By identifying each tangent space of the fibre of N(M) with the fibre itself, the tangent space $T_\sigma(N(M))$ of N(M) at $\sigma \in N(M)$ is expressible as a direct sum by

$$T_\sigma(N(M)) = T_x(M) \oplus F_x = L_x \oplus N_x \oplus F_x,$$

x being the point $p(\sigma)$ of M. There exists naturally an identification $j : N_x \longrightarrow F_x$.

Let there be given a connection ω^* in the vector bundle N(M). Taking a tangent vector X of the base space M at x, we denote by X^* the horizontal lift of X at each point σ of the fibre $p^{-1}(x)$ with respect to the connection ω^*. We define a linear operator J_σ applied to the tangent space $T_\sigma(N(M))$ of the manifold N(M) at a point x by

$$J_\sigma(X^*) = (fX)^*, \quad J_\sigma(Y^*) = j(Y), \quad J_\sigma(Z) = -(j^{-1}(Z))^*,$$

where X, Y and Z belong respectively to L_x, N_x and F_x, x being the point $p(\sigma)$. It is easily verified that the operator J_σ defined in each tangent space $T_\sigma(N(M))$ determine an almost complex structure J in the manifold N(M), i.e., $J^2 = -I$.

We shall now give the tensor representation J_i^h of the almost complex structure J. The fibre of N(M) being (n-r)-dimensional vector space R^{n-r}, the collection $p^{-1}(U) = U \times R^{n-r}$ of local product of N(M) over U's forms an open covering of N(M). In $p^{-1}(U)$, any element v of N(M) such that $p(v) \in U$ is expressed by (η^a, v^x), where (η^a) are coordinates of p(v) and $v = v^x f_x$. Any tangent vector of the bundle space N(M) is expressed by

$$\begin{pmatrix} V^a \\ V^x \end{pmatrix},$$

if the tangent space of R^{n-r} are identified with R^{n-r} itself. That is to say, (η^a, v^x) are local coordinates defined in each $p^{-1}(U)$ of N(M).

Let there be given a linear connection ω^* in N(M) and $\Gamma_c{}^x{}_y$ its components with respect to local coordinates (η^a) and (n-r)-frame $\{f_x\}$ in U of M. Then, in the tangent space of N(M) at any point (η^a, v^x) of $p^{-1}(U) \times R^{n-r}$, the horizontal plane is defined by a linear equation

$$V^x + \Gamma_a{}^x V^a \quad (\Gamma_a{}^x = \Gamma_a{}^x{}_y v^y),$$

and the vertical plane is defined by a linear equation

$$V^a = 0.$$

If, in each tangent space of N(M), we consider a frame consisting of 2n-r vectors $V_{(i)}$ with components $V^h_{(i)}$ such that

$$(V^h_{(b)}) = \begin{pmatrix} V^a_{(b)} \\ V^x_{(b)} \end{pmatrix} = \begin{pmatrix} \delta^a_b \\ -\Gamma_b{}^x \end{pmatrix}, \quad (V^h_{(y)}) = \begin{pmatrix} V^a_{(y)} \\ V^x_{(y)} \end{pmatrix} = \begin{pmatrix} 0 \\ \delta^x_y \end{pmatrix},$$

then $V_{(b)}$ are horizontal and $V_{(y)}$ are vertical, where the indices h,i,j,... run over the range {1,2,...,n,n+1,...,2n-r}. We now define in each tangent space of N(M) a linear operator J by

$$J(V_{(b)}) = \sum_{a=1}^{n} f^a_b V_{(a)} + \sum_{x=n+1}^{2n-r} f^x_b V_{(x)},$$

$$J(V_{(y)}) = - \sum_{a=1}^{n} f^a_y V_{(a)}.$$

We see that J has the components

$$(J^h_i) = \begin{pmatrix} \delta^a_b & 0 \\ -\Gamma_b{}^x & \delta^x_y \end{pmatrix} \begin{pmatrix} f^a_b & -f^a_y \\ f^x_b & 0 \end{pmatrix} \begin{pmatrix} \delta^a_b & 0 \\ -\Gamma_b{}^x & \delta^x_y \end{pmatrix}^{-1},$$

i.e.,

$$(*) \qquad (J^h_i) = \begin{pmatrix} f^a_b - \Gamma_b{}^y f^a_y & -f^a_y \\ f^x_b - f^e_b \Gamma_e{}^x + \Gamma_b{}^z f^e_z \Gamma_e{}^x & f^e_y \Gamma_e{}^x \end{pmatrix}.$$

We can easily verify that

$$\begin{pmatrix} f_b^a & -f_y^a \\ f_b^x & 0 \end{pmatrix}^2 = -\begin{pmatrix} \delta_b^a & 0 \\ 0 & \delta_y^x \end{pmatrix}.$$

Consequently, we obtain $J^2 = -I$. Hence J is an almost complex structure in the bundle space $N(M)$. Summing up, we have

THEOREM 2.1. If a manifold M admits an f-structure f of rank r, then there exist almost complex structures in the bundle space of the $(n-r)$-dimensional vector bundle $N(M)$ over M. Given a connection with components $\Gamma_c{}^x{}_y$ in $N(M)$, then an almost complex structure $J = (J_i^h)$ is determined by (*).

In the next place, we define a tensor field $S_{cb}{}^a$ of type (1,2) by

$$S_{cb}{}^a = N_{cb}{}^a + (\partial_c f_b^z - \partial_b f_c^z) f_z^a - (f_c^u \Gamma_b{}^z{}_u - f_b^u \Gamma_c{}^z{}_u) f_z^a,$$

where $N_{cb}{}^a$ is the Nijenhuis tensor of f, i.e.,

$$N_{cb}{}^a = f_c^e \partial_e f_b^a - f_b^e \partial_e f_c^a - (\partial_c f_b^e - \partial_b f_c^e) f_e^a.$$

We put

$$S_{cb}{}^x = f_c^e(\partial_e f_b^x - \partial_b f_e^x) - f_b^e(\partial_e f_c^x - \partial_c f_e^x) - (f_c^z f_b^e - f_b^z f_c^e)\Gamma_e{}^x{}_z,$$

$$S_c{}^a{}_y = f_y^e \partial_e f_c^a - f_c^e \partial_e f_y^a + f_e^a \partial_c f_y^e + f_c^e f_z^a \Gamma_e{}^z{}_y,$$

$$S_{cy}{}^x = f_y^e(\partial_e f_c^x - \partial_c f_e^x) + f_c^z f_y^e \Gamma_e{}^x{}_z - \Gamma_c{}^x{}_y,$$

$$S^a{}_{xy} = f_x^e \partial_e f_y^a - f_y^e \partial_e f_x^a - (f_x^e \Gamma_e{}^z{}_y - f_y^e \Gamma_e{}^z{}_x) f_z^a.$$

Then we have

$$S_{cb}{}^a + S_{bc}{}^a = 0, \quad S_{cb}{}^x + S_{bc}{}^x = 0, \quad S^a{}_{xy} + S^a{}_{yx} = 0.$$

In term of covariant differentiation ∇_c these are expressed respectively as follows:

$$S_{cb}{}^a = N_{cb}{}^a + (\nabla_c f_b^z - \nabla_b f_c^z) f_z^a,$$

$$S_{cb}{}^x = f_c^e(\nabla_e f_b^x - \nabla_b f_e^x) - f_b^e(\nabla_e f_c^x - \nabla_c f_e^x),$$

$$S_c{}^a{}_y = f_y^e \nabla_e f_e^a - f_c^e \nabla_e f_y^a + f_e^a \nabla_c f_y^e,$$

$$S_{cy}{}^x = f_y^e(\nabla_e f_c^x - \nabla_c f_e^x),$$

$$S^a{}_{xy} = f_x^e \nabla_e f_y^a - f_y^e \nabla_e f_x^a.$$

We denote by $H_{ji}{}^h$ the Nijenhuis tensor of the almost complex structure J of N(M), which is given by

$$H_{ji}{}^h = J_j^l \partial_l J_i^h - J_i^l \partial_l J_j^h - (\partial_j J_i^l - \partial_i J_j^l) J_l^h$$

with respect to local coordinates (η^a, v^x) in $P^{-1}(U) = U \times R^{n-r}$, where we have put $\eta^x = v^x$ and $\partial_i = \partial/\partial\eta^i$. We find

$$\begin{aligned} H_{cb}{}^a = {} & S_{cb}{}^a - (\Gamma_c{}^z S_b{}^a{}_z - \Gamma_b{}^z S_c{}^a{}_z) + \Gamma_c{}^z \Gamma_b{}^u S_{zu}{}^a \\ & - (f_c^e R_{eb}{}^z - f_b^e R_{ec}{}^z) f_z^a + (\Gamma_c{}^z f_z^e R_{eb}{}^u - \Gamma_b{}^z f_z^e R_{ec}{}^u) f_u^a, \end{aligned}$$

$$\begin{aligned} H_{cb}{}^x = {} & S_{cb}{}^x - (\Gamma_c{}^z S_{bz}{}^x - \Gamma_b{}^z S_{cz}{}^x) - S_{cb}{}^e \Gamma_e{}^x + (\Gamma_c{}^z S_b{}^e{}_z \\ & - \Gamma_b{}^z S_c{}^e{}_z) \Gamma_e{}^x - \Gamma_c{}^z \Gamma_b{}^y S_{zy}{}^e \Gamma_e{}^x + (R_{cb}{}^x - f_c^e f_b^d R_{ed}{}^x) \\ & + (\Gamma_c{}^z f_z^e f_b^d - \Gamma_b{}^z f_z^e f_c^d) R_{ed}{}^x - \Gamma_c{}^z f_z^e \Gamma_b{}^u f_u^d R_{ed}{}^x \\ & + (f_c^e R_{eb}{}^z - f_b^e R_{ec}{}^z) f_z^d \Gamma_d{}^x - (\Gamma_c{}^z f_z^e R_{eb}{}^u - \Gamma_b{}^z f_z^e R_{ec}{}^u) f_u^e \Gamma_d{}^x, \end{aligned}$$

$$H_{cy}{}^a = S_c{}^a{}_y + \Gamma_c{}^z S_{zy}{}^a + R_{ce}{}^z f_y^e f_z^a,$$

$$H_{cy}{}^{a} = S_{c}{}^{a}{}_{y} + \Gamma_{c}{}^{z}S_{zy}{}^{a} + R_{ce}{}^{z}f_{y}^{e}f_{z}^{a},$$

$$H_{cy}{}^{x} = S_{cy}{}^{x} - S_{c}{}^{e}{}_{y}\Gamma_{e}{}^{x} + \Gamma_{c}{}^{z}S_{yz}{}^{e}\Gamma_{e}{}^{x} + f_{c}^{e}f_{y}^{d}R_{ed}{}^{x}$$
$$+ \Gamma_{c}{}^{z}f_{z}^{d}f_{y}^{e}R_{ed}{}^{x} + f_{y}^{e}R_{ec}{}^{z}f_{z}^{d}\Gamma_{d}{}^{x},$$

$$H_{zy}{}^{a} = S^{a}{}_{zy},$$

$$H_{zy}{}^{x} = - S^{e}{}_{zy}\Gamma_{e}{}^{x} - f_{z}^{e}f_{y}^{d}R_{cd}{}^{x},$$

where $R_{cb}{}^{x} = R_{cby}{}^{x}v^{y}$ and $R_{cby}{}^{x}$ is the curvature tensor of the linear connection $\Gamma_{c}{}^{x}{}_{y}$, $R_{cby}{}^{x}$ being defined by

$$R_{cby}{}^{x} = \partial_{c}\Gamma_{b}{}^{x}{}_{y} - \partial_{b}\Gamma_{c}{}^{x}{}_{y} + \Gamma_{c}{}^{x}{}_{z}\Gamma_{b}{}^{z}{}_{y} - \Gamma_{b}{}^{x}{}_{z}\Gamma_{c}{}^{z}{}_{y}.$$

Noticing that

$$H_{jl}{}^{h}J_{i}^{l} + H_{ji}{}^{k}J_{k}^{h} = 0, \quad H_{jl}{}^{h}J_{i}^{l} - H_{li}{}^{h}J_{j}^{l} = 0,$$

we obtain

$$S_{cb}{}^{x} = S_{ce}{}^{a}f_{b}^{e}f_{a}^{x} - S_{ey}{}^{x}f_{c}^{e}f_{b}^{y},$$

$$S_{b}{}^{a}{}_{y} = S_{de}{}^{a}f_{b}^{d}f_{y}^{e} - S^{e}{}_{zy}f_{b}^{z}f_{e}^{a},$$

$$S_{cy}{}^{x} = S_{de}{}^{x}f_{c}^{d}f_{y}^{e} + S^{e}{}_{zy}f_{c}^{z}f_{e}^{x},$$

$$S^{a}{}_{zy} = -S_{cb}{}^{a}f_{z}^{c}f_{y}^{b}.$$

These identities show that

$$S_{cb}{}^{a} = 0 \text{ implies } S_{cb}{}^{x} = 0,\ S_{c}{}^{a}{}_{y} = 0,\ S_{cy}{}^{x} = 0,\ S^{a}{}_{zy} = 0$$

(cf. Nakagawa [1], Ishihara-Yano [1]).

When the almost complex structure J is complex in the bundle space N(M), we say that the given f-structure f is *normal* with respect to a connection ω^* given in the vector bundle N(M). f is normal if and only if $H_{ji}{}^{h}$ vanishes identically. If $H_{ji}{}^{h} = 0$, then

$$(2.1) \qquad S_{cb}{}^{a} = 0,\ S_{cb}{}^{x} = 0,\ S_{cy}{}^{a} = 0,\ S_{cy}{}^{x} = 0,\ S^{a}{}_{zy} = 0.$$

Therefore we have $R_{cb}{}^{x} = 0$, and hence

$$(2.2) \qquad R_{cby}{}^{x} = 0.$$

Conversely, if we assume that (2.1) and (2.2), then $H_{ji}{}^{h} = 0$. Therefore, we obtain

PROPOSITION 2.1. A necessary and sufficient condition for an f-structure f to be normal in M with respect to a connection ω^* given in the vector bundle N(M) is that the tensor fields

$$S_{cb}{}^{a},\ S_{cb}{}^{x},\ S_{c}{}^{a}{}_{y},\ S_{cy}{}^{x},\ S^{a}{}_{zy}$$

vanish identically and the connection ω^* is of zero curvature.

If $S_{cb}{}^{a}$ vanishes, then $S_{cb}{}^{x}$, $S_{c}{}^{a}{}_{y}$, $S_{cy}{}^{x}$, $S^{a}{}_{zy}$ are equal identically to zero. Thus we have

THEOREM 2.2. A necessary and sufficient condition for an f-structure f to be normal in M with respect to a connection ω^* given in the vector bundle N(M) is that the tensor field $S_{cb}{}^{a}$ vanishes identically and the connection ω^* is of zero curvature.

If an f-structure in M is normal with respect to a connection given in N(M), then the connection has zero curvature by means of Theorem 2.2. Thus, M assumed to be simply connected, the vector bundle N(M) is trivial, that is, it is a product bundle. Therefore, we have

PROPOSITION 2.2. If an f-structure in M is normal with respect to a connection given in the vector bundle N(M), then the vector bundle N(M) induced from N(M) by the covering projection $\pi : \tilde{M} \longrightarrow M$ is trivial, where $\tilde{M}$ is the universal covering space of M.

If the vector bundle N(M) is trivial, then it is naturally identified with the product space $M \times R^{n-r}$. Then there exists naturally a connection ω_0^* of zero curvature in N(M). In this case, a necessary and sufficient condition for f to be normal with respect to ω_0^* is

$$S_{cb}{}^a = N_{cb}{}^a + (\partial_c f_b^z - \partial_b f_c^z) f_z^a = 0$$

(cf. Nakagawa [1]).

3. FRAMED f-STRUCTURE

Let M be an n-dimensional manifold with an f-structure f of rank r. If there are n-r vector fields E_a spanning the distribution N at each point of M, and if there exist n-r differential form η^a satisfying

$$\eta^a(E_b) = \delta_b^a,$$

where a,b,c,... = 1,...,n-r, and if

$$f^2 = -I + \eta^a \otimes E_a,$$

then M is said to have a *globally framed* f-*structure* or, simply, a *framed structure*. We call such a manifold M a *framed manifold,* and denote it by $M(f,E_a,\eta^a)$. We see that

$$fE_a = 0, \quad \eta^a \cdot f = 0, \quad a = 1,\ldots,n-r.$$

A framed manifold $M(f,E_a,\eta^a)$ is called a *framed metric manifold* if a Riemannian metric g on M is distinguished such that (i) $\eta^a = g(E_a,\cdot)$, a = 1,...,n-r, and (ii) f is skew-symmetric with respect to g. It can be shown that a framed manifold carries a metric with these properties. We put F(X,Y) = g(fX,Y) and call it *fundamental* 2-*form* of the framed manifold.

A framed metric manifold $M(f,\eta^a,g)$ is said to be *covariant constant* if the covariant derivatives with respect to g of its structure tensors are zero.

For a framed manifold M, if we put

$$\tilde{f} = f + \eta^{2i} \otimes E_{2i-1} - \eta^{2i-1} \otimes E_{2i}, \quad i = 1,\dots,[(n-r)/2],$$

an almost contact structure $\tilde{f}$ is defined on M if dim M = n = 2m, and an almost contact structure $(\tilde{f}, E_{2m-r-1}, \eta^{2m-r-1})$ if n = 2m-1. If M is a framed metric manifold, then an almost complex structure $\tilde{f}$ is defined on M with n = 2m in terms of which the metric g is Hermitian. Setting $\tilde{F}(X,Y) = g(\tilde{f}X,Y)$, we obtain

$$\tilde{F} = F + 2\sum_i \eta^{2i} \wedge \eta^{2i-1}.$$

If the fundamental 2-form F and the η^a are closed forms, the almost Hermitian structure on M is almost Kaehlerian. It is Kaehlerian if $\tilde{f}$ has vanishing covariant derivative with respect to g, that is, if the structure tensors f and E_a are covariant constant with respect to the metric g.

THEOREM 3.1. A covariant constant even dimensional framed manifold M carries a Kaehlerian structure.

In the odd dimensional case the framed metric structure $M(f,\eta^a,g)$ gives rise to the almost contact metric structure $M(\tilde{f},\eta^{2m-r-1},g)$. Then

$$g(\tilde{f}X,\tilde{f}Y) = g(X,Y) - \eta^{2m-r-1}(X)\eta^{2m-r-1}(Y).$$

We put

$$\phi = \tilde{f}, \quad E = E_{2m-r-1}, \quad \eta = \eta^{2m-r-1}, \quad \Phi(X,Y) = g(\phi X,Y),$$

then

$$\Phi = F + 2\sum_i \eta^{2i} \wedge \eta^{2i-1}.$$

If the fundamental 2-form Φ and the 1-form η are closed, the almost contact structure on M is almost cosymplectic. It is cosymplectic if and only if the almost contact structure is normal.

THEOREM 3.2. A covariant constant odd dimensional framed manifold M carries a cosymplectic structure.

Proof. Since f and η^a have vanishing covariant derivatives with respect to the Riemannian metric g, so does ϕ. Thus we can easily verify that the torsion of ϕ vanishes. Therefore, $M(\phi,E,\eta)$ is normal. QED.

In the next place we consider the relation of the normality of the framed structure and the integrability (normality) of the induced almost complex (almost contact) structure. First of all, we have

LEMMA 3.1. Let $M(f,E_a,\eta^a)$ be a framed manifold. Then

$$X(\eta^a(Y)) = (L_X\eta^a)(Y) + \eta^a([X,Y]),$$

$$d\eta^a(E_b,X) = (L_{E_b}\eta^a)(X),$$

$$d\eta^a(fX,Y) = (L_{fX}\eta^a)(Y).$$

Since the f-structure is framed, there exists a connection of zero curvature in N(M). Then a framed f-structure f is normal if

$$S = N_f + d\eta^a \otimes E_a$$

vanishes.

LEMMA 3.2. Let $M(f,E_a,\eta^a)$ be a normal framed manifold. Then

$$L_{E_a}\eta^b = 0, \qquad [E_a,E_b] = 0,$$

$$L_{E_a} f = 0, \qquad d\eta^a(fX,Y) + d\eta^a(X,fY) = 0$$

for any vector fields X and Y and $a,b = 1,\ldots,n-r$.

Proof. Since the structure of M is normal, we have

$$N_f(X,Y) + d\eta^a(X,Y)E_a = 0,$$

from which

$$-f[fX,E_b] + f^2[X,E_b] + d\eta^a(X,E_b)E_a = 0,$$

that is,

$$f(L_{E_b} f)X + d\eta^a(X,E_b)E_a = 0.$$

Taking the interior product of the both sides of the equation above by η^c, we obtain $d\eta^c(X,E_b) = 0$ which is equivalent to $L_{E_a}\eta^b = 0$. Moreover, we obtain

$$f^2[E_a,E_b] + d\eta^c(E_a,E_b)E_c = f^2[E_a,E_b] = 0,$$

and hence $f[E_a,E_b] = 0$. Thus, $[E_a,E_b] = \lambda^c_{ab}E_c$ for some function λ^c_{ab} on M. Then we get $\eta^c([E_a,E_b]) = \lambda^c_{ab} = 0$.

We also have $fL_{E_a} f = 0$, so $(L_{E_a} f)X = \mu^b_a(X)E_b$ for some function $\mu^b_a(X)$. Consequently, we get $\eta^b((L_{E_a} f)X) = \mu^b_a(X)$. Thus, since

$$0 = (L_{E_a}(\eta^b\cdot f))X = ((L_{E_a}\eta^b)f)X + \eta^b((L_{E_a} f)X),$$

the $\mu^b_a(X)$ vanish, that is, $L_{E_a} f = 0$ for all a.

Next, we find

$$\eta^a([fX,fY]) + d\eta^a(X,Y) = 0,$$

so that

$$\begin{aligned} 0 &= \eta^a([fX,f^2Y]) + d\eta^a(X,fY) \\ &= -\eta^a([fX,Y]) + fX(\eta^a(Y)) + d\eta^a(X,fY). \end{aligned}$$

On the other hand, we have

$$fX(\eta^a(Y)) - Y(\eta^a(fX)) - \eta^a([fX,Y]) = d\eta^a(fX,Y).$$

This proves the last equation. QED.

THEOREM 3.3. Let $M(f,E_a,\eta^a)$ be a 2m-dimensional normal framed manifold of rank r, $a = 1,\ldots,2m-r$. Then the induced almost complex structure $\tilde{f} = f + \eta^{2i} \otimes E_{2i-1} - \eta^{2i-1} \otimes E_{2i}$ on M is integrable.

Proof. Let $N_{\tilde{f}}$ be the torsion tensor of $\tilde{f}$. By a straightforward computation, we find

$$\begin{aligned} N_{\tilde{f}}(X,Y) &= N_f(X,Y) + d\eta^a(X,Y)E_a + (d\eta^{2i}(fX,Y) + d\eta^{2i}(X,fY))E_{2i} \\ &\quad - (d\eta^{2i-1}(fX,Y) + d\eta^{2i-1}(X,fY))E_{2i} + \eta^{2i}(X)(L_{E_{2i-1}}\tilde{f})Y \\ &\quad - \eta^{2i}(Y)(L_{E_{2i-1}}\tilde{f})X - \eta^{2i-1}(X)(L_{E_{2i}}\tilde{f})Y + \eta^{2i-1}(Y)(L_{E_{2i}}\tilde{f})X \\ &\quad + 2(\eta^{2i} \wedge \eta^{2i-1})(X,Y)[E_{2i},E_{2i-1}]. \end{aligned}$$

Our theorem is now a consequence of Lemma 3.2. QED.

We state the following converse of Theorem 3.3.

THEOREM 3.4. Let $M(f,E_a,\eta^a)$ be an even dimensional framed manifold of rank r, $a = 1,\ldots,n-r$, where induced almost complex structure $\tilde{f} = \eta^{2i} \otimes E_{2i-1} - \eta^{2i-1} \otimes E_{2i}$ is integrable. Then, if (a) the $d\eta^a$ are of bidegree (1,1) with respect to f, i.e., $d\eta^a(fX,Y) + d\eta^a(X,fY) = 0$, (b) the vector fields E_a are holomorphic and (c) $[E_{2i-1},E_{2i}] = 0$, the f-structure is normal.

If $M(f,E_a,\eta^a)$ is an even dimensional normal framed manifold, then the $d\eta^a$ are of bidegree (1,1) with respect to the induced almost complex structure f. We have

$$\begin{aligned} d\eta^a(fX,Y) = d\eta^a(\tilde{f}X,Y) &- \eta^{2i}(X)d\eta^a(E_{2i-1},Y) \\ &+ \eta^{2i-1}(X)d\eta^a(E_{2i},Y). \end{aligned}$$

But, by Lemma 3.2,

$$d\eta^a(E_b,Y) = E_b(\eta^a(Y)) - \eta^a([E_b,Y])$$
$$= E_b(\eta^a(Y)) - \eta^a(L_{E_b} Y)$$
$$= E_b(\eta^a(Y)) - L_{E_b}(\eta^a(Y)) = 0.$$

THEOREM 3.5. Let $M(f,E_a,\eta^a)$ be a (2m-1)-dimensional normal framed manifold of rank r, $a = 1,\dots,2m-r-1$. Then, the induced almost contact structure $\tilde{f} = f + \eta^{2i} \otimes E_{2i-1} - \eta^{2i-1} \otimes E_{2i}$ on M is normal.

Proof. Let $\tilde{S}$ be the torsion tensor of $\tilde{f}$. Then

$$\tilde{S}(X,Y) = [\tilde{f}X,\tilde{f}Y] - \tilde{f}[\tilde{f}X,Y] - \tilde{f}[X,\tilde{f}Y] - [X,Y]$$
$$+ d\eta^{2m-r-1}(X,Y)E_{2m-r-1} + \eta^{2m-r-1}([X,Y])E_{2m-r-1}$$
$$= [fX,fY] - f[fX,Y] - f[X,fY] - [X,Y]$$
$$+ (\eta^a([X,Y]) + d\eta^a(X,Y))E_a - (\eta^{2m-r-1}([X,Y])$$
$$+ d\eta^{2m-r-1}(X,Y))E_{2m-r-1} + (\eta^{2m-r-1}([X,Y]$$
$$+ d\eta^{2m-r-1}(X,Y))E_{2m-r-1}$$
$$= N_f(X,Y) + d\eta^a(X,Y)E_a = 0.$$

Thus, $(\tilde{f},E_{2m-r-1},\eta^{2m-r-1})$ is normal. QED.

We also have the following converse.

THEOREM 3.6. Let $M(f,E_a,\eta^a)$ be an odd dimensional framed manifold of rank r where induced almost contact structure $\tilde{f} = f + \eta^{2i} \otimes E_{2i-1} - \eta^{2i-1} \otimes E_{2i}$ is normal. Then, if (a) the $d\eta^a$ are of bidegree (1,1) with respect to f, (b) $L_{E_a}\tilde{f}$ vanishes and (c) $[E_{2i-1},E_{2i}] = 0$, the f-structure is normal.

4. HYPERSURFACES OF FRAMED MANIFOLDS

Let $\bar{M}(\bar{f},\bar{E}_a,\bar{\eta}^a)$ be a framed manifold of dimension n and rank r, $a = 1,\ldots,n-r$. We consider an (n-1)-dimensional hypersurface M immersed in $\bar{M}$ such that: For each $x \in M$ the vectors $\bar{E}_a$, $a = 1,\ldots,n-r-1$ belong to the tangent space $T_x(M)$ and $\bar{E}_{n-r} \notin T_x(M)$. The vector field $\bar{E} = \bar{E}_{n-r}$ is then an affine normal to M, so we may write

$$\bar{f}X = fX + \theta(X)\bar{E}, \quad \bar{f}\bar{E} = 0,$$

where f and θ are tensor fields on M of type (1,1) and (0,1), respectively. If $\theta = 0$, the submanifold is an invariant hypersurface of $\bar{M}$. On the other hand, if $\theta \neq 0$, it provides a measure of the deviation of M from this property. Such a hypersurface will be called *noninvariant* or a *normal variation* of M. A hypersurface may, of course, be neither invariant or noninvariant. However, in the sequel, unless otherwise specified, M will be a noninvariant hypersurface of the framed manifold $\bar{M}$.

First of all, we have

$$-X + \bar{\eta}^a(X)\bar{E}_a = f^2X + \theta(fX)\bar{E}.$$

Since there are vector fields E on M such that

$$\bar{E}_x = E_x, \quad x = 1,\ldots,n-r-1,$$

we obtain

$$f^2 = -I + \eta^x \otimes E_x, \quad \eta^x = \bar{\eta}^x, \quad x = 1,\ldots,n-r-1,$$

$$c\theta = \eta, \quad \eta = \bar{\eta}^{n-r},$$

$c\theta$ is the 1-form on M defined by $c\theta(X) = \theta(fX)$.

Moreover, we obtain

$$\eta^x(E_y) = \bar{\eta}^x(E_y) = \delta^x_y, \quad \bar{f}\bar{E}_x = fE_x + \theta(E_x)\bar{E},$$

$$fE_x = 0, \qquad \theta(E_x) = 0, \quad x = 1,\dots,n-r-1.$$

THEOREM 4.1. A noninvariant hypersurface of a framed manifold admits a framed structure of the same rank as the ambient manifold. Moreover, it admits a 1-form θ determining an (n-r-1)-dimensional distribution complementary to the distribution determined by the Pfaffian system $\eta^x = 0$, $x = 1,\dots,n-r-1$.

If $\bar{M}$ is integrable, we obtain

THEOREM 4.2. A noninvariant hypersurface of a normal framed manifold $\bar{M}(\bar{f},\bar{E}_a,\bar{\eta}^a)$ is a normal framed manifold of the same rank r carring a 1-form whose differential has bidegree (1,1) with respect to the induced f-structure.

Proof. Given a symmetric affine connection $\bar{\nabla}$ on $\bar{M}$, an affine connection ∇ is defined on M with respect to the affine normal $\bar{E}$ by the Gauss formula

$$\bar{\nabla}_X Y = \nabla_X Y + h(X,Y)\bar{E},$$

where h is the second fundamental tensor of the immersion with respect to $\bar{E}$. We express $S_{\bar{f}}$ in the form

$$S_{\bar{f}}(\bar{X},\bar{Y}) = (\bar{\nabla}_{\bar{f}\bar{X}}\bar{f})\bar{Y} - (\bar{\nabla}_{\bar{f}\bar{Y}}\bar{f})\bar{X} + \bar{f}(\bar{\nabla}_{\bar{Y}}\bar{f})\bar{X} - \bar{f}(\bar{\nabla}_{\bar{X}}\bar{f})\bar{Y}$$
$$+ [(\bar{\nabla}_{\bar{X}}\bar{\eta}^a)(\bar{Y}) - (\bar{\nabla}_{\bar{Y}}\bar{\eta}^a)(\bar{X})]\bar{E}_a$$

for any vector fields $\bar{X}$ and $\bar{Y}$ on $\bar{M}$. For any vector fields X and Y on M, by a straightforward computation, we obtain

$$S_{\bar{f}}(X,Y) = S_f(X,Y) + [d\theta(fX,Y) + d\theta(X,fY)]\bar{E}.$$

From the equation above we have our assertion. QED.

Let $\bar{M}(\bar{f},\bar{\eta}^a,g)$ be an n-dimensional framed metric manifold with rank r. Let $\bar{F}$ be the fundamental 2-form of $\bar{M}$ defined by $\bar{F}(\bar{X},\bar{Y}) = g(\bar{f}\bar{X},\bar{Y})$. Let F be the fundamental 2-form of the induced f-structure on M. We then have

$$\bar{F}(X,Y) = g(\bar{f}X,Y) = g(fX,Y) + \theta(X)\bar{\eta}(Y)$$
$$= F(X,Y) + (\theta \wedge c\theta)(X,Y),$$

that is,

$$\bar{F} = F + \theta \wedge c\theta.$$

Since $\bar{f}$ is not of maximal rank, the tensor field

$$G = g - c\theta \otimes c\theta$$

is not a Riemannian metric. However, if $n = 2m+1$ and $r = 2m$, f is of maximal rank, so G defines a positive definite metric. In this case, it is easily checked that G is Hermitian with respect to f. In fact, if $\bar{F}$ is closed, G is an almost Kaehlerian metric and $F + \theta \wedge c\theta$ is the fundamental 2-form of the almost Kaehlerian manifold M(f,G). If the structure on $\bar{M}$ is integrable, then M is Kaehlerian.

THEOREM 4.3. In addition to the canonical framed metric structure (f,η^x,g) the noninvariant hypersurface $M(f,E_x,\eta^x)$ admits the framed metric structure (f,η^x,g^*), where $g^* = g + \theta \otimes \theta$.

Proof. We have

$$g(fX,Y) + \theta(X)\bar{\eta}(Y) = -g(X,fY) - \theta(Y)\bar{\eta}(X).$$

Hence we have

$$g(fX,Y) + \theta(X)c\theta(Y) = -g(X,fY) - \theta(Y)c\theta(X),$$

that is,

$$(g + \theta \otimes \theta)(fX,Y) = -(g + \theta \otimes \theta)(X,fY).$$

Moreover, we obtain

$$\eta^X(X) = g(X,E_X) = g^*(X,E_X) - \theta(X)\theta(E_X) = g^*(X,E_X).$$

QED.

EXERCISES

A. AUTOMORPHISMS: Let $M(f,E_a,\eta^a)$ and $M'(f',E'_a,\eta'^a)$ be framed manifolds of the same rank. A diffeomorphism μ of M onto M' is called an isomorphism of M onto M' if

$$\mu_*\cdot f = f'\cdot\mu_* \quad \text{and} \quad \mu_* E_a = E'_a.$$

If $M' = M$ and $f' = f$, $E'_a = E_a$, $\eta'^a = \eta^a$, $a = 1,\ldots,n-r$, then μ is said to be an automorphism of M. Then we have (Goldberg-Yano [2])

THEOREM. The group of automorphisms of a compact framed structure is a Lie group.

B. PRODUCT FRAMED MANIFOLDS: Let $M(f,E_a,\eta^a)$, $a = 1,\ldots,n-r$, and $\bar{M}(\bar{f},\bar{E}_x,\bar{\eta}^x)$, $x = \bar{1},\ldots,\bar{n}-\bar{r}$, be framed manifolds. An f-structure may be defined on the product manifold $M \times \bar{M}$ canonically as follows: For any $X \in T_p(M)$ and $\bar{X} \in T_{\bar{p}}(\bar{M})$, we put

$$\tilde{f}(X,\bar{X}) = (fX,\bar{f}\bar{X}).$$

Then

$$\tilde{f}^2 = -(I,\bar{I}) + (\eta^a \otimes E_a,0) + (0,\bar{\eta}^x \otimes \bar{E}_x),$$

where 0 is the zero vector and I, $\bar{I}$ the identity tensors of M, $\bar{M}$ respectively. Clearly, $\tilde{f}^3 + \tilde{f} = 0$ and $\tilde{f}$ has rank $r + \bar{r}$. We put

$$\tilde{E}_a = (E_a,0), \quad \tilde{E}_{n-r+x} = (0,\bar{E}_x), \quad \tilde{\eta}^a = (\eta^a,0), \quad \tilde{\eta}^{n-r+x} = (0,\bar{\eta}^x).$$

Then

$$\tilde{f}\tilde{E}_A = 0, \qquad \tilde{\eta}^A(E_B) = \delta^A_B,$$

where $A,B = 1,\dots,n+\bar{n}-r-\bar{r}$. Thus $M \times \bar{M}$ carries a framed structure (f,E_A,η^A) of rank $r+\bar{r}$. Then we have (Goldberg [3])

THEOREM. The direct product of two normal framed manifolds is a normal framed manifold.

For the product framed manifolds see also Millman [1] and Nakagawa [1].

C. QUASI-SYMPLECTIC MANIFOLDS: An even dimensional framed metric manifold $M(f,E_a,g)$ of rank r is called *quasi-symplectic* if the fundamental 2-form F is closed and parallel along the integral curves of the vector fields E_a. It is symplectic if $\dim M = 2n$ and $r = 2n$. A quasi-symplectic manifold with zero torsion will be called an integral quasi-symplectic manifold. Goldberg [4] proved

THEOREM. The betti number $b_{2q}(M)$ of a compact integral quasi-symplectic manifold M is different from zero for $q = 0,1,\dots,r/2$.

For the quasi-symplectic manifolds see also Goldberg-Yano [3].

D. HYPERSURFACES: Let M be a hypersurface of a framed manifold $\bar{M}(f,E_a,\eta^a)$. We suppose that $\bar{M}$ is covariant constant, that is, $\bar{\nabla}f = 0$ and $\bar{\nabla}\eta^a = 0$ for all a. If for any vector field X on M, $\bar{\nabla}_X E$ is proportional to E, then M is said to be totally flat, where E is the affine normal of M. Then we have (Goldberg [2])

THEOREM. A noninvariant hypersurface of a covariant constant framed manifold is a totally flat covariant constant framed manifold and the connection in the affine normal bundle is trivial. If the hypersurface is invariant, it is also totally geodesic.

CHAPTER VIII

PRODUCT MANIFOLDS

In this chapter we give the fundamental results concerning geometry of product manifolds. We also prove some theorems of submanifolds of product manifolds.

In §1, we discuss locally product manifolds. §2 is devoted to the study of locally decomposable Riemannian manifolds. We give some properties of Riemannian curvature tensors of locally decomposable Riemannian manifolds. Moreover, we define an almost product Riemannian manifold. In §3, we discuss submanifolds of almost product Riemannian manifolds which are invariant with respect to an almost product structure. In the last §4, we consider Kaehlerian product manifolds and its submanifolds.

For the theory of product manifolds we refer to Tachibana [2] and Yano [5].

1. LOCALLY PRODUCT MANIFOLDS

Let us consider an n-dimensional manifold M which is covered by such a system of coordinate neighborhoods (ξ^h) that in any intersection of two coordinate neighborhoods (ξ^h) and $(\xi^{h'})$ we have

$$\xi^{a'} = \xi^{a'}(\xi^a), \quad \xi^{x'} = \xi^{x'}(\xi^x),$$

with

$$|\partial_a \xi^{a'}| \neq 0, \qquad |\partial_x \xi^{x'}| \neq 0,$$

where the indices a,b,c,... run over the range 1,2,...,p and the indices x,y,z,... run over p+1,...,p+q=n. Then we say that the manifold M admits a *locally product structure* defined by the existence of such a system of coordinate neighborhoods called *separating coordinate system*. We call *locally product manifold* a manifold which admits a locally product structure.

Let $v^h = (v^a, v^x)$ are components of a contravariant vector, then $(v^a, 0)$ and $(0, v^x)$ are also components of contravariant vectors. Similarly, if $\omega_i = (\omega_b, \omega_y)$ are components of covariant vectors and also if, for example,

$$T_i^h = \begin{pmatrix} T_b^a & T_y^a \\ \\ T_b^x & T_y^x \end{pmatrix}$$

are components of a tensor, then

$$\begin{pmatrix} T_b^a & 0 \\ \\ 0 & 0 \end{pmatrix}, \quad \begin{pmatrix} 0 & T_y^a \\ \\ 0 & 0 \end{pmatrix}, \quad \dots\dots$$

are all components of tensors of the same type as T_i^h. For $v^h = (v^a, v^x)$ we see that $(v^a, -v^x)$ are also components of a contravariant vector. This process may be represented by $v^h \longrightarrow F_i^h v^i$, where

$$F_i^h = \begin{pmatrix} \delta_b^a & 0 \\ \\ 0 & -\delta_y^x \end{pmatrix}.$$

Then we have

$$F_j^i F_i^h = I_j^h,$$

where I denotes the identity tensor, $I_j^h = \delta_j^h$.

If we put

$$P_i^h = \begin{pmatrix} \delta_b^a & 0 \\ 0 & 0 \end{pmatrix}, \quad Q_i^h = \begin{pmatrix} 0 & 0 \\ 0 & \delta_y^x \end{pmatrix}$$

in a separating coordinate system, then P_i^h is an operator which projects any vector

$$v^h = (v^a, v^x) \quad \text{into} \quad P_i^h v^i = (v^a, 0)$$

and Q_i^h is an operator which projects

$$v^h = (v^a, v^x) \quad \text{into} \quad Q_i^h v^i = (0, v^x).$$

These operators may be expressed in terms of I_i^h and F_i^h as follows

$$P_i^h = \tfrac{1}{2}(I_i^h + F_i^h), \qquad Q_i^h = \tfrac{1}{2}(I_i^h - F_i^h),$$

from which

$$I_i^h = P_i^h + Q_i^h, \qquad F_i^h = P_i^h - Q_i^h.$$

If a tensor T_i^h has components of the type

$$T_i^h = \begin{pmatrix} T_b^a & 0 \\ 0 & T_y^x \end{pmatrix},$$

we say that T_i^h is *pure*, and if it has components of the type

$$T_i^h = \begin{pmatrix} 0 & T_y^a \\ T_b^x & 0 \end{pmatrix},$$

we say that T_i^h is *hybrid*. Similarly if a tensor field T_{ji} has components of the type

$$T_{ji} = \begin{pmatrix} T_{cb} & 0 \\ 0 & T_{zy} \end{pmatrix},$$

we say that T_{ji} is *pure*, and if it has components of the type

$$T_{ji} = \begin{pmatrix} 0 & T_{zb} \\ T_{cy} & 0 \end{pmatrix},$$

we say that T_{ji} is *hybrod*.

We denote by M^p the system of subspaces defined by ξ^x = const. and by M^q the system of subspaces defined by ξ^a = const. Then the manifold M is locally the product $M^p \times M^q$ of two manifolds. Now, suppose that a linear connection $\Gamma_{ji}^h(\xi)$ is given in the manifold M. The linear connection Γ has the following transformation law:

$$\Gamma_{j'i'}^{h'} = \frac{\partial\xi^{h'}}{\partial\xi^h}\left(\frac{\partial\xi^j}{\partial\xi^{j'}}\frac{\partial\xi^i}{\partial\xi^{i'}}\Gamma_{ji}^h + \frac{\partial^2\xi^h}{\partial\xi^{j'}\partial\xi^{i'}}\right).$$

We can see the quantities

$$\Gamma_{cy}^a, \quad \Gamma_{zb}^a, \quad \Gamma_{zy}^a, \quad \Gamma_{cb}^x, \quad \Gamma_{zb}^x, \quad \Gamma_{cy}^x$$

are all components of tensors.

We can prove the following propositions.

PROPOSITION 1.1. An arbitrary contravariant vector tangent to M^p, displaced parallelly along M^p, is still tangent to M^p, if and only if $\Gamma_{cb}^x = 0$.

If this condition is satisfied, we say that M^p is totally geodesic in M.

PROPOSITION 1.2. An arbitrary contravariant vector tangent to M^p, displaced parallelly along M^q, is still tangent to M^p, if and only if $\Gamma_{zb}^{x} = 0$.

PROPOSITION 1.3. An arbitrary contravariant vector tangent to M^p, displaced parallelly in any direction, is still tangent to M^p if and only if $\Gamma_{cb}^{x} = 0$ and $\Gamma_{zb}^{x} = 0$.

PROPOSITION 1.4. An arbitrary contravariant vector tangent to M^q, displaced parallelly along M^p, is still tangent to M^q, if and only if $\Gamma_{cy}^{a} = 0$.

PROPOSITION 1.5. An arbitrary contravariant vector tangent to M^q, displaced parallelly along M^q, is still tangent to M^q, if and only if $\Gamma_{zy}^{a} = 0$.

In this case M^q is totally geodesic in M.

PROPOSITION 1.6. An arbitrary contravariant vector tangent to M^q, displaced parallelly in any direction, is still tangent to M^q, if and only if $\Gamma_{cy}^{a} = 0$ and $\Gamma_{zy}^{a} = 0$.

In this case, M^q is parallel in M.

2. LOCALLY DECOMPOSABLE RIEMANNIAN MANIFOLDS

Suppose that, in an n-dimensional locally product manifold M, there is given a Riemannian metric

$$ds^2 = g_{ji}(\xi)d\xi^j d\xi^i$$

which satisfies

$$F_j^t F_i^s g_{ts} = g_{ji}.$$

Then we see that g_{ji} is pure, and consequently that

$$ds^2 = g_{cb}(\xi)d\xi^c d\xi^b + g_{zy}(\xi)d\xi^z d\xi^y,$$

or equivalently that the manifolds M^p and M^q are orthogonal. We call such a manifold a *locally product Riemannian manifold*. It is easily seen that g^{ih} is also pure. If we put $F_j^t g_{ti} = F_{ji}$, then $F_{ji} = F_{ij}$ and in fact

$$F_{ji} = \begin{pmatrix} g_{cb} & 0 \\ 0 & -g_{zy} \end{pmatrix},$$

the tensor F_{ji} is pure. If we put $P_{ji} = P_j^t g_{ti}$, $Q_{ji} = Q_j^t g_{ti}$, then we have also

$$P_{ji} = \begin{pmatrix} g_{cb} & 0 \\ 0 & 0 \end{pmatrix}, \qquad Q_{ji} = \begin{pmatrix} 0 & 0 \\ 0 & g_{zy} \end{pmatrix}$$

and

$$F_{ji} = P_{ji} - Q_{ji}.$$

Suppose that the matrix has the form

$$ds^2 = g_{cb}(\xi^a)d\xi^c d\xi^b + g_{zy}(\xi^x)d\xi^z d\xi^y,$$

that is, $g_{cb}(\xi)$ are functions of ξ^a only, $g_{cy} = 0$, and $g_{zy}(\xi)$ are functions of ξ^x only, then we call the manifold a *locally decomposable Riemannian manifold.*

For a locally decomposable Riemannian manifold M, we have

$$(2.1) \qquad \Gamma^x_{cb} = 0, \quad \Gamma^x_{zb} = 0, \quad \Gamma^a_{cy} = 0, \quad \Gamma^a_{zy} = 0.$$

Conversely, if (2.1) are satisfied in a locally product Riemannian manifold, we can easily prove that $g_{cb}(\xi)$ are functions of ξ^a only, $g_{cy} = 0$ and $g_{zy}(\xi)$ are functions of ξ^x only. Thus we have

THEOREM 2.1. A necessary and sufficient condition for a locally product Riemannian manifold M to be a locally decomposable Riemannian manifold is that (2.1) are satisfied in a separating coordinate system.

Combining the propositions in the previous section and this theorem, we have

THEOREM 2.2. A necessary and sufficient condition for a locally product Riemannian manifold $M^p \times M^q$ to be a locally decomposable Riemannian manifold is that M^p and M^q are both totally geodesic and are parallel.

On the other hand, we obtain

$$\nabla_j F_b^a = 0, \qquad \nabla_j F_y^a = -2\Gamma^a_{jy},$$
(2.2)
$$\nabla_j F_y^x = 0, \qquad \nabla_j F_b^x = 2\Gamma^x_{jb}.$$

Thus, from Theorem 2.1 and (2.2), we have

THEOREM 2.3. A necessary and sufficient condition for a locally product Riemannian manifold M to be a locally decomposable Riemannian manifold is that

$$\nabla_j F_i^h = 0,$$

or equivalently

$$\nabla_j F_{ih} = 0.$$

Since, in a locally decomposable Riemannian manifold M, the Christoffel symbols Γ^h_{ji} are all zero except Γ^a_{cb} and Γ^x_{zy}, and besides, Γ^a_{cb} are functions of ξ^a only and Γ^x_{zy} are functions of ξ^x only, the curvature tensor

$$R_{kji}{}^h = \partial_k \Gamma^h_{ji} - \partial_j \Gamma^h_{ki} + \Gamma^h_{kt}\Gamma^t_{ji} - \Gamma^h_{jt}\Gamma^t_{ki} \tag{2.3}$$

is pure in all its indices and also the successive covariant derivatives of $R_{kji}{}^{h}$, $\nabla_l R_{kji}{}^{h}$, $\nabla_m \nabla_l R_{kji}{}^{h}$,... are all pure in its indices.

Now suppose that g_{cb} and g_{zy} are both Einstein, then

$$R_{cb} = \lambda g_{cb}, \qquad R_{zy} = \mu g_{zy} \tag{2.4}$$

for certain constants λ and μ provided that $p,q > 2$, where R_{ji} is the Ricci tensor of M. The equations (2.4) may also written in the form

$$R_{ji} = \tfrac{1}{2}(\lambda+\mu)g_{ji} + \tfrac{1}{2}(\lambda-\mu)F_{ji}.$$

Conversely suppose that the Ricci tensor of a locally decomposable Riemannian manifold has the form

$$R_{ji} = ag_{ji} + bF_{ji}.$$

Then we have

$$R_{cb} = (a+b)g_{cb}, \qquad R_{zy} = (a-b)g_{zy}.$$

Thus, if $p,q > 2$, then g_{cb} and g_{zy} are both metrics of Einstein manifolds, and $a+b = \lambda$, $a-b = \mu$ are both constants. Thus we have

THEOREM 2.4. In a locally decomposable Riemannian manifold $M^p \times M^q$ $(p,q > 2)$, a necessary and sufficient condition that the two components are both Einstein is that the Ricci tensor of the manifold has the form

$$R_{ji} = ag_{ji} + bF_{ji},$$

a and b being necessarily constant.

We next suppose that the two components are both of constant curvature. Then we have

$$R_{dcba} = \lambda(g_{da}g_{cb} - g_{ca}g_{db}), \qquad R_{zyxw} = \mu(g_{zw}g_{yx} - g_{yw}g_{zx})$$

for certain constants λ and μ. The equations above may also written in the form

$$R_{kjih} = \tfrac{1}{4}(\lambda+\mu)[(g_{kh}g_{ji} - g_{jh}g_{ki}) + (F_{kh}F_{ji} - F_{jh}F_{ki})]$$
$$+ \tfrac{1}{4}(\lambda-\mu)[(F_{kh}g_{ji} - F_{jh}g_{ki}) + (g_{kh}F_{ji} - g_{jh}F_{ki})].$$

Conversely suppose that the curvature tensor of a locally decomposable Riemannian manifold has the form

$$R_{kjih} = a[(g_{kh}g_{ji} - g_{jh}g_{ki}) + (F_{kh}F_{ji} - F_{jh}F_{ki})]$$
$$+ b[(F_{kh}g_{ji} - F_{jh}g_{ki}) + (g_{kh}F_{ji} - g_{jh}F_{ki})].$$

Then we have

$$R_{dcba} = 2(a+b)(g_{da}g_{cb} - g_{ca}g_{db}),$$

$$R_{zyxw} = 2(a-b)(g_{zw}g_{yx} - g_{yw}g_{zx}).$$

Thus if $p,q > 2$, then g_{cb} and g_{zy} are both metrics of manifolds of constant curvature, and $2(a+b) = \lambda$, $2(a-b) = \mu$ are both constants. Thus we have

THEOREM 2.5. In a locally decomposable Riemannian manifold $M^p \times M^q$ $(p,q > 2)$, a necessary and sufficient condition that two components are both of constant curvature is that the curvature tensor of the manifold has the form

$$R_{kjih} = a[(g_{kh}g_{ji} - g_{jh}g_{ki}) + (F_{kh}F_{ji} - F_{jh}F_{ki})]$$
$$+ b[(F_{kh}g_{ji} - F_{jh}g_{ki}) + (g_{kh}F_{ji} - g_{jh}F_{ki})],$$

a and b being necessarily constant.

In the following we define an almost product manifold. Let M be an n-dimensional manifold with a tensor field F of type (1,1) such that

$$F^2 = I.$$

Then we say that M is an *almost product manifold* with *almost product structure* F. We put

$$P = \tfrac{1}{2}(I + F), \qquad Q = \tfrac{1}{2}(I - F).$$

Then

$$P + Q = I, \qquad P^2 = P, \qquad Q^2 = Q, \qquad PQ = QP = 0,$$

$$F = P - Q.$$

Thus P and Q define two complementary distributions P and Q globally. We easily see that the eigenvalues of F are +1 or −1. An eigenvector corresponding to the eigenvalue +1 is in P and an eigenvector corresponding to −1 is in Q. Thus, if F has eigenvalue +1 of multiplicity p and eigenvalue −1 of multiplicity q, then the dimension of P is p and that of Q is q.

Conversely, if there exist in M two globally complementary distributions P and Q of dimension p and q respectively, where $p + q = n$ and $p,q \geq 1$. Then we can define an almost product structure F on M by $F = P - Q$.

If an almost product manifold M admits a Riemannian metric g such that

$$g(FX,FY) = g(X,Y)$$

for any vector fields X and Y on M, then M is called an *almost product Riemannian manifold*.

3. SUBMANIFOLDS OF PRODUCT MANIFOLDS

Let $\bar{M}$ be an m-dimensional almost product Riemannian manifold with structure tensors (F,g). Let M be an n-dimensional Riemannian manifold isometrically immersed in $\bar{M}$. For any vector field X tangent to M we put

$$FX = fX + hX,$$

where fX is the tangential part of FX and hX the normal part of FX. For any vector field V normal to M we put

$$FV = tV + sV,$$

where tV is the tangential part of FV and sV the normal part of FV. We then have

$$f^2X = X - thX, \qquad hfX + shX = 0,$$

$$s^2V = V - htV, \qquad ftV + tsV = 0.$$

We easily see that

$$g(fX,Y) = g(X,fY), \quad g(fX,fY) = g(X,Y) - g(hX,hY).$$

If $FT_x(M) \subset T_x(M)$ for each $x \in M$, then M is said to be *F-invariant* in $\bar{M}$. Then h vanishes identically, and hence $f^2 = I$ and $g(fX,fY) = g(X,Y)$. Therefore, (f,g) is an almost product Riemannian structure on M. Conversely, if (f,g) is an almost product Riemannian structure on M, then h = 0 and M is F-invariant in $\bar{M}$. Consequently, we obtain (Adati [1])

THEOREM 3.1. Let M be a submanifold of an almost product Riemannian manifold $\bar{M}$. A necessary and sufficient condition for M to be F-invariant is that the induced structure (f,g) of M is an almost product Riemannian structure.

Let $\bar{M}$ be a locally decomposable Riemannian manifold, that is, $\bar{\nabla}_X F = 0$, where $\bar{\nabla}$ denotes the operator of covariant differentiation in $\bar{M}$. We denote by ∇ the operator of covariant differentiation in M with respect to the induced connection on M. If M is F-invariant in $\bar{M}$, then we easily see that $FT_x(M)^\perp \subset T_x(M)^\perp$ for each $x \in M$. Then we have

$$\bar{\nabla}_X FY = F\bar{\nabla}_X Y = F\nabla_X Y + FB(X,Y) = f\nabla_X Y + sB(X,Y),$$

$$\bar{\nabla}_X FY = \bar{\nabla}_X fY = (\nabla_X f)Y + f\nabla_X Y + B(X,fY).$$

Comparing the tangential and normal parts of these equations, we obtain $(\nabla_X f)Y = 0$ and $sB(X,Y) = B(X,fY)$. Thus M is locally decomposable.

THEOREM 3.2. Let M be an F-invariant submanifold of a locally decomposable Riemannian manifold $\bar{M}_1 \times \bar{M}_{-1}$. Then M is a locally decomposable Riemannian manifold $M_1 \times M_{-1}$, where M_1 is a submanifold of $\bar{M}_1$ and M_{-1} is a submanifold of $\bar{M}_{-1}$, M_1 and M_{-1} being both totally geodesic in M.

Proof. We put

$$T_1(x) = \{X \in T_x(M) : fX = X\}, \quad T_{-1}(x) = \{X \in T_x(M) : fX = -X\}.$$

Then the correspondence of $x \in M$ to $T_1(x)$ and that to $T_{-1}(x)$ define two distributions T_1 and T_{-1} in M respectively. Let $Y \in T_1$. Then for any vector field X tangent to M, we have $f\nabla_X Y = \nabla_X fY = \nabla_X Y$ and hence $\nabla_X Y \in T_1$. This shows that the distribution T_1 is parallel. Similarly, we see that T_{-1} is also parallel. Consequently, the integral manifolds of T_1 and T_{-1} are both totally geodesic in M. We denote them by M_1

and M_{-1} respectively. We now show that M_1 is in $\bar{M}_1$. Let $X \in T_1$. Then $\bar{Q}X = \frac{1}{2}(IX - FX) = \frac{1}{2}(X - fX) = 0$. Thus X belongs to the tangent space $T_x(\bar{M}_1)$. Therefore M_1 is a submanifold of $\bar{M}_1$. Similarly we see that M_{-1} is a submanifold of $\bar{M}_{-1}$. QED.

A submanifold M of an almost product Riemannian manifold $\bar{M}$ is said to be F-*anti-invariant* if $FT_x(M) \subset T_x(M)^{\perp}$ for each $x \in M$.

THEOREM 3.3. Let M be a submanifold of a locally decomposable Riemannian manifold $\bar{M}$. If M is anti-invariant with respect to F, then $A_{hX}Y = 0$. Moreover, if 2dim M = dim $\bar{M}$, then M is totally geodesic.

Proof. Let X and Y be vector fields tangent to M. Then

$$\bar{\nabla}_X FY = \bar{\nabla}_X hY = -A_{hY}X + D_X hY,$$

$$\bar{\nabla}_X FY = F\bar{\nabla}_X Y = h\nabla_X Y + tB(X,Y) + sB(X,Y).$$

From these equations we obtain

$$-g(A_{hY}X,Z) = g(tB(X,Y),Z) = g(B(X,Y),hZ).$$

Since B is symmetric, we have $A_{hX}Y = A_{hY}X$. Thus we obtain

$$-g(A_{hY}X,Z) = g(A_{hZ}X,Y) = g(A_{hY}X,Z).$$

Thus we have $A_{hX}Y = 0$.

If dim $\bar{M}$ = 2dim M, the normal space $T_x(M)^{\perp}$ is spanned by $\{hX : X \in T_x(M)\}$. Therefore $A_{hX}Y = 0$ means that M is totally geodesic. QED.

Let S^n be an n-dimensional sphere of radius 1, and consider $S^n \times S^n$ as an ambient manifold $\bar{M}$. We denote by $\bar{P}$ and $\bar{Q}$ the projection operators of the tangent space of $\bar{M}$ to each components S^n respectively. The almost product structure F of $\bar{M}$ satisfies TrF = 0, where TrF is the trace of F. The Riemannian curvature tensor of $\bar{M}$ is given by

$$(3.1) \quad \bar{R}(X,Y)Z = \tfrac{1}{2}[g(Y,Z)X-g(X,Z)Y+g(FY,Z)FX-g(FX,Z)FY]$$

for any vector fields X, Y and Z on $\bar{M}$. We can easily see that $\bar{M}$ is an Einstein manifold.

Let M be a hypersurface of $\bar{M}$. We denote by N the unit normal of M in $\bar{M}$. We can put

$$FX = fX + u(X)N, \qquad FN = U + \lambda N.$$

Then f, u, U and λ define a symmetric linear transformation of the tangent bundle of M, a 1-form, a vector field and a function on M respectively. Moreover, we easily see that

$$g(U,X) = u(X).$$

The Gauss and Weingarten formulas of M are given by

$$\bar{\nabla}_X Y = \nabla_X Y + g(AX,Y)N, \qquad \bar{\nabla}_X N = -AX,$$

where A is the second fundamental tensor of M with respect to N. The Gauss and Codazzi equations of M are given respectively by

$$(3.2) \quad R(X,Y)Z = \tfrac{1}{2}[g(Y,Z)X-g(X,Z)Y+g(fY,Z)fX-g(fX,Z)fY] + g(AY,Z)AX - g(AX,Z)AY,$$

and

$$(3.3) \quad (\nabla_X A)Y - (\nabla_Y A)X = \tfrac{1}{2}[u(X)fY - u(Y)fX].$$

Moreover, we obtain

$$f^2X = X - u(X)U, \quad u(fX) = -\lambda u(X), \quad u(U) = 1 - \lambda^2,$$

$$fU = -\lambda U, \quad \mathrm{Tr} f = -\lambda, \quad X\lambda = -2u(AX), \quad \nabla_X U = -fAX + \lambda AX.$$

We also have

$$(\nabla_X f)Y = g(AX,Y)U + u(Y)AX,$$

$$(\nabla_X u)(Y) = \lambda g(AX,Y) - g(AfX,Y).$$

We now assume that the hypersurface M has constant mean curvature. Then, by a straightforward computation, we have

$$(3.4) \quad \tfrac{1}{2}\Delta|A|^2 = -\lambda \mathrm{Tr}fA^2 + \mathrm{Tr}(fA)^2 + \tfrac{1}{2}(\mathrm{Tr}A)g(AU,U) - (\mathrm{Tr}Af)^2 + \tfrac{1}{2}\lambda(\mathrm{Tr}A)(\mathrm{Tr}fA) + g(AU,AU) - \tfrac{1}{2}(\mathrm{Tr}A)^2 - |A|^2(|A|^2-(n-1)) + (\mathrm{Tr}A)(\mathrm{Tr}A^3) + |\nabla A|^2.$$

On the other hand, we obtain

$$(3.5) \quad \mathrm{div}((\mathrm{Tr}fA)U - fAU) = g(AU,AU) - (\mathrm{Tr}fA)^2 + \lambda(\mathrm{Tr}A)(\mathrm{Tr}fA) - (\mathrm{Tr}A)g(AU,U) + \mathrm{Tr}(fA)^2 - \lambda \mathrm{Tr}fA^2 + (n-1)(1-\lambda^2),$$

$$(3.6) \quad \mathrm{div}((\mathrm{Tr}A)U) = -(\mathrm{Tr}A)(\mathrm{Tr}fA) + \lambda(\mathrm{Tr}A)^2.$$

From these equations we obtain (Ludden-Okumura [1])

THEOREM 3.4. Let M be a hypersurface of $S^n \times S^n$. If the mean curvature vector of M is constant, then

$$\tfrac{1}{2}\Delta|A|^2 - \mathrm{div}((\mathrm{Tr}fA)U - fAU) - \tfrac{1}{2}\mathrm{div}((\mathrm{Tr}A)U) = \frac{3}{2}(\mathrm{Tr}A)g(AU,U) - \tfrac{1}{2}(\lambda-1)(\mathrm{Tr}A)(\mathrm{Tr}fA) - \tfrac{1}{2}(1+\lambda)(\mathrm{Tr}A)^2 - |A|^2(|A|^2-(n-1)) + (\mathrm{Tr}A)(\mathrm{Tr}A^3) - (n-1)(1-\lambda^2) + |\nabla A|^2.$$

In the following we shall give some applications of Theorem 3.4 (Ludden-Okumura [1]).

THEOREM 3.5. A compact minimal hypersurface M of $S^n \times S^n$ $(n > 1)$ satisfying

$$\int_M [|A|^2 - (n-1)]|A|^2 *1 \geq \int_M |\nabla A|^2 *1$$

is an F-invariant hypersurface.

Proof. From the assumption TrA = 0 and hence

$$\int_M [|A|^2(n-1) - |A|^4 - (n-1)(1-\lambda^2) + |\nabla A|^2]*1 = 0.$$

From this we see that $\lambda^2 = 1$. Therefore, M is an F-invariant hypersurface of $S^n \times S^n$. QED.

In view of Theorem 3.2 we can prove

LEMMA 3.1. A complete F-invariant hypersurface M of $S^n \times S^n$ is a Riemannian product manifold $M' \times S^n$, where M' is a hypersurface of S^n.

Since M' is totally geodesic in $M' \times S^n$, the second fundamental form of M' in S^n has the quite similar properties to A. From this and Theorem 5.8 of Chapter II, we have

THEOREM 3.6. The $S^m((m/(n-1))^{1/2}) \times S^{n-m-1}(((n-m-1)/(n-1))^{1/2}) \times S^n$ in $S^n \times S^n$ are the only compact F-invariant minimal hypersurfaces of $S^n \times S^n$ satisfying $|A|^2 = n-1$.

4. SUBMANIFOLDS OF KAEHLERIAN PRODUCT MANIFOLDS

In this section we study submanifolds of Kaehlerian product manifolds (see Yano-Kon [7]).

Let $\bar{M}^m$ be a Kaehlerian manifold of complex dimension m (of real dimension 2m) and $\bar{M}^n$ be a Kaehlerian manifold of complex dimension n (of real dimension 2n). We denote by J_m and J_n almost complex structures of $\bar{M}^m$ and $\bar{M}^n$ respectively. We consider the Kaehlerian product

$\bar{M} = \bar{M}^m \times \bar{M}^n$ and put

$$JX = J_m\bar{P}X + J_n\bar{Q}X$$

for any vector field X on $\bar{M}$, where $\bar{P}$ and $\bar{Q}$ denote the projection operators. Then we see that

$$J_m\bar{P} = \bar{P}J, \quad J_n\bar{Q} = \bar{Q}J, \quad FJ = JF,$$

$$J^2 = -I, \quad g(JX,JY) = g(X,Y), \quad \bar{\nabla}_X J = 0,$$

F being an almost product structure on M. Thus J is a Kaehlerian structure on $\bar{M}$. If $\bar{M}^m$ is of constant holomorphic sectional curvature c_1 and $\bar{M}^n$ is of constant holomorphic sectional curvature c_2, then the Riemannian curvature tensor $\bar{R}$ of $\bar{M}$ is given by

$$
\begin{aligned}
(4.1) \quad \bar{R}(X,Y)Z = {} & \frac{1}{16}(c_1+c_2)[g(Y,Z)X-g(X,Z)Y+g(JY,Z)JX-g(JX,Z)JY \\
& +2g(X,JY)JZ+2g(FY,Z)FX-g(FX,Z)FY+g(FJY,Z)FJX \\
& -g(FJX,Z)FJY+2g(FX,JY)FJZ] \\
& +\frac{1}{16}(c_1-c_2)[g(FY,Z)X-g(FX,Z)Y+g(Y,Z)FX-g(X,Z)FY \\
& +g(FJY,Z)JX-g(FJX,Z)JY+g(JY,Z)FJX-g(JX,Z)FJY \\
& +2g(FX,JY)JZ+2g(X,JY)JFZ]
\end{aligned}
$$

for any vector fields X, Y and Z on $\bar{M}$.

We now consider an F-invariant submanifold M of a Kaehlerian product manifold $\bar{M} = \bar{M}^m \times \bar{M}^n$. Suppose that M is a Kaehlerian submanifold of $\bar{M}$. Since M is F-invariant, M is a Riemannian product manifold $M^p \times M^q$ and M^p is a submanifold of $\bar{M}^m$ and M^q is a submanifold of $\bar{M}^n$. We now show that M^p is a Kaehlerian submanifold of $\bar{M}^m$ and M^q is that of $\bar{M}^n$. Let $X \in T_x(M^p)$. Then

$$JX = J_m\bar{P}X + J_n\bar{Q}X = J_mX \in T_x(\bar{M}^m) \cap T_x(M) = T_x(M^p).$$

Therefore M^p is a Kaehlerian submanifold of $\bar{M}^m$. Similarly, M^q is a Kaehlerian submanifold of $\bar{M}^n$. Thus we have

THEOREM 4.1. Let M be an F-invariant Kaehlerian submanifold of a Kaehlerian product manifold $\bar{M} = \bar{M}^m \times \bar{M}^n$. Then M is a Kaehlerian product manifold $M^p \times M^q$, where M^p is a Kaehlerian submanifold of $\bar{M}^m$ and M^q is a Kaehlerian submanifold of $\bar{M}^n$.

We next assume that M is an F-invariant, anti-invariant submanifold of $\bar{M}$. Then M is the product $M^p \times M^q$. Let $X \in T_x(M^p)$. Then we have

$$JX = J_m\bar{P}X + J_n\bar{Q}X = J_mX \in T_x(M)^\perp.$$

Since $\bar{Q}JX = J_n\bar{Q}X = 0$, we see that $\bar{Q}J_mX = 0$. This means that J_mX is in $T_x(\bar{M}^m)$. Thus M^p is anti-invariant in $\bar{M}^m$. Similarly, M^q is anti-invariant in $\bar{M}^n$.

THEOREM 4.2. Let M be an F-invariant submanifold of a Kaehlerian product manifold $\bar{M} = \bar{M}^m \times \bar{M}^n$. If M is anti-invariant in $\bar{M}$, then M is a Riemannian product manifold $M^p \times M^q$, where M^p is an anti-invariant submanifold of $\bar{M}^m$ and M^q is an anti-invariant submanifold of $\bar{M}^n$.

Let M be a (p+q)-dimensional anti-invariant submanifold of a Kaehlerian product manifold $\bar{M} = \bar{M}^m(c_1) \times \bar{M}^n(c_2)$. Then the Guass and Codazzi equations are respectively given by

$$(4.2)\quad R(X,Y)Z = \frac{1}{16}(c_1+c_2)[g(Y,Z)X-g(X,Z)Y+g(FY,Z)FX-g(FX,Z)FY]$$
$$+ \frac{1}{16}(c_1-c_2)[g(FY,Z)X-g(FX,Z)Y+g(Y,Z)FX-g(X,Z)FY]$$
$$+ A_{B(Y,Z)}X - A_{B(X,Z)}Y,$$

$$(4.3)\quad (\nabla_X B)(Y,Z) = (\nabla_Y B)(X,Z).$$

From (4.2) we see that the Ricci tensor S of M is given by

$$(4.4) \qquad S(X,Y) = \frac{1}{16}(c_1+c_2)[(p+q-2)g(X,Y)+\mathrm{Tr}Fg(FX,Y)]$$
$$+ \frac{1}{16}(c_1-c_2)[(p+q-2)g(FX,Y)+\mathrm{Tr}Fg(X,Y)]$$
$$+ \sum_i [g(B(e_i,e_i),B(X,Y))-g(B(X,e_i),B(Y,e_i))],$$

where $\{e_i\}$ is an orthonormal frame of M and hence the scalar curvature r of M is given by

$$(4.5) \qquad r = \frac{1}{16}(c_1+c_2)[(p+q)(p+q-2) + (\mathrm{Tr}F)^2]$$
$$+ \frac{1}{16}(c_1-c_2)[(p+q-2)\mathrm{Tr}F + (p+q)\mathrm{Tr}F]$$
$$+ \sum_{i,j} [g(B(e_i,e_i),B(e_j,e_j)) - g(B(e_i,e_j),B(e_i,e_j))].$$

From (4.2) we have

PROPOSITION 4.1. Let M be a (p+q)-dimensional F-invariant, anti-invariant submanifold of a Kaehlerian product manifold $\bar{M}^m(c_1) \times \bar{M}^n(c_2)$. If M is totally geodesic, then $M = M^p(\frac{1}{4}c_1) \times M^q(\frac{1}{4}c_2)$, where $M^p(\frac{1}{4}c_1)$ is a real space form of constant curvature $\frac{1}{4}c_1$ and $M^q(\frac{1}{4}c_2)$ is a real space form of constant curvature $\frac{1}{4}c_2$.

If M is minimal, then $\Sigma B(e_i,e_i) = 0$. Then (4.2), (4.4) and (4.5) imply the following

PROPOSITION 4.2. Let M be a real 2n-dimensional F-invariant, anti-invariant minimal submanifold of $\bar{M}^n(c) \times \bar{M}^n(c)$. Then M is totally geodesic if and only if M satisfies one of the following conditions:

(a) M is a Riemannian product manifold of two $M^n(\frac{1}{4}c)$;

(b) $S = \frac{1}{4}(n-1)cg$;

(c) $r = \frac{1}{2}n(n-1)c$.

If, for any vectors X, Y and Z tangent to M, $\bar{R}(X,Y)Z$ is also tangent to M, i.e., $\bar{R}(X,Y)T_x(M) \subset T_x(M)$ for each $x \in M$, then M is said to be *curvature invariant*.

THEOREM 4.3. Let M be a real (p+q)-dimensional F-invariant submanifold of a Kaehlerian product manifold $\bar{M}^m(c) \times \bar{M}^n(c)$ $(c \neq 0)$. If M is curvature invariant, then M is a Riemannian product manifold $M^p \times M^q$ such that

(a) M^p and M^q are invariant in $\bar{M}^m(c)$ and $\bar{M}^n(c)$ respectively;

(b) M^p and M^q are anti-invariant in $\bar{M}^m(c)$ and $\bar{M}^n(c)$ respectively;

(c) M^p is invariant in $\bar{M}^m(c)$ and M^q is anti-invariant in $\bar{M}^n(c)$;

(d) M^p is anti-invariant in $\bar{M}^m(c)$ and M^q is invariant in $\bar{M}^n(c)$.

Proof. Since $c \neq 0$, (4.1) implies that

$$g(JY,Z)JX-g(JX,Z)JY+2g(X,JY)JZ+g(FJY,Z)FJX-g(FJX,Z)FJY$$
$$+2g(FX,JY)FJZ \in T_x(M)$$

for any vector fields X, Y and Z tangent to M. Putting Y = Z in the equation above, we obtain

$$g(X,JY)JY + g(X,JFY)JFY \in T_x(M).$$

On the other hand, M is of the form $M^p \times M^q$. If $X \in T_x(M^p)$, then $FX = X$ and if $X \in T_x(M^q)$, then $FX = -X$. Now let $Y \in T_x(M^p)$. Then $g(X,JY)JY \in T_x(M)$ for each point x of M. Therefore $JY \in T_x(M)$ or $g(X,JY) = 0$. If $JY \in T_x(M)$, then $JY = J_mY \in T_x(M^p)$ and hence M^p is invariant in $\bar{M}^m(c)$. If $g(X,JY) = 0$ for X, $Y \in T_x(M^p)$, then M^p is anti-invariant in $\bar{M}^m(c)$. Similarly we see that M^q is invariant or anti-invariant in $\bar{M}^n(c)$. QED.

In Theorem 4.3, if M is totally geodesic, M is obviously curvature invariant. Thus we have

THEOREM 4.4. Let M be a real r-dimensional complete F-invariant submanifold of $CP^m \times CP^n$. If M is totally geodesic, then M is $CP^p \times CP^q$ (2p+2q=r), or $RP^p \times RP^q$ (p+q=r), or $CP^p \times RP^q$ (2p+q=r) or $RP^p \times CP^q$ (p+2q=r).

We now study invariant submanifold (Kaehlerian submanifold) M of a Kaehlerian product manifold $\bar{M}^m(c) \times \bar{M}^n(c)$ $(c \neq 0)$. We now assume that M is curvature invariant. Then, for any vectors X, Y and Z tangent to M, (4.1) implies

$$(4.6)\quad g(FY,Z)FX-g(FX,Z)FY+g(FJY,Z)FJX -g(FJX,Z)FJY+2g(FX,JY)FJZ \in T_x(M).$$

Putting Y = Z in (4.6), we find

$$(4.7)\quad g(FY,Y)FX-g(FX,Y)FY+3g(FX,JY)FJY \in T_x(M).$$

Replacing Y by JY in (4.7), we see that

$$(4.8)\quad g(FY,Y)FX-g(FX,JY)FJY+3g(FX,Y)FY \in T_x(M).$$

From (4.7) and (4.8) we have

$$(4.9)\quad g(FY,Y)FX+g(FX,Y)FY+g(FX,JY)FJY \in T_x(M).$$

Therefore (4.7) and (4.9) imply

$$(4.10)\quad g(FY,Y)FX + 2g(FX,JY)FJY \in T_x(M),$$

from which

$$(4.11)\quad g(FY,Y)FX + 2g(FX,Y)FY \in T_x(M).$$

Putting X = Y in (4.11), we get

$$g(FY,Y)FY \in T_x(M).$$

Thus we see that $FY \in T_x(M)$ or $g(FY,Y) = 0$. If $g(FY,Y) = 0$, then (4.11) implies that $g(FX,Y)FY \in T_x(M)$ and hence $FY \in T_x(M)$ or $FY \in T_x(M)^\perp$. Consequently we see that for any $Y \in T_x(M)$, $FY \in T_x(M)$ or $FY \in T_x(M)^\perp$. Thus we have $FT_x(M) \subset T_x(M)$ or $FT_x(M) \subset T_x(M)^\perp$. This means that M is F-invariant or F-anti-invariant. Therefore we have

THEOREM 4.5. Let M be a Kaehlerian submanifold of a Kaehlerian product manifold $\bar{M}^m(c) \times \bar{M}^n(c)$ $(c \neq 0)$. If M is curvature invariant, then M is F-invariant or F-anti-invariant.

THEOREM 4.6. Let M be a Kaehlerian submanifold of a Kaehlerian product manifold $\bar{M}^n(c) \times \bar{M}^n(c)$ $(c \neq 0)$. If M is totally geodesic, then M is F-invariant or F-anti-invariant.

EXERCISES

A. DECOMPOSABLE VECTORS: Consider a contravariant vector field $v^i(\xi)$ in a locally product manifold M. When v^a are functions of ξ^b only, and v^x functions of ξ^y only in a separating coordinate system, we call v^h a contravariant decomposable vector field in M. Then we have the following theorems (see Tachibana [1], Yano [5]).

THEOREM 1. A necessary and sufficient condition for a contravariant vector field v^h in a locally product manifold to be decomposable is that the derivative of F_i^h with respect to v^h vanishes.

THEOREM 2. A Killing vector field in a compact locally decomposable Riemannian manifold is decomposable.

THEOREM 3. A conformal Killing vector in a compact locally decomposable Riemannian manifold is a Killing vector and is decomposable.

B. INTEGRABILITY CONDITIONS: Let M be an almost product manifold with almost product structure F. Then the integrability condition of the distributions P and Q are stated as follows:

THEOREM 1. In order that the distribution P (resp. Q) be completely integrable, it is necessary and sufficient that

$$N_{ji}{}^h - N_{ji}{}^k F_k^h = 0 \quad (\text{resp. } N_{ji}{}^h + N_{ji}{}^k F_k^h = 0).$$

Consequently, in order that both of the distributions P and Q be completely integrable, it is necessary and sufficient that

$$N_{ji}{}^h = 0,$$

where $N_{ji}{}^h$ is the Nijenhuis tensor formed with F_i^h, i.e.,

$$N_{ji}{}^h = F_j^l(\partial_l F_i^h - \partial_i F_l^h) - F_i^l(\partial_l F_j^h - \partial_j F_l^h).$$

We also have

THEOREM 2. In order that an almost product manifold be integrable, it is necessary and sufficient that it is possible to introduce a symmetric affine connection with respect to which the structure tensor F is covariantly constant.

(See Fukami [1], Lichnerowicz [1], [2], [3], de Rham [1], Walker [1], Willmore [1], Yano [3].)

C. HYPERSURFACES OF $S^n \times S^n$: Matsuyama [1] proved the following theorems.

THEOREM 1. A compact minimal hypersurface with non-negative sectional curvature of $S^n \times S^n$ $(n > 1)$ satisfying

$$\int_M [|A|^4 - (n-1)|A|^2]*1 \geq 0$$

is an F-invariant hypersurface.

THEOREM 2. $S^{n-1}(r) \times S^n(1)$ and $S^m(s) \times S^{n-m-1}((1-s^2)^{1/2}) \times S^n(1)$ are the only compact F-invariant hypersurfaces of $S^n \times S^n$ with constant mean curvature and non-negative sectional curvature.

D. SUBMANIFOLDS OF $CP^n \times CP^n$: Let M be a Kaehlerian hypersurface of $CP^n \times CP^n$. We denote by v, Jv an orthonormal basis of the normal space of M in $CP^n \times CP^n$, and by H the second fundamental tensor with respect to v. Then the second fundamental form with respect to Jv is given by JH. We put $S = 2\mathrm{Tr}H^2$, that is, the square of the length of the second fundamental form of M. Then we have (Matsuyama [2])

THEOREM 1. A compact Kaehlerian hypersurface of $CP^n \times CP^n$ satisfying

$$\int_M [\frac{2n+1}{2n-1}S^2 - (n+1)S]*1 \geq 4\int_M |\nabla A|^2 *1$$

is an F-invariant hypersurface.

THEOREM 2. $CP^{n-1} \times CP^n$ and $Q^{n-1} \times CP^n$ are the only compact F-invariant Kaehlerian hypersurfaces of $CP^n \times CP^n$ with constant scalar curvature, where Q^{n-1} is the complex quadric.

For anti-invariant submanifolds we have (Yano-Kon [7])

THEOREM 3. Let M be a real (m+n)-dimensional ($m \geq n$) compact F-invariant, anti-invariant minimal submanifold of $CP^n \times CP^n$. If $|A|^2 = n(n+1)/(2m-1)$, then $m = n = 2$ and M is of the form $S^1 \times S^1 \times RP^2$ or $RP^2 \times S^1 \times S^1$.

CHAPTER IX

SUBMERSIONS

In this chapter we study the submersions and its applications to the theory of submanifolds.

In §1, following O'Neill [2] we give the fundamental equations and some examples of submersions. §2 is devoted to the study of almost Hermitian submersions (Watson [1]). In the last §3, we discuss the relation between Sasakian manifolds and Kaehlerian manifolds which is a special case of the submersion with totally geodesic fibres. We also consider the relation between submanifolds of Sasakian manifolds and submanifolds of Kaehlerian manifolds. Then the various notions in submanifolds of Sasakian manifolds which correspond to that of Kaehlerian manifolds will be clear.

1. FUNDAMENTAL EQUATIONS OF SUBMERSIONS

Let M and B be Riemannian manifolds. A surjective mapping $\pi : M \longrightarrow B$ is called a *Riemannian submersion* (O'Neill [2]) if:

(S1) π has maximal rank;

that is, each derivative map π_* of π is onto; hence, for each $x \in B$, $\pi^{-1}(x)$ is a submanifold of M of dimension dim M - dim B. The submanifolds $\pi^{-1}(x)$ are called *fibres*, and a vector field on M is *vertical* if it is always tangent to the fibres, *horizontal* if always orthogonal to fibres;

(S2) π_* preserves lengths of horizontal vectors.

For a submersion $\pi : M \longrightarrow B$, let H and V denote the projections of the tangent spaces of M onto the subspaces of horizontal and vertical vectors, respectively. We denote by ∇ the operator of covariant differentiation of M. Following O'Neill [2] we define a tensor field T of type (1,2) for arbitrary vector fields E and F on M by

$$T_E F = H\nabla_{VE}(VF) + V\nabla_{VE}(HF).$$

We shall make frequent use of the following three properties of T:

(a) T_E is a skew-symmetric operator on the tangent space of M, and it reverses the horizontal and vertical subspaces;

(b) T is vertical, that is, $T_E = T_{VE}$;

(c) For vertical vector fields, T has the symmetric property, that is, for vertical vector fields V and W, $T_V W = T_W V$.

This last fact, well known for second fundamental forms, follows immediately from the integrability of the vertical distribution.

Next we define a tensor field A of type (1,2) on M by

$$A_E F = V\nabla_{HE}(HF) + H\nabla_{HE}(VF).$$

A has the following properties:

(a') A_E is a skew-symmetric operator on the tangent space of M, and it reverses the horizontal and vertical subspaces;

(b') A is horizontal, that is, $A_E = A_{HE}$;

(c') For horizontal vector fields X, Y, A has the alternation property $A_X Y = -A_Y X$.

The last property (c') will be proved in the proof of Lemma 1.2.

A *basic vector field* on M is a horizontal vector field X which is π-related to a vector field X_* on B, that is, $\pi_* X_p = X_{*\pi(p)}$ for all $p \in M$. Every vector field X_* on B has a unique *horizontal lift* X to M, and X is basic. Thus $X \longleftrightarrow X_*$ is a one-to-one correspondence between basic vector fields on M and arbitrary vector fields on M.

In the following we prepare some lemmas which give the basic formulas for submersions.

LEMMA 1.1. Let X and Y be vector fields on M. Then

(1) $G(X,Y) = g(X_*,Y_*)\cdot\pi$, where G is the metric tensor field on M and g the metric tensor field on B;

(2) $\mathcal{H}[X,Y]$ is the basic vector field corresponding to $[X_*,Y_*]$;

(3) $\mathcal{H}\nabla_X Y$ is the basic vector field corresponding to $\nabla^*_{X_*}Y_*$, where ∇^* is the operator of covariant differentiation of B.

Proof. The first assertion (1) follows from (S2), the second from the identity $\pi_*[X,Y] = [X_*,Y_*]$. We shall prove (3). For any basic vector fields X, Y and Z on M we have

$$2G(\nabla_X Y,Z) = XG(Y,Z) + YG(X,Z) - ZG(X,Y)$$

$$- G(X,[Y,Z]) - G(Y,[X,Z]) + G(Z,[X,Y]).$$

But, for example, $XG(Y,Z) = X(g(Y_*,Z_*)\cdot\pi) = X_*g(Y_*,Z_*)\cdot\pi$. Thus we have

$$g(\nabla^*_{X_*}Y_*,Z_*)\cdot\pi = G(\nabla_X Y,Z).$$

Therefore, $\nabla_X Y$ is π-related to $\nabla^*_{X_*}Y_*$. Thus we have (3). QED.

LEMMA 1.2. If X and Y are horizontal vector fields on M, then

$$A_X Y = \tfrac{1}{2}\mathcal{V}[X,Y].$$

Proof. First of all we have

$$\mathcal{V}[X,Y] = \mathcal{V}\nabla_X Y - \mathcal{V}\nabla_Y X = A_X Y - A_Y X.$$

Thus it sufficies to prove the alternation property (c'), or equivalently, to show that $A_X X = 0$. We may assume that X is basic, hence that $0 = VG(X,X) = 2G(\nabla_V X,X)$ for any vertical vector field V. Since V is π-related to the zero vector field, we see that $[V,X] = \nabla_V X-\nabla_X V$ is vertical. Hence

$$G(\nabla_V X, X) = G(\nabla_X V, X) = -G(V, \nabla_X X) = -G(V, A_X X).$$

Since $A_X X$ is vertival, the result follows. QED.

We denote by $\hat{\nabla}$ the Riemannian connection along a fibre with respect to the induced metric. We see that $\hat{\nabla}_V W = V\nabla_V W$ for any vertical vector fields V and W. We easily see the following

LEMMA 1.3. Let X and Y be horizontal vector fields, and V and W vertical vector fields on M. Then

(1) $\nabla_V W = T_V W + \hat{\nabla}_V W$, (2) $\nabla_V X = H\nabla_V X + T_V X$,

(3) $\nabla_X V = A_X V + V\nabla_X V$, (4) $\nabla_X Y = H\nabla_X Y + A_X Y$,

(5) if X is basic, $H\nabla_V X = A_X V$.

LEMMA 1.4. Let X, Y be horizontal vector fields and V, W be vertical vector fields on M. Then

(1) $(\nabla_V A)_W = -A_{T_V W}$, $(\nabla_X A)_W = -A_{A_X W}$,

(2) $(\nabla_X T)_Y = -T_{A_X Y}$, $(\nabla_V T)_Y = -T_{T_V Y}$.

Proof. We will only prove (1). Let E be an arbitrary vector field on M. Then

$$(\nabla_V A)_W E = \nabla_V (A_W E) - A_{\nabla_V W} E - A_W (\nabla_V E).$$

Since A is horizontal, we see that $A_W = 0$. On the other hand, we have

$$A_{\nabla_V W} E = A_{H\nabla_V W} E = A_{T_V W} E.$$

Thus we obtain (1). QED.

Furthermore, we prepare the following lemmas.

LEMMA 1.5. If X is a horizontal vector field and U, V, W are vertical vector fields, then

$$G((\nabla_U A)_X V, W) = G(T_U V, A_X W) - G(T_U W, A_X V).$$

LEMMA 1.6. Let X and Y be horizontal vector fields and V and W be vertical vector fields. Then we have

(1) $G((\nabla_E A)_X Y, V)$ is alternate in X and Y;

(2) $G((\nabla_E T)_V W, X)$ is symmetric in V and W.

LEMMA 1.7. Let X, Y and Z be horizontal vector fields and V be a vertical vector field. Then we have

$$\mathfrak{S} G((\nabla_Z A)_X Y, V) = \mathfrak{S} G(A_X Y, T_V Z),$$

where $\mathfrak{S}$ denotes the cyclic sum of over the horizontal vector fields X, Y and Z.

Proof. Since this is a tensor equation, we can assume that X, Y and Z are basix, and even that all three brackets [X,Y],.. are vertical. Thus, Lemma 1.2 implies that $\frac{1}{2}[X,Y] = A_X Y$. Hence

$$\tfrac{1}{2}G([[X,Y],z],V) = G([A_X Y,Z],V) = G(\nabla_{A_X Y} Z,V) - G(\nabla_Z(A_X Y),V).$$

On the other hand, we have

$$\begin{aligned} G(\nabla_{A_X Y} Z,V) &= G(T_{A_X Y} Z,V) = -G(Z,T_{A_X Y} V) \\ &= -G(Z,T_V(A_X Y)) = G(T_V Z,A_X Y). \end{aligned}$$

Thus, using the Jacobi's identity, we find

$$\mathfrak{S} G(\nabla_Z(A_X Y),V) = \mathfrak{S} G(T_V Z,A_X Y).$$

Thus it remains only to show that $\mathfrak{S}G(\nabla_Z(A_XY),V) = \mathfrak{S}G((\nabla_ZA)_XY,V)$. But

$$G(\nabla_Z(A_XY),V) - G((\nabla_ZA)_XY,V) = G(A_{\nabla_ZX}Y,V) + G(A_X(\nabla_ZY),V).$$

The first term on the right hand side of this equation equals to

$$-G(A_Y(\mathcal{H}\nabla_ZX),V),$$

and since we assume that $[X,Z] = 0$, this becomes $-G(A_Y(\mathcal{H}\nabla_XZ),V)$, from which the projection $\mathcal{H}$ may now be deleted. From this we have our assertion. QED.

For a submersion $\pi : M \longrightarrow B$ we now derive the equations analogous to the Gauss and Codazzi equations of an immersion.

Since we can consider the fibres as submanifolds of M, we have the following theorem. Let R be the Riemannian curvature tensor of M and $\hat{R}$ the Riemannian curvature tensor of the fibre. Then

THEOREM 1.1. If U, V, W, F are vertical vector fields and X is horizontal, then

$$G(R(U,V)W,F) = G(\hat{R}(U,V)W,F) + G(T_UW,T_VF) - G(T_VW,T_UF),$$

$$G(R(U,V)W,X) = G((\nabla_UT)_VW,X) - G((\nabla_VT)_UW,X).$$

Let R* be the Riemannian curvature tensor of B. Since there is no danger of ambiguity, we denote the horizontal lift of R* by R* as well. Explicitely, if X, Y, Z and W are horizontal vectors of M, we set

$$G(R^*(X,Y)Z,W) = g(R^*(\pi_*X,\pi_*Y)\pi_*Z,\pi_*W).$$

THEOREM 1.2. If X, Y, Z, H are horizontal vector fields and V is vertical, then

$$G(R(X,Y)Z,H) = G(R^*(X,Y)Z,H) + 2G(A_XY,A_ZH)$$
$$- G(A_YZ,A_XH) + G(A_XZ,A_YH),$$

$$G(R(X,Y)Z,V) = -G((\nabla_Z A)_X Y,V) - G(A_XY,T_VZ)$$
$$+ G(A_YZ,T_VX) - G(A_XZ,T_VY).$$

Proof. Since the two equations are tensor equation, we can assume that X, Y and Z are basic vector fields whose brakets are vertical. Then $[X,Y] = 2A_XY$. We write the basic vector field $\mathcal{H}\nabla_YZ$ as $\nabla^*{}_YZ$. Then $\nabla_YZ = \nabla^*{}_YZ + A_YZ$. From Lemma 1.3 we obtain

$$\nabla_X\nabla_YZ = \nabla^*{}_X\nabla^*{}_YZ + A_X\nabla^*{}_YZ + A_XA_YZ + \mathcal{V}\nabla_XA_YZ,$$

$$\nabla_{[X,Y]}Z = 2A_ZA_XY + 2T_{A_XY}Z.$$

Therefore we have

$$R(X,Y)Z = \nabla_X\nabla_YZ - \nabla_Y\nabla_XZ - \nabla_{[X,Y]}Z$$
$$= \nabla^*{}_X\nabla^*{}_YZ - \nabla^*{}_Y\nabla^*{}_XZ + A_XA_YZ - A_YA_XZ - 2A_ZA_XY - 2T_{A_XY}Z$$
$$+ \mathcal{V}\nabla_XA_YZ - \mathcal{V}\nabla_YA_XZ + A_X\nabla^*{}_YZ - A_Y\nabla^*{}_XZ$$
$$= R^*(X,Y)Z + A_XA_YZ - A_YA_XZ - 2A_ZA_XY - 2T_{A_XY}Z$$
$$+ \mathcal{V}(\nabla_XA_YZ - \nabla_YA_XZ) + A_X\nabla^*{}_YZ - A_Y\nabla^*{}_XZ,$$

where we have used the fact that $\mathcal{H}[X,Y] = 0$, i.e., $\pi_*[X,Y] = 0$. Taking the inner product in the equation above with H, we have the first equation. Taking the inner product with V, we obtain

$$G(R(X,Y)Z,V) = -2G(T_{A_XY}Z,V) + G(\nabla_XA_YZ,V) - G(\nabla_YA_XZ,V)$$
$$+ G(A_X\nabla_YZ,V) - G(A_Y\nabla_XZ,V).$$

On the other hand, in the proof of Lemma 1.7, we have

$$G(T_{A_X Y}Z,V) = G(T_V Z,A_X Y).$$

Moreover, we find

$$G((\nabla_X A)_Y Z,V) - G((\nabla_Y A)_X Z,V)$$

$$= G(\nabla_X A_Y Z,V) - G(A_Y \nabla_X Z,V) - G(\nabla_Y A_X Z,V) + G(A_X \nabla_Y Z,V),$$

because of [X,Y] is vertical. Thus we have

$$G(R(X,Y)Z,V) = -2g(T_V Z,A_X Y) + G((\nabla_X A)_Y Z,V) - G((\nabla_Y A)_X Z,V).$$

From this and Lemmas 1.6, 1.7 we have the second equation. QED.

By the similar argument as that of the preceding theorem we have the following

THEOREM 1.3. If X and Y are horizontal vector fields, and V and W are vertical, then

$$G(R(X,V)Y,W) = G((\nabla_V A)_X Y,W) - G((\nabla_X T)_V W,Y)$$

$$+ G(A_X V,A_Y W) - G(T_V X,T_W Y).$$

We denote by K, K_* and $\hat{K}$ the sectional curvatures of M, B and the fibre, respectively. Then we have

THEOREM 1.4. Let X and Y be orthonormal horizontal vectors and V and W be orthonormal vertical vectors. Then

$$K(V,W) = \hat{K}(V,W) + G(T_V W,T_V W) - G(T_V V,T_W W),$$

$$K(X,V) = G((\nabla_X T)_V V,X) + G(A_X V,A_X V) - G(T_V X,T_V X),$$

$$K(X,Y) = K_*(\pi_* X,\pi_* Y) - 3G(A_X Y,A_X Y).$$

The first equation of Theorem 1.4 is one formulation of the Gauss equation of the fibre.

Example 1.1. Let C^{n+1} be a complex (n+1)-dimensional number space with the natural almost complex structure J. We consider a (2n+1)-dimensional unit sphere S^{2n+1}. Let N be the outward unit normal of S^{2n+1} in C^{n+1}. Then the integral curves of the tangent vector field JN are great circles in S^{2n+1} that are the fibres of a bundle mapping $\pi : S^{2n+1} \longrightarrow CP^n$, CP^n being the complex projective space (see Example 2.6 of Chapter III). The usual Riemannian structure on CP^n is characterized by the fact that π is a submersion. Since the fibres are totally geodesic in S^{2n+1}, the tensor T vanishes. If X and Y are horizontal vector fields, then

$$A_X Y = G(X,JY)JN, \qquad A_X JN = JX.$$

Thus we have

$$G(A_X Y, JN) = G(X,JY).$$

Example 1.2. Let G be a Lie group furnished with two-sided invariant Riemannian structure. If K is a closed subgroup, then the usual Riemannian structure on G/K is characterized by the fact that the natural mapping $\pi : G \longrightarrow G/K$ is a submersion. The fibres, left cosets of G mod K, are totally geodesic and hence T = 0. Let X and Y be left invariant horizontal vector fields on G, that is, X and Y be in the orthogonal complement of the Lie algebra of K in the Lie algebra of G. By Lemma 1.1, $A_X Y = \frac{1}{2}\mathcal{V}[X,Y]$ is in the Lie algebra of K, and it is known that $K(p) = \frac{1}{4}|[X,Y]|^2$ for plane section p spanned by orthonormal vectors X and Y. Then for plane section $p_* = \pi_*(p)$ tangent to G/K, Theorem 1.4 implies that

$$K_*(p_*) = \tfrac{1}{4}|[X,Y]|^2 + \frac{3}{4}|\mathcal{V}[X,Y]|^2 = \tfrac{1}{4}|\mathcal{H}[X,Y]|^2 + |\mathcal{V}[X,Y]|^2.$$

Example 1.3. Let F(B) be the frame bundle over B with structure group O(n). We identify the elements of the Lie algebra o(n) with skew-symmetric matrices, and use the inner product

$$\langle a,b\rangle = -\mathrm{Trace}(ab) = \sum_{i,j} a_{ij}b_{ij}.$$

Then there exists a natural Riemannian structure on F(B) such that the projection $\pi : F(B) \longrightarrow B$ is a submersion. To define it, let ω be the Riemannian connection form on F(B) taking values in o(n), and let $H = \mathrm{Kernel}\,\omega$ be the Riemannian connection on F(B). If v is a vertical vector and $\omega(h) = 0$, define

$$|v| = |\omega(v)|, \qquad \langle v,h\rangle = 0, \qquad |h| = |\pi_*(h)|.$$

Clearly, π is a submersion with H as its horizontal distribution. By a straightforward computation we see that the fibres are totally geodesic. Next, we compute A. If X and Y are horizontal vector fields on F(B), then $\omega([X,Y]) = -\Omega(X,Y)$, where Ω is the curvature form of B on F(B). Hence, by Lemma 1.2, if x and y are horizontal vectors, $A_x y$ is the vertical vector such that $\omega(A_x y) = -\frac{1}{2}\Omega(x,y)$. If x, y and a vertical vector v are all tangent to F(B) at $f = (f_1,\ldots,f_n)$, then

$$\langle A_x v,y\rangle = -\langle v,A_x y\rangle = -\langle\omega(v),\omega(A_x y)\rangle = \tfrac{1}{2}\langle\omega(v),\Omega(x,y)\rangle$$

$$= \tfrac{1}{2}\sum_{i,j}\omega_{ij}(v)\Omega_{ij}(x,y) = \tfrac{1}{2}\sum_{i,j}\omega_{ij}(v)\langle R^*(f_i,f_j)x_*,y_*\rangle.$$

This implies that $A_x v$ is the horizontal vector at f.

2. ALMOST HERMITIAN SUBMERSIONS

Let M be a complex m-dimensional almost Hermitian manifold with Hermitian metric G and almost complex structure J, and B be a complex n-dimensional almost Hermitian manifold with Hermitian metric g and almost complex structure J'. We suppose that there exists a submersion

$\pi : M \longrightarrow B$ such that π is an almost complex mapping, i.e., $\pi_* J = J'\pi_*$. Then we say that π is an *almost Hermitian submersion*.

We denote the vertical and horizontal distributions in the tangent bundle of M by $V(M)$ and $H(M)$, respectively. Then $T(M) = V(M) \oplus H(M)$. The orthogonal projection mappings are denoted by $V : T(M) \longrightarrow V(M)$ and $H : T(M) \longrightarrow H(M)$ respectively.

PROPOSITION 2.1. Let $\pi : M \longrightarrow B$ be an almost Hermitian submersion. Then the horizontal and vertical distributions are J-invariant, i.e., $JH(M) = H(M)$ and $JV(M) = V(M)$.

Proof. Since π is almost complex, JW is vertical for $W \in V(M)$. Let X be in $H(M)$. Then $G(JX,W) = -G(X,JW) = 0$ and therefore JX is in $H(M)$. QED.

From Proposition 2.1 we see that any fibre of the submersion $\pi : M \longrightarrow B$ is a complex submanifold of M. We notice here that if X is a basic vector field, then JX is the basic vector field associated to $J'X_*$.

We prove the following (Watson [1])

THEOREM 2.1. Let $\pi : M \longrightarrow B$ be an almost Hermitian submersion. If M is quasi-Kaehlerian, nearly Kaehlerian, Kaehlerian, or Hermitian, then B has the same property.

Proof. Let Φ and Φ' be the fundamental 2-forms of M and B respectively. We first claim $\Phi = \pi^*\Phi'$ on basic vector fields. If X and Y are basic vector fields on M, and X_* and Y_* are their associated vector fields on B, then

$$\Phi(X,Y) = G(X,JY) = g(X_*,J'Y_*)\cdot\pi = \pi^*\Phi'(X,Y).$$

Since π^* commutes with d on differential forms, we also see that $d\Phi = \pi^*(d'\Phi')$.

If M is almost Kaehlerian, then $\pi^*d'\Phi' = 0$. Since π^* is a linear isometry, we obtain $d'\Phi' = 0$ and therefore B is almost Kaehlerian.

Suppose that M is nearly Kaehlerian. It is easy to see that the basic vector field associated to $\nabla'_{X_*}J'X_*$ for any vector field X_* on B is $\mathcal{H}\nabla_X JX$ which vanishes on M. Thus B is nearly Kaehlerian. Similarly, we see that if M is quasi-Kaehlerian, then B is quasi-Kaehlerian. Moreover, the basic vector field on M associated to the Nijenhuis tensor $N'(X_*,Y_*)$ on B is $\mathcal{H}N(X,Y)$. Therefore, if M is Hermitian, B is also Hermitian. Furthermore, when M is Kaehlerian, B is Kaehlerian.

QED.

Now we can begin to examin how the almost Hermitian structure on M places restriction on T and A.

LEMMA 2.1. Let $\pi : M \longrightarrow B$ be a quasi-Kaehlerian submersion, V and W vertical vectors, and X and Y horizontal vectors. Then

(a) $T_V JW = T_{JV}W$, (b) $T_{JV}X = -T_V X$,

(c) $A_X JX = 0$, (d) $A_X JY = -A_Y JX$.

Proof. (a) follows from the similar result on the second fundamental tensor of complex submanifolds. To see (b), note that

$$G(T_{JV}X,W) = -G(T_{JV}W,X) = -G(T_V JW,X) = -G(JT_V X,W).$$

Thus we have $T_{JV}X = -JT_V X$. Since M is quasi-Kaehlerian, we obtain

$$\nabla_X JX - \nabla_{JX}X = J\nabla_X X + J\nabla_{JX}JX.$$

Taking the vertical part of this equation, we find

$$A_X JX - A_{JX}X = JA_X X + JA_{JX}JX = 0.$$

Therefore, we obtain $A_X JX = 0$.

Assertion (d) follows from (c) by the standard polarization trick.

QED.

LEMMA 2.2. Let $\pi : M \longrightarrow B$ be a nearly Kaehlerian submersion. Then, for all vertical vectors V and W,

(a) $T_V JW = JT_V W$, (b) $T_{JV} W = JT_V W$,

(c) $T_V JX = JT_V X$ for all horizontal X.

Proof. For (a) and (b) it will be sufficient to show that $T_V JV = JT_V V$. By a straightforward computation we have

$$\begin{aligned} T_V JV &= \nabla_V JV - \hat{\nabla}_V JV \\ &= (\nabla_V J)V + J\nabla_V V - (\hat{\nabla}_V J)V - J\hat{\nabla}_V V \\ &= J(\nabla_V V - \hat{\nabla}_V V) = JT_V V. \end{aligned}$$

Assertion (c) is an easy calculation. QED.

Clearly, we have the following

LEMMA 2.3. Let $\pi : M \longrightarrow B$ be a Kaehlerian submersion. If V is a vertical vector field and E is any vector field, then

$$T_V JE = JT_V E.$$

We prove the following theorem (Watson [1]).

THEOREM 2.2. Let $\pi : M \longrightarrow B$ be an almost Hermitian submersion with M, an almost semi-Kaehlerian manifold. Then B is almost semi-Kaehlerian if and only if the fibres of π are minimal submanifolds of M.

Proof. Let $\{E_1,\ldots,E_{m-n},JE_1,\ldots,JE_{m-n},F_1,\ldots,F_n,JF_1,\ldots,JF_n\}$ be a local J-basis on M whose horizontal vector fields are basic. Then

$$0 = \delta\Phi(X) = \overline{\delta}\Phi(X) - \sum_{i=1}^{n} [\nabla_{F_i} \Phi(F_i,X) + \nabla_{JF_i} \Phi(JF_i,X)],$$

where X is a basic vector field and $\bar{\delta}\Phi(X)$ is given by

$$\bar{\delta}\Phi(X) = -\sum_{j=1}^{m-n} [\nabla_{E_j}\Phi(E_j,X) + \nabla_{JE_j}\Phi(JE_j,X)].$$

On the other hand, we obtain

$$\begin{aligned}
\nabla_{F_i}\Phi(F_i,X) &= -G(\nabla_{F_i}JF_i,X) + G(J\nabla_{F_i}F_i,X) \\
&= -G(H\nabla_{F_i}JF_i,X) + G(HJ\nabla_{F_i}F_i,X) \\
&= -g(\nabla'_{F_{i*}}J'F_{i*},X_*)\cdot\pi + g(J'\nabla'_{F_{i*}}F_{i*},X_*)\cdot\pi \\
&= \nabla'_{F_{i*}}\Phi'(F_{i*},X_*).
\end{aligned}$$

Similarly, we have

$$\nabla_{JF_i}\Phi(JF_i,X) = \nabla'_{J'F_{i*}}\Phi'(J'F_{i*},X_*).$$

Therefore

$$0 = \bar{\delta}\Phi(X) + \delta'\Phi'(X_*).$$

Since we have

$$\begin{aligned}
\bar{\delta}\Phi(X) &= \sum_{j=1}^{m-n} [G((\nabla_{E_j}J)E_j,X) + G((\nabla_{JE_j}J)JE_j,X)] \\
&= \sum_{j=1}^{m-n} G(T_{E_j}JE_j - JT_{E_j}E_j - T_{JE_j}E_j - JT_{E_j}E_j,X) \\
&= -2\sum_{j=1}^{m-n} G(JT_{E_j}E_j,X),
\end{aligned}$$

$\bar{\delta}\Phi(X) = 0$ if and only if a fibre of π is minimal. Therefore we have our assertion. QED.

THEOREM 2.3. Let M be semi-Kaehlerian, and $\pi : M \longrightarrow B$ an almost Hermitian submersion. Then B is semi-Kaehlerian if and only if the fibres of π are minimal.

We next consider the integrability of the horizontal distribution of a Kaehlerian submersion (see Johnson [1], Watson [1]).

THEOREM 2.4. The horizontal distribution of a Kaehlerian submersion is completely integrable.

Proof. Let $\pi : M \longrightarrow B$ be a Kaehlerian submersion, X a basic horizontal vector field on M, Y horizontal and V vertical. Then

$$G(A_{JX}Y,V) = G(A_X JY,V) = -G(JY,\mathcal{H}\nabla_V X)$$

$$= G(Y,\mathcal{H}\nabla_V JX) = G(Y,A_{JX}V) = -G(A_{JX}Y,V).$$

Therefore $A_{JX}Y = 0$, and hence $A_X Y = 0$. From this we see that the distribution is completely integrable. Moreover, the integral manifold is totally geodesic. QED.

In the sequel, we consider the Riemannian curvature tensors of an almost Hermitian submersion.

The *holomorphic bisectional curvature* of an almost Hermitian manifold M is defined for any pair of unit vectors E and F on M by (Goldberg-Kobayashi [1])

$$B(E,F) = G(R(E,JE)JF,F).$$

Then the holomorphic sectional curvature of M is given by $H(E) = B(E,E)$. We have the following theorems.

THEOREM 2.5. Let $\pi : M \longrightarrow B$ be an almost Hermitian submersion. Let X and Y be horizontal unit vectors, and V and W be unit vertical vectors. Then

$$B(V,W) = \hat{B}(V,W) + G(T_V JW,T_{JV}W) - G(T_V W,T_{JV}JW),$$

$$B(X,V) = G((\nabla_V A)_X JX,JV) - G(A_X JV,A_{JX}V) + G(A_X V,A_{JX}JV)$$

$$- G((\nabla_{JV}A)_X JX,V) + G(T_{JV}X,T_V JX) - G(T_V X,T_{JV}JX),$$

$$B(X,Y) = B'(X_*,Y_*) - 2G(A_X JX, A_Y JY) + G(A_{JX}Y, A_X JY) + G(A_Y X, A_{JX} JY).$$

THEOREM 2.6. If $\pi : M \longrightarrow B$ is a quasi-Kaehlerian submersion, then

$$B(V,W) = \hat{B}(V,W) + |T_V JW|^2 + |T_V W|^2,$$

$$B(X,V) = G((\nabla_V A)_X JX, JV) - G(A_X JV, A_{JX} V) + G(A_X V, A_{JX} JV) - G((\nabla_{JV} A)_X JX, V) - 2G(T_V X, T_{JV} JX),$$

$$B(X,Y) = B'(X_*,Y_*) + |A_X JY|^2 + |A_X Y|^2.$$

From this we have

THEOREM 2.7. Let $\pi : M \longrightarrow B$ be a quasi-Kaehlerian submersion. Then we have the following:

(a) $B(V,W) \geq \hat{B}(V,W)$, and equality holds if and only if the fibres of π are totally geodesic;

(b) $B(X,Y) \geq B'(X_*,Y_*)$, and equality holds if and only if the horizontal distribution is completely integrable.

We also have the following

THEOREM 2.8. If $\pi : M \longrightarrow B$ is a nearly Kaehlerian submersion, then

$$B(V,W) = \hat{B}(V,W) + 2|T_V W|^2,$$

$$B(X,V) = G((\nabla_V A)_X JX, JV) - G(A_X JV, A_{JX} V) + G(A_X V, A_{JX} JV) - G((\nabla_{JV} A)_X JX, V) + 2|T_V X|^2.$$

THEOREM 2.9. If $\pi : M \longrightarrow B$ is a Kaehlerian submersion, then

$$B(X,V) = 2|T_V X|^2, \qquad B(X,Y) = B'(X_*,Y_*).$$

From this we obtain

THEOREM 2.10. Let $\pi : M \longrightarrow B$ is an almost Hermitian submersion. If M is of constant holomorphic bisectional curvature b, then B is of constant holomorphic bisectional curvature b.

THEOREM 2.11. If $\pi : M \longrightarrow B$ is an almost Hermitian submersion, then

$$H(V) = \hat{H}(V) + |T_V JV|^2 - G(T_V V, T_{JV} JV),$$

$$H(X) = H'(X_*) - 3|A_X JX|^2.$$

THEOREM 2.12. If $\pi : M \longrightarrow B$ is a quasi-Kaehlerian submersion, then

$$H(V) = \hat{H}(V) + |T_V V|^2 + |T_V JV|^2,$$

$$H(X) = H'(X_*).$$

THEOREM 2.13. If $\pi : M \longrightarrow B$ is a nearly Kaehlerian submersion, then

$$H(V) = \hat{H}(V) + 2|T_V V|^2.$$

3. SUBMERSIONS AND SUBMANIFOLDS

Let $\bar{M}$ be a (2m+1)-dimensional Sasakian manifold with structure tensors (ϕ,ξ,η,G) such that there exists a fibering $\bar{\pi} : \bar{M} \longrightarrow \bar{M}/\xi = \bar{N}$, where $\bar{N}$ denote the set of orbits of ξ and is a real 2m-dimensional Kaehlerian manifold (see §6 of Chapter V). This is a special case of a submersion with totally geodesic fibres. We denote by (J,g) the

Kaehlerian structure of $\bar{N}$. We denote by * the horizontal lift with respect to the connection η. Then we have

$$(JX)^* = \phi X^*, \qquad G(X^*,Y^*) = g(X,Y)$$

for any vector fields X and Y on $\bar{N}$, where we write $g(X,Y)\cdot\pi$ by $g(X,Y)$ to simplify the notation. We denote by $\bar{\nabla}$ (resp. $\bar{\nabla}'$) the operator of covariant differentiation with respect to G (resp. g). Then we obtain (see Example 1.1)

$$(\bar{\nabla}'_X Y)^* = -\phi^2\bar{\nabla}_{X^*}Y^* = \bar{\nabla}_{X^*}Y^* - G(\bar{\nabla}_{X^*}Y^*,\xi)\xi$$
$$= \bar{\nabla}_{X^*}Y^* - G(Y^*,\phi X^*)\xi.$$

Let $\bar{R}$ and $\bar{R}'$ be the Riemannian curvature tensors of $\bar{M}$ and $\bar{N}$ respectively. Then we have

PROPOSITION 3.1. The Riemannian curvature tensors $\bar{R}$ and $\bar{R}'$ satisfy

$$(\bar{R}'(X,Y)Z)^* = \bar{R}(X^*,Y^*)Z^* + G(Z^*,\phi Y^*)\phi X^*$$
$$- G(Z^*,\phi X^*)\phi Y^* - 2G(Y^*,\phi X^*)\phi Z^*$$

for any vector fields X, Y and Z on M.

Proof. First of all we obtain

$$(\bar{\nabla}'_X\bar{\nabla}'_Y Z)^* = \bar{\nabla}_{X^*}(\bar{\nabla}'_Y Z)^* - G((\bar{\nabla}'_Y Z)^*,\phi X^*)\xi$$
$$= \bar{\nabla}_{X^*}\bar{\nabla}_{Y^*}Z^* - G(\bar{\nabla}_{X^*}Z^*,\phi Y^*)\xi - G(Z^*,\phi\bar{\nabla}_{X^*}Y^*)\xi$$
$$+ G(Z^*,\phi Y^*)\phi X^* - G(\bar{\nabla}_{Y^*}Z^*,\phi X^*)\xi.$$

From this we have the similar expression of $(\bar{\nabla}'_Y\bar{\nabla}'_X Z)^*$. Moreover, we have

$$(\bar{\nabla}'_{[X,Y]}Z)^* = \bar{\nabla}_{[X^*,Y^*]}Z^* + 2G(Y^*,\phi X^*)\phi Z^* - G(Z^*,\phi[X^*,Y^*])\xi.$$

From these equations we have the equation of proposition. QED.

Let $\bar{S}$ and $\bar{S}'$ be the Ricci tensors of $\bar{M}$ and $\bar{N}$ respectively. Then Proposition 3.1 implies

$$\bar{S}'(X,Y) = \bar{S}(X^*,Y^*) + 2G(X^*,Y^*). \tag{3.1}$$

Therefore, the scalar curvature $\bar{r}$ of $\bar{M}$ and the scalar curvature $\bar{r}'$ of $\bar{N}$ satisfy

$$\bar{r}' = \bar{r} + 2m. \tag{3.2}$$

Moreover, the sectional curvatures of $\bar{M}$ and $\bar{N}$ determined by orthonormal vectors X and Y on $\bar{N}$ satisfy

$$\bar{K}'(X,Y) = \bar{K}(X^*,Y^*) + 3G(X^*,\phi Y^*)^2. \tag{3.3}$$

Especially, we have

$$\bar{K}'(X,JX) = \bar{K}(X^*,\phi X^*) + 3.$$

Thus we have

THEOREM 3.1. $\bar{M}$ is of constant ϕ-sectional curvature c if and only if $\bar{N}$ is of constant holomorphic sectional curvature (c+3)

Let M be an (n+1)-dimensional submanifold tangent to the structure vector field ξ of $\bar{M}$ and N be an n-dimensional submanifold of $\bar{N}$. Throughout in this section we assume that the diagram

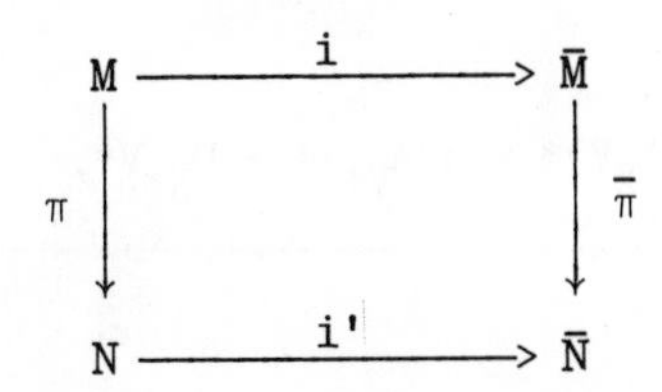

commutes. Let ∇ (resp. ∇') be the operator of covariant differentiation in M (resp. N). We denote by the same G and g the induced metric tensor fields on M and N respectively. We denote by B (resp. B') the second fundamental form of the immersion i (resp. i') and the associated second fundamental tensors of B and B' will be denoted by A and A' respectively. Let X and Y be vector fields tangent to N. Then we have the following Gauss formulas:

$$\bar{\nabla}'_X Y = \nabla'_X Y + B'(X,Y) \quad \text{and} \quad \bar{\nabla}_{X^*} Y^* = \nabla_{X^*} Y^* + B(X^*,Y^*).$$

Therefore we obtain

$$(\nabla'_X Y)^* + (B'X,Y))^* = -\phi^2 \nabla_{X^*} Y^* + B(X^*,Y^*).$$

Comparing the tangential and normal parts of this equation, we have respectively

$$(\nabla'_X Y)^* = -\phi^2 \nabla_{X^*} Y^* = \nabla_{X^*} Y^* - G(Y^*,\phi X^*)\xi, \tag{3.4}$$

$$(B'(X,Y))^* = B(X^*,Y^*). \tag{3.5}$$

Let D and D' be the operators of covariant differentiation with respect to the linear connection induced in the normal bundles of M and N respectively. For any vector field X tangent to N and vector field V normal to N, we obtain

$$\bar{\nabla}'_X V = -A'_V X + D'_X V \quad \text{and} \quad \bar{\nabla}_{X^*} V^* = -A_{V^*} X^* + D_{X^*} V^*.$$

Thus we have

$$-(A'_V X)^* + (D'_X V)^* = \phi^2 A_{V^*} X^* + D_{X^*} V^*,$$

from which

$$(A'_V X)^* = -\phi^2 A_{V^*} X^* = A_{V^*} X^* - \eta(A_{V^*} X^*)\xi, \tag{3.6}$$

(3.7) $$(D'_X V)^* = D_{X^*}V^*.$$

For any vector field X tangent to M we put

$$\phi X = PX + FX,$$

where PX is the tangential part of ϕX and FX the normal part of ϕX. Similarly, for any vector field V normal to M, we put

$$\phi V = tV + fV,$$

where tV is the tangential part of ϕV and fV the normal part of ϕV (see §3 of Chapter IV). We can define the operators P', F', t' and f' on N corresponding respectively to P, F, t and f (see §4 of Chapter VI). Then we have

$$(P'X)^* = PX^*, \qquad (F'X)^* = FX^*$$

for any vector field X tangent to N. Moreover, we obtain

$$(t'V)^* = tV^*, \qquad (f'V)^* = fV^*$$

for any vector field V normal to N. Thus we have

PROPOSITION 3.2. (1) M is a contact CR submanifold of $\bar{M}$ if and only if N is a CR submanifold of $\bar{N}$;

(2) M is a generic submanifold of $\bar{M}$ if and only if N is a generic submanifold of $\bar{N}$;

(3) M is an anti-invariant submanifold of $\bar{M}$ tangent to ξ if and only if N is an anti-invariant submanifold of $\bar{N}$;

(4) M is an invariant submanifold of $\bar{M}$ if and only if N is an invariant submanifold (a complex submanifold) of $\bar{N}$.

We now study the relation between covariant differentiations of the second fundamental forms B and B'. From (3.4), (3.5) and (3.7) we have

$$((\nabla_X B')(Y,Z))^* - G(Y^*,\phi X^*)B(\xi,Z^*) - G(Z^*,\phi X^*)B(Y^*,\xi)$$
$$= (\nabla_{X^*}B)(Y^*,Z^*).$$

On the other hand, we see that

$$\nabla_X \xi = -PX, \qquad B(X,\xi) = -FX, \qquad B(\xi,\xi) = 0.$$

Thus we obtain

$$(3.8) \qquad (\nabla_{X^*}B)(Y^*,Z^*) = [(\nabla_X B')(Y,Z)+g(Y,P'X)F'Z+g(Z,P'X)F'Y]^*.$$

Moreover, from (3.7) of Chapter IV, we obtain

$$(\nabla_{X^*}B)(Y^*,\xi) = (\nabla_{X^*}F)Y^* - B(Y^*,PX^*)$$
$$= fB(X^*,Y^*) - B(X^*,PY^*) - B(Y^*,PX^*),$$

from which

$$(3.9) \qquad (\nabla_{X^*}B)(Y^*,\xi) = [f'B'(X,Y)-B'(X,P'Y)-B'(Y,P'X)]^*.$$

From (3.8) and (3.9) we see that B is parallel if and only if B' satisfies

$$(\nabla_X B')(Y,Z) = g(X,P'Y)F'Z + g(X,P'Z)F'Y$$

and

$$f'B'(X,Y) = B'(X,P'Y) + B'(Y,P'X).$$

This is the proof of Lemma 3.1 of Chapter IV.

In the next place, we consider the normal connections of M and N. We denote by $K^{\perp}$ and $R^{\perp}$ the curvature tensors of the normal bundles of M and N respectively. We give the relation of $K^{\perp}$ and $R^{\perp}$.

Let X and Y be vector fields tangent to N and U and V be vector fields normal to N. Then (3.7) implies

$$(D'_X D'_Y V)^* = D_{X^*} D_{Y^*} V^*, \qquad (D'_Y D'_X V)^* = D_{Y^*} D_{X^*} V^*.$$

Since $[X,Y]^* = [X^*,Y^*] - 2G(Y^*,PX^*)\xi$, we find

$$(D'_{[X,Y]} V)^* = D_{[X^*,Y^*]} V^* - 2G(Y^*,PX^*) D_{\xi} V^*.$$

From these equations we have

$$G(K^{\perp}(X^*,Y^*)V^*,U^*) = [g(R^{\perp}(X,Y)V,U) + 2g(Y,P'X)g(f'V,U)]^*. \tag{3.10}$$

On the other hand, the Ricci equation of M implies

$$\begin{aligned} G(K^{\perp}(X^*,\xi)V^*,U^*) &= G([A_{U^*},A_{V^*}]X^*,\xi) \\ &= G(A_{V^*}X^*,tU^*) - G(A_{U^*}X^*,tV^*) \\ &= -[g(F'A'_V X,U) + g(B'(X,t'V),U)]^*. \end{aligned}$$

From this and (3.9) of Chapter IV we obtain

$$G(K^{\perp}(X^*,\ )V^*,U^*) = g((\nabla_X f')V,U)^*. \tag{3.11}$$

Consequently, the normal connection of M is flat if and only if

$$R^{\perp}(X,Y)V = 2g(X,P'Y)f'V \quad \text{and} \quad (\nabla_X f')V = 0.$$

This gives the proof of Lemma 3.2 of Chapter IV.

PROPOSITION 3.3. Let M and N be invariant submanifolds of $\bar{M}$ and $\bar{N}$ respectively. Then the second fundamental form B of M is η-parallel if and only if the second fundamental form B' of N is parallel.

Proof. From the assumption we see that $F' = 0$. Thus (3.8) shows that $(\nabla_{X^*}B)(Y^*,Z^*) = ((\nabla_X B')(Y,Z))^*$. From this we have our assertion.

QED.

When M is anti-invariant in $\bar{M}$, the normal connection of M is pseudo-flat if and only if $K^{\perp}(X^*,Y^*)V^* = 0$ (see Exercise E of Chapter VI). Thus, from (3.10), we have

PROPOSITION 3.4. If M and N are anti-invariant submanifolds of $\bar{M}$ and $\bar{N}$ respectively, then the normal connection of M is pseudo-flat if and only if the normal connection of N is flat.

When M is a generic submanifold of $\bar{M}$, f vanishes identically, and hence f' on N vanishes identically. Therefore, we have

PROPOSITION 3.5. If M and N are generic submanifolds of $\bar{M}$ and $\bar{N}$ respectively, then the normal connection of M is flat if and only if the normal connection of N is flat.

LEMMA 3.1. Let μ and μ' be the mean curvature vectors of M and N respectively. Then

$$(\mu')^* = \frac{n+1}{n}\,\mu.$$

Proof. If we take an orthonormal basis $\{e_i\}$ for $T_x(N)$. Then $\{e_i^*,\xi\}$ forms an orthonormal basis for $T_y(M)$ ($\pi(y) = x$). Thus (3.5) implies

$$(\mu')^* = \frac{1}{n}(\mathrm{Tr}B')^* = \frac{1}{n}[\Sigma B(e_i^*,e_i^*) + B(\xi,\xi)] = \frac{1}{n}\mathrm{Tr}B = \frac{n+1}{n}\,\mu.$$

In view of Lemma 3.1 we obtain

PROPOSITION 3.6. M is a minimal submanifold of $\bar{M}$ if and only if N is a minimal submanifold of $\bar{N}$.

From (3.7) and Lemma 3.1 we have

$$D_{X*}\mu = \frac{n}{n+1} D_{X*}(\mu')* = \frac{n}{n+1}(D'_X\mu')*.$$

Therefore we have

PROPOSITION 3.7. If the mean curvature vector μ of M is parallel, then the mean curvature vector μ' of N is also parallel.

We consider the converse of this proposition. We prove the following

LEMMA 3.2. Let M be a submanifold tangent to ξ of a Sasakian manifold $\bar{M}$. Then

$$D_\xi\mu = -f\mu.$$

Proof. We first notice that the second fundamental form B satisfies $(\nabla_X B)(\xi,Y) = (\nabla_\xi B)(X,Y)$. Then we have

$$\begin{aligned}(n+1)D_\xi\mu &= \sum(\nabla_\xi B)(e_i,e_i) = \sum(\nabla_{e_i} B)(\xi,e_i)\\ &= \sum[D_{e_i} B(\xi,e_i) - B(\nabla_{e_i}\xi,e_i)]\\ &= -\sum(\nabla_{e_i} F)e_i\\ &= -\sum[-B(e_i,Pe_i) + fB(e_i,e_i)]\\ &= -f\mu.\end{aligned}$$

In the above we have denoted by the same e_i local, orthonormal vector fields on M which extend e_i of the orthonormal basis $\{e_i\}$ and which are covariant constant with respect to ∇ at x of M. QED.

From Lemma 3.2 we have

PROPOSITION 3.8. Let M and N be generic submanifolds of $\bar{M}$ and $\bar{N}$ respectively. Then the mean curvature vector μ of M is parallel if and only if the mean curvature vector μ' of N is parallel.

When M is invariant in $\bar{M}$, the mean curvature vector μ of M is pseudo-parallel if and only if $D_{X^*}\mu = 0$ (see Exercise E of Chapter VI). Thus we have

PROPOSITION 3.9. Let M and N be anti-invariant submanifolds of $\bar{M}$ and $\bar{N}$ respectively. Then the mean curvature vector μ of M is pseudo-parallel if and only if the mean curvature vector μ' of N is parallel.

Let R and R' be the Riemannian curvature tensors of M and N respectively. From Proposition 3.1 we obtain

$$
\begin{aligned}
&(R'(X,Y)Z)^* - (A'_{B'(Y,Z)}X)^* + (A'_{B'(X,Z)}Y)^* \\
&\qquad ((\nabla_X B')(Y,Z))^* - ((\nabla_Y B)(X,Z))^* \\
&= R(X^*,Y^*)Z^* - A_{B(Y^*,Z^*)}X^* + A_{B(X^*,Z^*)}Y^* + (\nabla_{X^*}B)(Y^*,Z^*) \\
&\quad - (\nabla_{Y^*}B)(X^*,Z^*) + G(Z^*,PY^*)\phi X^* - G(Z^*,PX^*)\phi Y^* \\
&\quad - 2G(Y^*,PX^*)\phi Z^*.
\end{aligned}
$$

Taking the tangential part of this equation and using (3.6), we obtain

PROPOSITION 3.10. The Riemannian curvature tensors R and R' satisfy

$$
\begin{aligned}
(R'(X,Y)Z)^* &= R(X^*,Y^*)Z^* + G(Y^*,tB(X^*,Z^*))\xi - G(X^*,tB(Y^*,Z^*))\xi \\
&\quad + G(Z^*,PY^*)PX^* - G(Z^*,PX^*)PY^* - 2G(Y^*,PX^*)PZ^*.
\end{aligned}
$$

We denote by S and S' the Ricci tensors of M and N respectively. Then we have

$$
S'(X,Y) = S(X^*,Y^*) + 2G(PX^*,PY^*). \tag{3.12}
$$

Let $\{e_i\}$ be an orthonormal basis of N. Then, by (3.12), the scalar curvature r of M and the scalar curvature r' of N satisfy

$$(3.13) \qquad r' = r + \sum G(Pe_i^*, Pe_i^*), \qquad r' = r + n - \sum G(Fe_i^*, Fe_i^*).$$

Therefore we obtain the following

PROPOSITION 3.11. (1) M and N are anti-invariant submanifolds if and only if $r = r'$;

(2) M and N are invariant submanifolds if and only if $r = r' - n$.

We now compute the square of the length of the second fundamental forms. We obtain

$$\begin{aligned} G(A,A) &= \sum_{i,j} G(B(e_i^*, e_j^*), B(e_i^*, e_j^*)) + 2\sum_i G(B(e_i^*, \xi), B(e_i^*, \xi)) \\ &= \sum_{i,j} g(B'(e_i, e_j), B'(e_i, e_j)) + 2\sum_i G(Fe_i^*, Fe_i^*) \\ &= g(A', A') + 2\sum_i g(F'e_i, F'e_i), \end{aligned}$$

that is,

$$(3.14) \qquad \begin{aligned} |A|^2 &= |A'|^2 + 2\sum_i g(F'e_i, F'e_i) \\ &= |A'|^2 + 2n - 2\sum_i g(F'e_i, F'e_i). \end{aligned}$$

From (3.14) we have

PROPOSITION 3.12. (1) M and N are invariant submanifolds if and only if $|A|^2 = |A'|^2$;

(2) M and N are anti-invariant submanifolds if and only if $|A|^2 = |A'|^2 + 2n$.

As an application of submersions with totally geodesic fibre, we consider submanifolds of a complex projective space (see Lawson [1]).

THEOREM 3.2. Let N be an n-dimensional compact minimal submanifold of CP^m. If the scalar curvature r' of N satisfies the inequality

$$r' \geq n(n+2) - \frac{n+1}{2-1/p} \quad (p = 2m-n),$$

then M is a totally geodesic complex projective space $CP^{n/2}$.

Proof. We consider the fibering $\bar{\pi} : S^{2m+1} \longrightarrow CP^m$. By Proposition 3.6 the submanifold N of S^{2m+1} is minimal. Thus the scalr curvature r of N is given by

$$r = n(n+1) - G(A,A).$$

From this and (3.13) we obtain

$$G(A,A) = n(n+2) - r' - \sum_i G(Fe_i^*,Fe_i^*) \leq n(n+2) - r'.$$

Therefore, by the assumption, we have

$$G(A,A) \leq \frac{n+1}{2-1/p}.$$

Hence Theorem 5.4 of Chapter II implies that $A = 0$ or $|A|^2 = (n+1)/(2-1/p)$. If $A = 0$, then M is totally geodesic in S^{2m+1} and hence N is a totally geodesic complex projective space $CP^{n/2}$. If $|A|^2 = (n+1)/(2-1/p)$, then $F = 0$, which shows that M is an invariant submanifold of S^{2m+1}. Moreover, from Theorem 5.8 of Chapter II, M is a Veronese surface in S^4 or a Clifford minimal hypersurface in S^{n+2}. But the ambient manifold S^{2m+1} is odd dimensional and any hypersurface of S^{2m+1} is not invariant. From these considerations we have our assertion.

QED.

EXERCISES

A. POSITIVE RICCI CURVATURE: Nash [1] construct complete metrices of positive Ricci curvature on a large class of fibre bundles.

THEOREM. Let $\pi : M \longrightarrow B$ be a vector bundle over B, a compact manifold admitting a metric of positive Ricci curvature. If the fibre dimension is greater than two, M admits a complete metric of positive Ricci curvature.

B. FIBRE BUNDLES AND SUBMERSIONS: A mapping $f : M \longrightarrow B$ of Riemannian manifolds is said to be *totally geodesic* if for each geodesic x_t in M the image $f(x_t)$ is a geodesic in B. Vilms [1] proved

THEOREM 1. A Riemannian submersion is totally geodesic if and only if the fibres are totally geodesic submanifolds and the horizontal subbundle is integrable.

THEOREM 2. Let $\pi : M \longrightarrow B$ be a fibre bundle with standard fibre F and Lie structure group G. Assume the bundle admits a connection in the sense of Ehresman [1]. Endow B and F with Riemannian metrices, and assume F is G invariant. Then there exists a natural metric on M such that π is a Riemannian submersion with totally geodesic fibres.

We also have (Hermann [1], Muto [1], Nagano [1])

THEOREM 3. Let $\pi : M \longrightarrow B$ be a Riemannian submersion, and assume M to be connected. If M is complete, so is B, and π is locally trivial fibre space. If, in addition, the fibres are totally geodesic, then π is a fibre bundle with structure group the Lie group of isometries of the fibre.

For a geodesic we have (Hermann [1], O'Neill [3])

THEOREM 4. Let $\pi : M \longrightarrow B$ be a Riemannian submersion. If x is a geodesic of M which is horizontal at one point, then it is always horizontal, and hence $\pi \cdot x$ is a geodesic of B.

C. EQUIVALENCE PROBLEM: Let π_1 and π_2 be Riemannian submersions from a complete M onto B. Assume the fibres of π_1 and π_2 are totally geodesic. π_1 and π_2 are said to be *equivalent* provided there exists an isometry f of M which induces an isometry f' of B, so that the following diagram commutes:

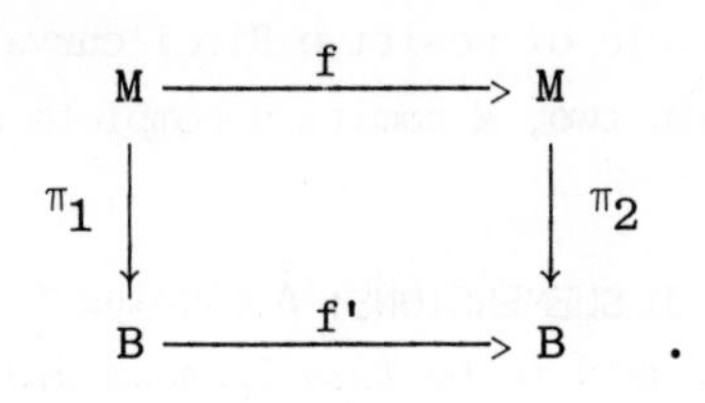

Escobales [1] proved

THEOREM 1. Let π_1 and π_2 be Riemannian submersions from M onto B satisfying the above hypotheses. Suppose f is an isometry of M satisfying the following two conditions:

(1) $f_{*p} : \mathcal{H}_{1p} \longrightarrow \mathcal{H}_{2f(p)}$ is an isometry from the horizontal distribution $\mathcal{H}_{1p}$ of π_1 at p onto the horizontal distribution $\mathcal{H}_{2f(p)}$ of π_2 at f(p);

(2) For E, F ε $T_p(M)$, $f_*(A_{1E}F) = A_{2f_*E}f_*F$, where A_i are the integrability tensors of π_i.

Then f induces an isometry f' of B so that π_1 and π_2 are equivalent.

O'Neill [2] gives the following

THEOREM 2. Let π_1 and π_2 be submersions of a Riemannian manifold M and B. If π_1 and π_2 have the same tensors T and A, and if their derivation maps agree at one point of M, then $\pi_1 = \pi_2$.

D. SUBMERSIONS WITH TOTALLY GEODESIC FIBRE: Escobales [1] gives the classification those B for which there is a Riemannian submersion $\pi : S^n \longrightarrow B$, where S^n is a square and the fibres are totally geodesic.

THEOREM 1. Let $\pi : S^n \longrightarrow B$ be a Riemannian submersion with connected totally geodesic fibres, and assume $1 \leq \dim$ fibre $\leq n-1$. Then, as a fibre bundle, π is one of the following types:

(a) $S^1 \longrightarrow S^{2m+1} \xrightarrow{\pi} CP^m$ for $n \geq 2$

(b) $S^3 \longrightarrow S^{4m+3} \xrightarrow{\pi} HP^m$ for $n \geq 2$

(c) $S^1 \longrightarrow S^3 \xrightarrow{\pi} S^2(\frac{1}{2})$

(d) $S^3 \longrightarrow S^7 \xrightarrow{\pi} S^4(\frac{1}{2})$

(e) $S^7 \longrightarrow S^{15} \xrightarrow{\pi} S^8(\frac{1}{2})$.

In case (a) and (b), B is isometric to complex and quaternion projective space with sectional curvature $1 \leq K_* \leq 4$. In case (c), (d) and (e), B is isometric to a sphere of curvature 4.

Moreover, if $_1$ and $_2$ are say two submersions both in class (a), (b) or (c), then π_1 is equivalent to π_2.

Escobales [2] proved the following

THEOREM 2. Any submersion $\pi : CP^n \longrightarrow B$ with connected complex totally geodesic fibres and with $2 \leq \dim$ fibre $\leq 2n-2$ must fall into one of the following two classes:

(a) $S^2 \longrightarrow CP^{2m+1} \xrightarrow{\pi} HP^m$

(b) $CP^3 \longrightarrow CP^7 \xrightarrow{\pi} S^8(\frac{1}{2})$.

In fact, $1 \leq K_* \leq 4$, where K_* denotes the curvature of HP^m, and $S^8(\frac{1}{2})$ denotes the sphere of radius $\frac{1}{2}$. Moreover, class (a) is not empty. Finally, if $n \geq 2$, any two submersions in class (a) are equivalent.

E. SUBMERSION WITH MINIMAL FIBRE: When a submersion has minimal fibres we obtain

THEOREM 1. Let $\pi : M \longrightarrow B$ be a Riemannian submersion with minimal fibres. Then a closed hypersurface P of B has constant mean curvature in B if and only if $\pi^{-1}(P)$ has constant mean curvature in M.

We also have (see Escobales [2])

THEOREM 2. Let $\pi : M \longrightarrow B$ be a Riemannian submersion with totally geodesic fibres. Assume P is a totally geodesic submanifold of B. Then $\pi^{-1}(P)$ is totally geodesic provided $A_X Y = 0$ whenever Y is horizontal and tangent to $\pi^{-1}(P)$ and X is normal to $\pi^{-1}(P)$.

For the submersion with totally umbilical fibres see Bishop [1].

F. REAL SUBMANIFOLDS AND SUBMERSIONS: Let $\pi : M^{n+1} \longrightarrow N^n$ be a submersion with 1-dimensional fibre. Then the fundamental tensor P of the submersion is a skew symmetric tensor of type (1,1) on N defined by

$$\nabla_{X^*} Y^* = (\nabla'_X Y)^* + g(PY,X)^* V,$$

where V is a unit vertical vector. Then Okumura [1] proved the following

THEOREM. Let N^n be a hypersurface of $CP^{(n+1)/2}$ and $\pi : M^{n+1} \longrightarrow N^n$ the submersion which is compatible with the fibration $S^1 \longrightarrow S^{n+2} \longrightarrow CP^{(n+1)/2}$. In order that the second fundamental tensor A' of N^n commutes with the fundamental tensor P of the submersion π, it is necessary and sufficient that the second fundamental tenor A of M^{n+1} is parallel.

G. QUATERNION KAEHLERIAN MANIFOLDS AND SASAKIAN 3-STRUCTURE: Let $\pi : M \longrightarrow B$ be a Riemannian submersion. Then we have the following (see Ishihara-Konishi [1])

THEOREM. (1) If M admits a K-contact 3-structure, then B admits an almost quaternion structure;

(2) If M admits a Sasakian 3-structure, then B admits a quaternion Kaehlerian structure;

(3) If M admits a Sasakian structure with constant curvature c, then c = 1 and the induced quaternion Kaehlerian structure of B is of constant Q-sectional curvature 4.

BIBLIOGRAPHY

The following bibliography includes all those references that are quoted in the text. The bibliography also contains books and papers on the topics closely related to the content of our treatises.

Adati, T.

[1] Submanifolds of an almost product Riemannian manifold, Kodai Math. J. 4 (1981), 327-343.

Ako, M.

[1] Submanifolds of a Kaehlerian manifold, Kōdai Math. Sem. Rep. 20 (1968), 1-11.

Alexander, S. and Maltz, R.

[1] Isometric immersions of Riemannian product in Euclidean space, J. Differential Geometry 11 (1976), 47-57.

Ambrose, W. and Singer, I. M.

[1] A theorem on holonomy, Trans. Amer. Math. Soc. 75 (1953), 428-443.

Auslander, L.

[1] Four dimensional compact locally Hermitian manifolds, Trans. Amer. Math. Soc. 84 (1957), 379-391.

Bejancu, A.

[1] CR submanifolds of a Kaehler manifold I, Proc. Amer. Math. Soc. 69 (1978), 135-142.

[2] CR submanifolds of a Kaehler manifold II, Trans. Amer. Math. Soc. 250 (1979), 333-345.

[3] Umbilical CR-submanifolds of a Kaehler manifold, Rendiconti di Mat. (3) 15 serie VI (1980), 431-446.

Bejancu, A., Kon, M. and Yano, K.

[1] CR-submanifolds of a complex space form, J. Differential Geometry 16 (1981), 137-145.

Berger, M.

[1] Sur les variétés d'Einstein compactes, C. R. IIIe Reunion Math. Expression Latine, Namur (1965), 35-55.

Bishop, R. L.

[1] Clairaut submersion, Differential Geometry, in Honor of K. Yano, Kinokumiya, Tokyo, 1972, 21-31.

Bishop, R. L. and Crittenden, R. J.

[1] Geometry of Manifolds, Academic Press, New York, 1964.

Bishop, R. L. and Goldberg, S. I.

[1] On the topology of positively curved Kaehler manifolds II, Tôhoku Math. J. 17 (1965), 310-318.

Blair, D. E.

[1] Contact Manifolds in Riemannian Geometry, Lecture Notes in Math. 509, Springer Verlag, 1976.

Blair, D. E. and Chen, B. Y.

[1] On CR-submanifolds of Hermitian manifolds, Israel J. Math. 34 (1979), 353-363.

Bochner, S.

[1] Vector field and Ricci curvature, Bull. Amer. Math. Soc. 52 (1946), 776-797.

Boothby, W. M. and Wang, H. C.

[1] On contact manifolds, Ann. of Math. 68 (1958), 721-734.

Calabi, E.

[1] Isometric imbedding of complex manifolds, Ann. of Math. 58 (1953), 1-23.

Calabi, E. and Eckmann, B.

[1] A class of compact, complex manifolds which are not algebraic, Ann. of Math. 58 (1953), 494-500.

Cartan, E.

[1] Lecons sur les Invariants Intégraux, Hermann Paris, 1922.

Cecil, T. E. and Ryan, P. J.

[1] Focal sets and real hypersurfaces in complex projective space, Trans. Amer. Math. Soc. 269 (1982), 481-499.

Chen, B. Y.

[1] Geometry of Submanifolds, Marcel Dekker, INC. New York,1973.

[2] CR-submanifolds of a Kaehlerian manifold. I, J. Differential Geometry 16 (1981), 305-322.

[3] CR-submanifolds of a Kaehler manifolds. II, J. Differential Geometry 16 (1981), 493-509.

Chen, B. Y. and Ogiue, K.

[1] Some characterizations of complex space forms, Duke Math. J. 44 (1973), 797-799.

[2] On totally real submanifolds, Trans. Amer. Math. Soc. 193 (1974), 257-266.

[3] Some characterizations of complex space forms in terms of Chern classes, Quart. J. Math. 26 (1975), 459-464.

Chern, S. S.

[1] Einstein hypersurfaces in a Kaehlerian manifold of constant holomorphic curvature, J. Differential Geometry 1 (1967), 21-31.

Chern, S. S., Do Carmo, M. and Kobayashi, S.

[1] Minimal submanifolds of a sphere with second fundamental form of constant length, Functional analysis and related fields, Proc. Conf. in Honor of Marshall Stone, Springer, Berlin, 1970.

Chevalley, C.

[1] Theory of Lie Groups, Princeton Univ. Press, 1946.

Ehresmann, C.

[1] Les connexions infinitésimales duns un espace fibre differentiable, Coll. de Topologie, Bruxelles, 1950, 29-55.

Ejiri, N.

[1] Totally real minimal immersions of n-dimensional real space forms into n-dimensional complex space forms, Proc. Amer. Math. Soc. 84 (1982), 243-246.

Endo, H.

[1] Invariant submanifolds in a K-contact Riemannian manifold, Tensor N. S. 28 (1974), 154-156.

Erbacher, J.

[1] Reduction of the codimension of an isometric immersion, J. Differential Geometry 5 (1971), 333-340.

[2] Isometric immersions of constant mean curvature and triviality of the normal connection, Nagoya Math. J. 45 (1971), 139-165.

Escobales, Jr., R. H.

[1] Riemannian submersions with totally geodesic fibers, J. Differential Geometry 10 (1975), 253-276.

[2] Riemannian submersions from complex projective space, J. Differential Geometry 13 (1978), 93-107.

Fialkow, A.

[1] Hypersurfaces of a space of constant curvature, Ann. of Math. 39 (1938), 762-785.

Fukami, T.

[1] Affine connections in almost product manifolds with some structure, Tôhoku Math. J. 11 (1959), 430-446.

Fukami, T. and Ishihara, S.

[1] Almost Hermitian structure on S^6, Tôhoku Math. J. 7 (1955), 151-156.

Funabashi, S.

[1] Totally real submanifolds of a quaternionic Kaehlerian manifold, Kōdai Math. Sem. Rep. 29 (1978), 261-270.

[2] Totally complex submanifolds of a quaternionic Kaehlerian manifold, Kodai Math. J. 2 (1979), 314-336.

Goldberg, S. I.

[1] Curvature and Homology, Academic Press, New York, 1962.

[2] Normal variations of invariant hypersurfaces of framed manifolds, Canad. Math. Bull. 15 (1972), 513-521.

[3] Framed manifolds, Differential Geometry, in honor of K. Yano, Kinokuniya, Tokyo, 1972, 121-132.

Goldberg, S. I.

[4] A generaralizatiion of Kaehler geometry, J. Differential Geometry 6 (1972), 343-355.

[5] On conformally flat space with definite Ricci curvature II, Kōdai Math. Sem. Rep. 27 (1976), 445-448.

Goldberg, S. I. and Kobayashi, S.

[1] Holomorphic bisectional curvature, J. Differential Geometry 1 (1967), 225-233.

Goldberg, S. I. and Okumura, M.

[1] Conformally flat manifolds and a pinching problem on the Ricci tensor, Proc. Amer. Math. Soc. 58 (1976), 234-236.

Goldberg, S. I. and Yano, K.

[1] Noninvariant hypersurfaces of almost contact manifolds, J. Math. Soc. Japan 22 (1970), 25-34.

[2] On normal globally framed f-manifolds, Tohoku Math. J. 22 (1970), 362-370.

[3] Globally framed f-manifolds, Illinois J. Math. 15 (1971), 456-474.

Gray, A.

[1] Vector cross products on manifolds, Trans. Amer. Math. Soc. 141 (1969), 465-504.

[2] Nearly Kaehler manifolds, J. Differential Geometry 4 (1970), 283-309.

[3] The structure of nearly Kaehlerian manifolds, Math. Ann. 223 (1976), 233-248.

[4] Compact Kaehler manifolds with nonnegative sectional curvature, Inventions Math. 41 (1977), 33-43.

Gray, J. W.

[1] Some global properties of contact structures, Ann. of Math. 69 (1959), 421-450.

Harada, M.

[1] On Sasakian submanifolds, Tôhoku Math. J. 25 (1973), 103-109.

Harada, M.

[2] On Kaehler manifolds satisfying the axiom of antiholomorphic 2-spheres, Proc. Amer. Math. Soc. 43 (1974), 186-189.

Hatakeyama, Y.

[1] Some notes on differentiable manifolds with almost contact structure, Tôhoku Math. J. 15 (1963), 176-181.

Hawley, N. S.

[1] Constant holomorphic curvature, Canad. J. Math. 5 (1953), 53-56.

Helgason, S.

[1] Differential Geometry and Symmetric Spaces, Academic Press, New York, 1962.

Hermann, R.

[1] A sufficient condition that a mapping of Riemannian manifolds be a fiber bundle, Proc. Amer. Math. Soc. 11 (1960), 236-242.

[2] On the differential geometry of foliations, Ann. of Math. 72 (1960), 445-457.

Hodge, W. V. D.

[1] Theory and Applications of Harmonic Integrals, Cambridge Univ. Press, London, 1941; 2 nd ed. 1951.

Hong, S. L.

[1] Isometric immersions of manifolds with plane geodesics into Euclidean space, J. Differential Geometry 8 (1973), 259-278.

Hopf, H.

[1] Zur Topologie der komplexen Mannigfeltigkeiten, Studies and Essays presented to R. Courant, Interscience, New York, 1948, 167-185.

Hopf, H. and Rinow, W.

[1] Über den Begriff der vollständigen differentialgeometrischen Flächen, Comm. Math. Helv. 3 (1931), 209-225.

Igusa, J.

[1] On the structure of a certain class of Kaehler manifolds, Amer. J. Math. 76 (1954), 669-678.

Ikawa, T. and Kon, M.

[1] Sasakian manifolds with vanishing contact Bochner curvature tensor and constant scalar curvature, Coll. Math. 37 (1977), 113-122.

Ishihara, I.

[1] Kaehler submanifolds satisfying a certain condition on normal bundle, Atti della Accademia Nazionale dei Lincei LXII (1977), 30-35.

[2] Anti-invariant submanifolds of a Sasakian space form, Kodai Math. J. 2 (1979), 171-186.

[2] Anti-invariant submanifolds satisfying a certain condition on normal connection, Kodai Math. J. 2 (1979), 371-383.

Ishihara, S.

[1] Normal structure f satisfying $f^3 + f = 0$, Kōdai Math. Sem. Rep. 18 (1966), 36-47.

[2] Quaternion Kaehlerian manifolds and fibred Riemannian spaces with Sasakian 3-structure, Kōdai Math. Sem. Rep. 25 (1973), 321-329.

[3] Quaternion Kaehlerian manifolds, J. Differential Geometry 9 (1974), 483-500.

Ishihara, S. and Konishi, M.

[1] Differential Geometry of Fibred Spaces, Publications of the study group of geometry, Tokyo, Vol. 8, 1973.

Ishihara, S. and Yano, K.

[1] On integrability conditions of a structure f satisfying $f^3 + f = 0$, Quart. J. Math. 15 (1964), 217-222.

Itoh, T.

[1] Positively curved complex submanifolds immersed in a complex space form, Proc. Amer. Math. Soc. 72 (1978), 341-345.

Itoh, T. and Ogiue, K.

[1] Isotropic immersions, J. Differential Geometry 8 (1973), 305-316.

Johnson, D. L.

[1] Kaehler submersions and holomorphic connections, J. Differential Geometry 15 (1980), 71-79.

Kashiwada, T.

[1] A note on a Riemannian space with Sasakian 3-structure, Nat. Sci. Rep. Ochanomizu Univ. 22 (1971), 1-2.

[2] On the induced almost contact structure of a Brieskorn manifold, Geometriae Dedicata 6 (1977), 415-418.

Kenmotsu, K.

[1] Invariant submanifolds in a Sasakian manifold, Tôhoku Math. J. 21 (1969), 495-500.

[2] On Sasakian immersions, Sem. on Contact Manifolds, Pub. of the study group of geometry 4 (1970), 42-59.

[3] Local classification of invariant η-Einstein submanifolds of codimension 2 in a Sasakian manifold with constant ϕ-sectional curvature, Tôhoku Math. J. 22 (1970), 270-272.

[4] Some remarks on minimal submanifolds, Tôhoku Math. J. 22 (1970), 240-248.

Ki, U. H. and Pak, J. S.

[1] Generic submanifolds of an even-dimensional Euclidean space, J. Differential Geometry 16 (1981), 293-303.

Ki, U. H., Pak, J. S. and Kim, Y. H.

[1] Generic submanifolds of complex projective spaces with parallel mean curvature vector, Kodai Math. J. 4 (1981), 137-155.

Kobayashi, S.

[1] Principal fibre bundles with 1-dimensional toroidal group, Tôhoku Math. J. 8 (1956), 29-45.

[2] Hypersurfaces of complex projective space with constant scalar curvature, J. Differential Geometry 1 (1967), 369-370.

Kobayashi, S. and Nomizu, K.

[1] Foundations of Differential Geometry, Vol. I, Interscience Publishers, 1963.

[2] Foundations of Differential Geometry, Vol. II, Interscience Publishers, 1969.

Kon, M.

[1] Invariant submanifolds of normal contact metric manifolds, Kōdai Math. Sem. Rep. 25 (1973), 330-336.

[2] Kaehler immersions with trivial normal connection, TRU Math. 9 (1973), 29-33.

[3] On some invariant submanifolds of normal contact metric manifolds, Tensor N. S. 28 (1974), 133-138.

[4] On invariant submanifolds in a Sasakian manifold of constant ϕ-sectional curvature, TRU Math. 10 (1974), 1-9.

[5] On some complex submanifolds in Kaehler manifolds, Canad. J. Math. 26 (1974), 1442-1449.

[6] Totally real minimal submanifolds with parallel second fundamental form, Atti della Accademia Nazionale dei Lincei 57 (1974), 187-189.

[7] Complex submanifolds with constant scalar curvature in a Kaehler manifold, J. Math. Soc. Japan 27 (1975), 76-81.

[8] Invariant submanifolds in Sasakian manifolds, Math. Ann. 219 (1975), 277-290.

[9] Kaehler immersions with vanishing Bochner curvature tensors, Kōdai Math. Sem. Rep. 27 (1976), 329-333.

[10] Remarks on anti-invariant submanifolds of a Sasakian manifold, Tensor N. S. 30 (1976), 239-246.

[11] Totally real submanifolds in a Kaehler manifold, J. Differential Geometry 11 (1976), 251-257.

[12] On hypersurfaces immersed in S^{2n+1}, Ann. Fac. Sci. de Kinshasa 4 (1978), 1-24.

[13] Pseudo-Einstein real hypersurfaces in complex space forms, J. Differential Geometry 14 (1979), 339-354.

[14] Real minimal hypersurfaces in a complex projective space, Proc. Amer. Math. Soc. 79 (1980), 285-288.

[15] Generic minimal submanifolds of a complex projective space, Bull. London Math. Soc. 12 (1980), 355-360.

Kubo, Y.

[1] Kaehlerian manifolds with vanishing Bochner curvature tensor, Kōdai Math. Sem. Rep. 28 (1976), 85-89.

Kuo, Y. Y.

[1] On almost contact 3-structure, Tôhoku Math. J. 22 (1970), 325-332.

Kuo, Y. Y. and Tachibana, S.

[1] On the distribution appeared in contact 3-structure, Taita J. of Math. 2 (1970), 17-24.

Lawson, Jr., H. B.

[1] Rigidity theorems in rank-1 symmetric spaces, J. Differential Geometry 4 (1970), 349-357.

Lichnerowicz, A.

[1] Théoremès de reductiblite des variétés kählériennes et applications, C. R. Acad. Sci. Paris 231 (1950), 1280-1282.

[2] Sur les forms harmoniques des variétés riemanniennes localement réductibles, C. R. Acad. Sci. Paris 232 (1951), 1634-1636.

[3] Sur la reductibilité des espaces homogénes riemanniens, C. R. Acad. Sci. Paris 243 (1956), 646-672.

[4] Géométrie des Groupes de Transformations, Dunod, Paris, 1958.

Ludden, G. D.

[1] Submanifolds of cosymplectic manifolds, J. Differential Geometry 4 (1970), 237-244.

Ludden, G. D. and Okumura, M.

[1] Some integral formulas and their applications for hypersurfaces of $S^n \times S^n$, J. Differential Geometry 9 (1974), 617-631.

Ludden, G. D., Okumura, M. and Yano, K.

[1] A totally real surface in CP^2 that is not totally geodesic, Proc. Amer. Math. Soc. 53 (1975), 186-190.

[2] Totally real submanifolds of complex manifolds, Atti della Accademia Nazionale dei Lincei LVIII (1975), 346-353.

Matsumoto, M.

[1] On 6-dimensional almost Tachibana spaces, Tensor N. S. 23 (1972), 250-252.

[2] A remark on 10-dimensional almost Tachibana spaces, Bull. of St. Marianna Univ. 9 (1980), 1-11.

[3] On 10-dimensional almost Tachibana spaces, Bull. of St. Marianna Univ. 10 (1981), 91-100.

Matsumoto, M. and Chūman, G.

[1] On the C-Bochner curvature tensor, TRU Math. 5 (1969), 21-30.

Matsumoto, M. and Tanno, S.

[1] Kaehlerian spaces with parallel or vanishing Bochner curvature tensor, Tensor N. S. 27 (1973), 291-294.

Matsushima, Y.

[1] Differentiable Manifolds, Marcel Dekker, Inc. New York, 1972.

Matsuyama, Y.

[1] Invariant hypersurfaces of $S^n \times S^n$ with constant mean curvature, Hokkaido Math. J. 5 (1976), 210-217.

[2] Complex hypersurfaces of $P_n(C) \times P_n(C)$, Tôhoku Math. J. 28 (1976), 541-552.

[3] Minimal submanifolds in S^N and R^N, Math. Z. 175 (1980), 275-282.

Millman, R. S.

[1] f-structures with parallelizable kernel on manifolds, J. Differential Geometry 9 (1974), 531-535.

Moore, J. D.

[1] Isometric immersions of Riemannian products, J. Differential Geometry 5 (1971), 159-168.

Morimoto, A.

[1] On normal almost contact structures, J. Math. Soc. Japan 15 (1963), 420-436.

[2] On normal almost contact structure with a regularity, Tôhoku Math. J. 16 (1964), 90-104.

Morrow, J. and Kodaira, K.

[1] Complex Manifolds, Holt, Rinehart and Winston, Inc. New York, 1971.

Muto, Y.

[1] On some properties of a fibred Riemannian manifold, Sci. Rep. Yokohama Nat. Univ. Sect. I, 1 (1952), 1-14.

Myers, S. B.

[1] Riemannian manifolds with positive mean curvature, Duke Math. J. 8 (1941), 401-404.

Nagano, T.

[1] On fibered Riemannian manifolds, Sci. Papers College Gen. Ed. Univ. Tokyo 10 (1960), 17-27.

Nakagawa, H.

[1] On framed f-manifolds, Kōdai Math. Sem. Rep. 18 (1966), 293-306.

[2] Einstein-Kaehler manifolds immersed in a complex projective space, Canad. J. Math. 28 (1976), 1-8.

Nakagawa, H. and Ogiue, K.

[1] Complex space forms immersed in complex space forms, Trans. Amer. Math. Soc. 219 (1976), 289-297.

Nakagawa, H. and Takagi, R.

[1] On locally symmetric Kaehler submanifolds in a complex projective space, J. Math. Soc. Japan 28 (1976), 638-667.

[2] Kaehler submanifolds with RS = 0 in a complex projective space, Hokkaido Math. J. 5 (1976), 67-70.

Nakagawa, H. and Yokote, I.

[1] Compact hypersurfaces in an odd dimensional sphere, Kōdai Math. Sem. Rep. 25 (1973), 225-245.

Nash, J. C.

[1] Positive Ricci curvature on fibre bundles, J. Differential Geometry 14 (1979), 241-254.

Newlander, A. and Nirenberg, L.

[1] Complex analytic coordinates in almost complex manifolds, Ann. of Math. 65 (1957), 391-404.

Nomizu, K.

[1] Lie Groups and Differential Geometry, Publ. Math. Soc. Japan, 1956.

Nomizu, K.

[2] On hypersurfaces satisfying a certain condition of the curvature tensor, Tôhoku Math. J. 20 (1968), 46-59.

[3] On the rank and curvature of non-singular complex hypersurfaces in a complex projective space, J. Math. Soc. Japan 21 (1969), 266-269.

[4] Some results in E. Cartan's theory of isoparametric families of hypersurfaces, Bull. Amer. Math. Soc. 79 (1973), 1184-1188.

[5] Conditions for constancy of the holomorphic sectional curvature, J. Differential Geometry 8 (1973), 335-339.

Nomizu, K. and Smyth, B.

[1] Differential geometry of complex hypersurfaces II, J. Math. Soc. Japan 20 (1968), 499-521

[2] A formula of Simons' type and hypersurfaces with constant mean curvature, J. Differential Geometry 3 (1969), 367-377.

Ogawa, Y.

[1] A condition for a compact Kaehlerian space to be locally symmetric, Nat. Sci. Rep. Ochanomizu Univ. 28 (1977), 21-23.

Ogiue, K.

[1] Positively curved complex submanifolds immersed in a complex projective space, J. Differential Geometry 7 (1972), 603-606.

[2] Positively curved complex submanifolds immersed in a complex projective space II, Hokkaisdo Math. J. 1 (1972),16-20.

[3] On Kaehler immersions, Canad. J. Math. 24 (1972), 1178-1182.

[4] n-dimensional complex space forms immersed in {n+n(n+1)/2}-dimensional complex space forms, J. Math. Soc. Japan 24 (1972), 518-526.

[5] Differential geometry of Kaehler submanifolds, Advances in Math. 13 (1974), 73-114.

[6] Positively curved totally real minimal submanifolds immersed in a complex projective space, Proc. Amer. Math. Soc. 56 (1976), 264-266.

Okumura, M.

[1] On some real hypersurfaces of a complex projective space, Trans. Amer. Math. Soc. 212 (1975), 355-364.

O'Neill, B.

[1] Isotropic and Kaehler immersions, Canad. J. Math. 17 (1965), 907-915.

[2] The fundamental equations of a submersion, Michigan Math. J. 13 (1966), 459-469.

[3] Submersions and geodesics, Duke Math. J. 34 (1967), 459-469.

Palais, R. S.

[1] A Global Formulation of the Lie Theory of Transformation Groups, Mem. Amer. Math. Soc. 22, 1957.

Reinhart, B.

[1] Foliated manifolds with bundle-like metrics, Ann. of Math. 69 (1959), 119-131.

de Rham, G.

[1] Sur la réductibilité d'un espace de Riemann, Comm. Math. Helv. 26 (1952), 328-344.

[2] Variétés Differentiables, Hermann, Paris, 1955.

Ryan, P. J.

[1] Homogeneity and some curvature conditions for hypersurfaces, Tôhoku Math. J. 21 (1969), 363-388.

[2] A class of complex hypersurfaces, Coll. Math. 26 (1972), 175-182.

Sakamoto, K.

[1] Complex submanifolds with certain condition, Kōdai Math. Sem. Rep. 27 (1976), 334-344.

[2] Planar geodesic immersions, Tôhoku Math. J. 29 (1977), 25-56.

Sasaki, S.

[1] A characterization of contact transformations, Tôhoku Math. J. 16 (1964), 285-290.

[2] Almost contact manifolds, Lecture Notes, Tôhoku Univ. Vol. 1, 1965, vol. 2, 1967, vol. 3, 1968.

Sasaki, S. and Hsu, C. J.

[1] On a property of Brieskorn manifolds, Tôhoku Math. J. 28 (1976), 67-78.

Sawaki, S., Watanabe, Y. and Sato, T.

[1] Notes on a K-space of constant holomorphic sectional curvature, Kōdai Math. Sem. Rep. 26 (1975), 438-445.

Schouten, J. A. and Struik, D. J.

[1] On some properties of general manifolds relating to Einstein's theory of gravitation, Amer. J. Math. 43 (1921), 213-216.

Schur, F.

[1] Über den Zusammenhang der Räume konstanter Krümmungsmasses mit den projectiven Räumen, Math. Ann. 27 (1886), 537-567.

Sekigawa, K.

[1] Almost Hermitian manifolds satisfying some curvature conditions, Kodai Math. J. 2 (1979), 384-405.

Simons, J.

[1] Minimal varieties in riemannian manifolds. Ann. of Math. 88 (1968), 62-105.

Smyth, B.

[1] Differential geometry of complex hypersurfaces, Ann. of Math. 85 (1967), 246-266.

[2] Homogeneous complex hypersurfaces, J. Math. Soc. Japan 20 (1968), 643-647.

[3] Submanifolds of constant mean curvature, Math. Ann. 205 (1973), 265-280.

Spencer, D. C.

[1] Some remarks on pertubation of structure, Analytic Functions, Princeton University Press, 1960, 67-87.

Steenrod, N.

[1] Topology of Fibre Bundles, Princeton Univ. Press, 1951.

Sternberg, S.

[1] Lectures on Differential Geometry, Printice-Hall, Englewood Cliffs, New Jersey, 1964.

Stong, R. E.

[1] The rank of an f-structure, Kōdai Math. Sem. Rep. 29 (1977), 207-209.

Tachibana, S.

[1] On almost-analytic vectors in certain almost-Hermitian manifolds, Tôhoku Math. J. 11 (1959), 351-363.

[2] Some theorems on locally product Riemannian spaces, Tôhoku Math. J. 12 (1960), 281-292.

[3] On infinitesimal conformal and projective transformations of compact K-space, Tôhoku Math. J. 13 (1961), 386-392.

Tachibana, S. and Yu, W. N.

[1] On a Riemannian space admitting more than one Sasakian Structure, Tôhoku Math. J. 22 (1970), 536-540.

Takagi, H.

[1] An example of Riemannian manifolds satisfying $R(X,Y)R = 0$ but not $\nabla R = 0$, Tôhoku Math. J. 24 (1972), 105-108.

Takagi, R.

[1] Real hypersurfaces in a complex projective space with constant principal curvatures, J. Math. Soc. Japan 27 (1975), 43-53.

[2] Real hypersurfaces in a complex projective space with constant principal curvatures II, J. Math. Soc. Japan 27 (1975), 507-516.

[3] A class of hypersurfaces with constant principal curvatures in a sphere, J. DIfferential Geometry 11 (1976), 225-233.

Takahashi Takao

[1] Deformations of Sasakian structures and its application to the Brieskorn manifolds, Tôhoku Math. J. 30 (1978), 37-43.

Takahashi Toshio

[1] Sasakian ϕ-symmetric spaces, Tôhoku Math. J. 29 (1977), 91-113.

Takahashi Toshio and Tanno, S.

[1] K-contact Riemannian manifolds isometrically immersed in a space of constant curvature, Tôhoku Math. J. 23 (1971), 535-539.

Takahashi Tsunero

[1] Minimal immersions of Riemannian manifolds, J. Math. Soc. Japan 18 (1966), 380-385.

[2] Hypersurface with parallel Ricci tensor in a space of constant holomorphic sectional curvature, J. Math. Soc. Japan 19 (1967), 199-204.

[3] Complex hypersurfaces with RS = 0 in C^{n+1}, Tôhoku Math. J. 25 (1973), 527-533.

Takamatsu, K.

[1] Some properties of 6-dimensional K-spaces, Kōdai Math. Sem. Rep. 23 (1971), 215-232.

Takamatsu, K. and Sato, T.

[1] A K-space of constant holomorphic sectional curvature, Kōdai Math. Sem. Rep. 27 (1976), 116-127.

Tanno, S.

[1] Locally symmetric K-contact Riemannian manifolds, Proc. Japan Acad. 43 (1967), 581-583.

[2] Isometric immersions of Sasakian manifolds in spheres, Kōdai Math. Sem. Rep. 21 (1969), 448-458.

[3] Hypersurfaces satisfying a certain condition on the Ricci tensor, Tôhoku Math. J. 21 (1969), 297-303.

[4] Sasakian manifolds with constant ϕ-holomorphic sectional curvature, Tôhoku Math. J. 21 (1969), 501-507.

[5] 2-dimensional complex submanifolds immersed in complex projective spaces, Tôhoku Math. J. 24 (1972), 71-78.

[6] Compact complex submanifolds immersed in complex projective spaces, J. Differential Geometry 8 (1973), 629-641.

[7] Constancy of holomorphic sectional curvature in almost Hermitian manifolds, Kōdai Math. Sem. Rep. 25 (1973), 190-201.

Tanno, S. and Takahashi Toshio

[1] Some hypersurfaces of a sphere, Tôhoku Math. J. 22 (1970), 212-219.

Tashiro, Y. and Tachibana, S.

[1] On Fubinian and C-Fubinian manifolds, Kōdai Math. Sem. Rep. 15 (1963), 176-183.

Thomas, T. Y.

[1] On closed spaces of constant mean curvature, Amer. J. Math. 58 (1936), 702-704.

[2] Extract from a letter by E. Cartan concerning my note: On closed spaces of constant mean curvature, Amer. J. Math. 59 (1937), 793-794.

Verheyen, P. and Verstraelen, L.

[1] Conformally flat totally real submanifolds of complex projective spaces, Soochow J. Math. 6 (1980), 137-143.

Waerden, R.

[1] Untermannigfaltigkeiten mit paralleler zweiter Fundamental form in Euklidishen Räumen und Sphären, Manuscripta Math. 10 (1973), 91-102.

Walker, A. G.

[1] Connexions for parallel distributions in the large I, II, Quart. J. Math. 6 (1955), 301-308; 9 (1958), 221-231.

Watanabe, Y. and Takamatsu, K.

[1] On a K-space of constant holomorphic sectional curvature, Kōdai Math. Sem. Rep. 25 (1973), 297-306.

Watson, B.

[1] Almost Hermitian submersion, J. Differential Geometry 11 (1976), 147-165.

Wells, Jr., R. O.

[1] Compact real submanifolds of a complex manifold with nondegenerate holomorphic tangent bundles, Math. Ann. 179 (1969), 123-129.

[2] Function theory on differentiable manifolds, Contribution to analysis, Academic Press, 1974, 407-441.

Weyl, H.

[1] Reine Infinitesimalygeometrie, Math. Z. 26 (1918), 384-411.

[2] Zur Infiniteimalgeometrie: Einordnung der projektiven und der konformen Auffassung, Göttingen Nachr. 1921, 99-112.

Willmore, T. J.

[1] Connexion for systems of parallel distributions, Quart. J Math. 7 (1956), 269-276.

Yamaguchi, S., Chūman, G. and Matsumoto, M.

[1] On a special almost Tachibana space, Tensor N. S. 24 (1972), 351-354.

Yamaguchi, S. and Kon, M.

[1] Kaehler manifolds satisfying the axiom of anti-invariant 2-spheres, Geometriae Dedicata 7 (1978), 403-406.

Yano, K.

[1] On harmonic and Killing vector fields, Ann. of Math. 55 (1952), 38-45.

[2] The Theory of Lie Derivatives and its Applications, North-Holland, Amsterdam, 1957.

[3] Affine connections in an almost product space, Kōdai Math. Sem. Rep. 11 (1959), 1-24.

[4] On a structure defined by a tensor field f of type (1,1) satisfying $f^3 + f = 0$, Tensor N. S. 14 (1963), 99-109.

[5] Differential Geometry on Complex and Almost Complex Spaces, Pergamon Press, New York, 1965.

[6] Integral Formulas in Riemannian Geometry, Marcel Dekker, Inc, New York, 1970.

[7] Manifolds and submanifolds with vanishing Weyl or Bochner curvature tensor, Proc. Symposia in Pure Math. 27 (1975), 253-262.

[8] Differential geometry of anti-invariant submanifolds of a Sasakian manifold, Bollettino U. M. I. 12 (1975), 279-296.

Yano, K. and Ako, M.

[1] An affine connection in an almost quaternion manifold, J. Differential Geometry 8 (1973), 341-347.

Yano, K. and Bochner, S.

[1] Curvature and Betti Numbers, Annala of Math. Studies, No. 32 Princeton Univ. Press, 1953.

Yano, K. and Ishihara, S.

[1] Differential geometry of fibred spaces, Kōdai Math. Sem. Rep. 257-288.

[2]Fibred spaces with invariant Riemannian metric, Kōdai Math. Sem. Rep. 19 (1967), 317-360.

Yano, K. and Ishihara, S.

[3] Submanifolds with parallel mean curvature vector, J. Differential Geometry 6 (1971), 95-118.

Yano, K. and Kon, M.

[1] Anti-invariant Submanifolds, Marcel Dekker Inc. New York, 1976.

[2] Totally real submanifolds of complex space forms I, Tôhoku Math. J. 28 (1976), 215-225.

[3] Totally real submanifolds of complex space forms II, Kōdai Math. Sem. Rep. 27 (1976), 385-399.

[4] Anti-invariant submanifolds of Sasakian space forms II, J. Korean Math. Soc. 13 (1976), 1-14.

[5] Anti-invariant submanifolds of Sasakian space forms I, Tôhoku Math. J. 29 (1977), 9-23.

[6] Submanifolds of an even-dimensional sphere, Geometriae Dedicata 6 (1977), 131-139.

[7] Submanifolds of Kaehlerian product manifolds, Memorie Accademia Nazionale dei Lincei CCCLXXVI (1979), 267-292.

[8] CR-sous-varietes d'un espace projectif complexe, C. R. Acad. Paris 288 (1979), 515-517.

[9] Generic submanifolds with semidefinite second fundamental form of a complex projective space, Kyungpook Math. J. 20 (1980), 47-51.

[10] Generic submanifolds of Sasakian manifolds, Kodai Math. J. 3 (1980), 163-196.

[11] Generic submanifolds, Annali Mat. CXXIII (1980), 59-92.

[12] Differential geometry of CR-submanifolds, Geometriae Dedicata 10 (1981), 369-391.

[13] CR submanifolds of a complex projective space, J. Differential Geometry 16 (1981), 431-444.

[14] Generic minimal submanifolds with flat normal connection, E. B. Christoffel, Aachen Birkhäuser Verlag, Basel, 1981, 592-599.

Yano, K. and Kon, M.

[15] On some minimal submanifolds of an odd dimensional sphere with flat normal connection, Tensor N. S. 36 (1982), 175-179.

[16] Contact CR submanifolds, Kodai Math. J. 5 (1982), 238-252.

[17] Generic minimal submanifolds with flat normal connection II, Tensor N. S. 36 (1982), 267-269.

[18] CR Submanifolds of Kaehlerian and Sasakian Manifolds, Birkhäuser Verlag, Boston, 1983.

Yano, K., Kon, M. and Ishihara, I.

[1] Anti-invariant submanifolds with flat normal connection, J. Differential Geometry 13 (1978), 577-588.

Yano, K. and Okumura, M.

[1] On (f,g,u,v,λ)-structures, Kōdai Math. Sem. Rep. 22 (1970), 401-423.

Yau, S. T.

[1] Submanifolds with constant mean curvature II, Amer. J. Math. 97 (1975), 76-100.

INDEX

A

Adapted frame of f-structure, 385
Affine
connection, 18
transformation, 20
Almost
complex manifold, 107
complex mapping, 109
complex structure, 107
contact manifold, 252
contact metric manifold, 254
contact metric structure, 254
contact structure, 252
Hermitian manifold, 124
Hermitian submersion, 449
Kaehlerian manifold, 128
product manifold, 423
product Riemannian manifold, 423
product structure, 423
quaternion manifold, 159
quaternion metric manifold, 160
quaternion metric structure, 160
quaternion structure, 159
semi-Kaehlerian manifold, 129
symplectic manifold, 129
Tachibana manifold, 129
Alternate mapping, 6
Analytic
function, 3
manifold, 2

Analytic
structure, 2
vector field, 114
Anti-invariant submanifold, 199, 329
Associated
second fundamental form, 66
vector field to η, 256

B

Base space, 48
Basic vector field, 440
Bianchi's
1st identity, 31
2nd identity, 31
Brieskorn manifold, 291
Bundle space, 48

C

Canonical
complex structure, 118
flat connection, 57
form, 58
global basis, 159
local basis, 159
Chart, 1
Clifford
minimal hypersurface, 90
torus, 90
Codazzi equation, 68
Codifferential, 42
Complete
manifold, 30
vector field, 14

C

Complex
- analytic mapping, 3
- conjugation, 105
- Lie algebra, 118
- Lie group, 119
- projective space, 120
- quadric, 142
- r-form, 110
- space form, 134
- structure, 105
- tangent space, 109
- tangent vector, 109
- torus, 120

Complexification, 105
Component of vector, 4
Concircular curvature tensor, 54
Conjugate vector, 105
Conformal
- change, 40
- Killing vector, 41
- transformation, 41

Conformally flat space, 40
Connection, 56
Connection form, 57
Constant
- curvature, 35
- holomorphic sectional curvature, 134
- Q-sectional curvature, 169

Contact
- Bochner curvature tensor, 308
- CR submanifold, 353
- distribution, 269
- form, 255

Contact
manifold, 255
manifold in the wider sense, 258
metric manifold, 256
metric structure, 256
structure, 255
structure in the wider sense, 257
totally umbilical submanifold, 374
transformation, 257, 272
Contraction, 9
Contravariant tensor, 6
Coordinate neighborhood, 2
Cosymplectic manifold, 372
Covariant
constant framed metric manifold, 402
differentiation, 18
tensor, 6
Covering space, 50
CR product, 239
CR structure, 217
CR submanifold, 216
Curvature
tensor, 25
tensor of the normal bundle, 69
transformation, 31

D

Derivation, 16
Derivative, 16
D-homothetic deformation, 281
Diffeomorphism, 13
Differentiable
curve, 3
function, 3

Differentiable
manifold, 2
mapping, 2
structure, 2
transformation, 13
vector field, 4
Differential form, 5
Distance, 28
Distribution, 17
Divergence, 43

E

Effictive action, 47
Einstein manifold, 38
Elliptic space, 39
Equation
of Codazzi, 68
of Gauss, 68
of Ricci, 69
Equivalence submersion, 468
η-Einstein manifold, 285
η-parallel
Ricci tensor, 321
second fundamental form, 314
Euclidean n-space, 39
Exponential mapping, 22
Exterior differentiation, 11

F

F-anti-invariant submanifold, 426
Fibre, 48, 439
Fibre bundle, 50
F-invariant submanifold, 424
First normal space, 77

F

Flat
connection, 58
normal connection, 70
space, 39
Framed
f-structure, 402
manifold, 402
metric manifold, 402
Free action, 47
f-structure, 379
Fubini-Study metric, 141
Fundamental
2-form, 125, 256, 402
vector field, 48

G

Gauss
equation, 68
formula, 66
Gaussian curvature, 37
Generic submanifold, 216, 353
Geodesic, 21
Geodesic hypersphere, 249
Globally
framed f-structure, 402
ϕ-symmetric manifold, 308
Great hypersphere, 40

H

Harmonic form, 42

H

Hermitian
manifold, 124
metric, 124
Hodge manifold, 291
Holomorphic
bisectional curvature, 453
form, 116
mapping, 3
sectional curvature, 134
vector field, 117
Homogeneous coordinate system, 120
Homomorphism, 49
Homothetic, 40
Horizontal
lift, 57, 440
subspace, 56
vector, 56, 439
Hybrid tensor, 145, 417
Hyperbolic space, 39
Hypersphere, 39

I

Imbedded submanifold, 12
Imbedding, 12, 49
Immersed submanifold, 12
Immersion, 12
Induced
connection, 65
metric, 62
Infinitesimal
affine transformation, 41

Infinitesimal
automorphism, 114
isometry, 41
Inhomogeneous coordinate system, 120
Injection, 49
Integrable, 113, 160, 391
Integral
curve, 13
manifold, 18
Invariant
submanifold, 180, 312
affine connection, 20
Involutive distribution, 17
Isometric
imbedding, 30
immersion, 30
Isometry, 30
Isomorphic, 50
Isoperimetric section, 101

K

Kaehlerian
manifold, 129
metric, 128
submanifold, 180
K-contact
manifold, 267
structure, 267
Killing vector, 41
K-space, 129

L

Laplacian, 44
Left invariant vector field, 47

L

Levi-Civita connection, 29
Lie
  algebra, 47
  differentiation, 14
  group, 46
  transformation group, 47
Lift, 57
Local
  coordinate, 2
  coordinate system, 2
  1-parameter group, 13
Locally
  decomposable Riemannian manifold, 419
  flat space, 38
  ϕ-symmetric manifold, 308
  product manifold, 415
  product structure, 415
  product Riemannian manifold, 419
  symmetric space, 38

M

Maximal integral manifold, 18
Mean curvature vector, 67
Metric connection, 29
Minimal
  section, 101
  submanifold, 67
Mixed tensor algebra, 9
Multilinear mapping, 6

N

Nearly Kaehlerian manifold, 129
Noninvariant hypersurface, 408
Non-trivial
contact CR submanifold, 353
CR submanifold, 216
Normal
almost contact structure, 263
connection, 62
contact metric manifold, 272
contact metric structure, 272
f-structure, 233, 401
neighborhood, 22
variation, 408

O

1-parameter group, 13
Orbit, 13

P

Parallel
f-structure, 333
normal vector, 67
second fundamental form, 67
Partially integrable, 220
Period function, 286
ϕ-section, 277
ϕ-sectional curvature, 277
Planer geodesic immersion, 102
Principal fibre bundle, 47
Product manifold, 3
Projection, 48

P

Projective curvature tensor, 54
Proper CR submanifold, 216
Pseudo-
Einstein hypersurface, 376
Einstein real hypersurface, 249
flat normal connection, 373
parallel mean curvature vector, 373
umbilical hypersurface, 375
Pure tensor, 145, 416, 417

Q

Quasi-Kaehlerian manifold, 129
Q-sectional curvature, 169
Quaternion
Kaehlerian manifold, 161
Kaehlerian structure, 161

R

Rank
of mapping, 12
of complex hypersurface, 191
Reduced bundle, 49
Reduction, 49
Regular
contact structure, 286
coordinate neighborhood, 286
Restriction, 48
Ricci
equation, 69
tensor, 37
Right invariant vector field, 47

R

Riemannian
connection, 29
manifold, 27
metric, 27
submersion, 439

S

Sasakian
manifold, 272
space form, 281
structure, 272
3-structure, 307
Scalar curvature, 37
Second fundamental form, 65
Semi-
flat normal connection, 224
Kaehlerian manifold, 129
Separable coordinate system, 415
Skew-derivarion, 16
Small hypersphere, 40
Space
form, 35
of constant curvature, 35
of constant holomorphic sectional curvature, 134
Standard CR product, 240
Strict contact transformation, 258, 272
Structure group, 48
Subbundle, 49
Submersion, 439
Symmetric space, 38

S

Symplectic
manifold, 129, 287
structure, 287

T

Tangent
space, 4
vector, 3
Tensor field, 6
Torsion tensor, 25
Total
differential, 5
space, 48
Totally
geodesic mapping, 467
geodesic submanifold, 67
real submanifold, 199, 250
umbilical submanifold, 67
Transformation, 13
Transitive function, 49
Transversal form, 392
Trivial
normal connection, 70
principal bibre bundle, 48

U

Umbilical section, 67
Universal covering space, 50

V

Vector, 3
Vector field, 4
Veronese surface, 90
Vertical
  subspace, 56
  vector, 56, 439

W

Weingarten formula, 66
Weyl conformal curvature tensor, 40